Lecture Notes in Civil Engineering

Volume 121

Lecture Notes in Civil Engineering (LNCE) publishes the latest developments in Civil Engineering - quickly, informally and in top quality. Though original research reported in proceedings and post-proceedings represents the core of LNCE, edited volumes of exceptionally high quality and interest may also be considered for publication. Volumes published in LNCE embrace all aspects and subfields of, as well as new challenges in, Civil Engineering. Topics in the series include:

- Construction and Structural Mechanics
- Building Materials
- Concrete, Steel and Timber Structures
- Geotechnical Engineering
- Earthquake Engineering
- Coastal Engineering
- Ocean and Offshore Engineering; Ships and Floating Structures
- Hydraulics, Hydrology and Water Resources Engineering
- Environmental Engineering and Sustainability
- Structural Health and Monitoring
- Surveying and Geographical Information Systems
- Indoor Environments
- Transportation and Traffic
- Risk Analysis
- Safety and Security

To submit a proposal or request further information, please contact the appropriate Springer Editor:

- Pierpaolo Riva at pierpaolo.riva@springer.com (Europe and Americas);
- Swati Meherishi at swati.meherishi@springer.com (Asia - except China, and Australia, New Zealand);
- Wayne Hu at wayne.hu@springer.com (China).

All books in the series now indexed by Scopus and EI Compendex database!

More information about this series at http://www.springer.com/series/15087

Arnab Jana · Pradipta Banerji
Editors

Urban Science and Engineering

Proceedings of ICUSE 2020

Springer

Editors
Arnab Jana
Centre for Urban Science and Engineering
Indian Institute of Technology Bombay
Mumbai, Maharashtra, India

Pradipta Banerji
Department of Civil Engineering
Indian Institute of Technology Bombay
Mumbai, Maharashtra, India

ISSN 2366-2557 ISSN 2366-2565 (electronic)
Lecture Notes in Civil Engineering
ISBN 978-981-33-4113-5 ISBN 978-981-33-4114-2 (eBook)
https://doi.org/10.1007/978-981-33-4114-2

This Springer imprint is published by the registered company Springer Nature Singapore Pte Ltd.
The registered company address is: 152 Beach Road, #21-01/04 Gateway East, Singapore 189721, Singapore

Organization

Program Chairs

Arnab Jana, Indian Institute of Technology Bombay, India
Pradipta Banerji, Indian Institute of Technology Bombay, India

Program Committee

Pradip Kalbar, IIT Bombay, India

Preface

This book presents selected papers from the 1st International Conference on Urban Science and Engineering (ICUSE 2020), which cover three themes: A. Sustainable infrastructure, mobility and planning, B. Urban water and sanitation, and C. Smart buildings engineering. Selected papers represent diverse topics ranging from urban planning, transportation planning and engineering, building sciences, building management, urban economics, urban water management, sanitation, and solid waste management. The book is interdisciplinary in scope and addresses a host of different areas relevant to urban research, making it of interest to scientists, policymakers, students, economists, environmental activists, and social scientists alike.

Mumbai, India
July 2020

Arnab Jana
Pradipta Banerji

Contents

About the Editors

Dr. Arnab Jana is Assistant Professor at the Centre for Urban Science and Engineering and Associate Faculty at Centre for Policy Studies, Indian Institute of Technology Bombay, India. He works in the area of urban infrastructure policy and planning, primarily focusing on public health policy, application of ICT in urban and regional planning, sustainability & environmental issues. He has been Visiting Faculty at Hiroshima University, Japan, in 2016. Dr. Jana holds Ph.D. in Urban Engineering from the University of Tokyo, Japan. He was awarded the prestigious Monbukagakusho Scholarship by the Ministry of Education, Culture, Sports, Science, and Technology (MEXT), Government of Japan. He is also an alumnus of Asian Program of Incubation of Environmental Leadership (APIEL), University of Tokyo. Dr. Jana holds a Master's degree in City Planning from IIT Kharagpur. He has guided several doctoral and postgraduate students and has authored numerous peer-reviewed articles in peer-reviewed journals and conferences and books.

Dr. Pradipta Banerji is Professor of Civil Engineering and current Head, Centre for Urban Science and Engineering at Indian Institute of Technology Bombay (IIT), India. He is also Former Director of IIT Roorkee. He has more than 35 years of academic and research experience. He has a B.Tech. degree from IIT Delhi, and M.S. and Ph.D. degrees from UC, Berkeley. He is a member, Fellow, and Chair of several committees and professional societies across the world. He has developed many popular courses for working professionals in the field of structural engineering and urban sustainability. He has supervised theses of around 70 Master's and Doctoral students at IIT Bombay, and has more than 100 research publications, books, and reports. He has delivered some of the most prestigious national lectures of professional societies in India, and several Keynote lectures in international conferences and symposia. He is the India representative on a few international academic and research networks.

Assessment of Social Cost for Limited Stop Bus Transit Operations in Durg–Raipur

Sachin Kumar Sahu and Vivek Agnihotri

Abstract The existing intercity public bus transit system can be improved by accelerated transit operation and reduce the travel time. The introduction of accelerated operations for bus transit will not only reduce the travel time of the passengers but will also help in increasing the comfort factor of passengers, this will lead to the increase in ridership of the bus transit and will make the bus services more efficient The research aims to identify suitable accelerated operations for the public transit corridor and evaluate these operations based on travel time and travel cost. A methodology is demonstrated for determining the suitable accelerated operations, i.e., skip-stop, zonal, and express operations and to compare the social cost involved with the operations to the existing normal operations. The study assumes the travel demand of the corridor to be constant. A social cost function is created to calculate the social cost of the operations. Based on the analysis, possible scenarios are assumed, and new travel times are discussed along with savings in travel time.

Keywords Bus transit · Limited-stop operations · Social cost · Travel time

1 Introduction

Transportation in urban influences the form of cities and their livability. It also shapes the economic, social, and environmental characteristics of any city. Trends indicate that the use of the public transportation system decreases considerably with an increase in vehicle ownership. In developing countries, the role of public transit is greater than that of the second part of the world due to its catchment of service to a greater number of people. Public transportation also offers better capacities that a highway cannot provide in rapidly growing cities. In Tier-II Indian cities, a big share of the working population travels intercity for work, education, business health purpose during their day-to-day lives. In developing Indian states like Chhattisgarh, public transportation systems are not that efficient as in bigger cities of other Indian

S. K. Sahu (✉) · V. Agnihotri
Department of Architecture, National Institute of Technology Raipur, Raipur, CG, India
e-mail: ar.sachinsahu@hotmail.com

A. Jana and P. Banerji (eds.), *Urban Science and Engineering*, Lecture Notes in Civil Engineering 121, https://doi.org/10.1007/978-981-33-4114-2_1

states. This increases private vehicle ownership. The state capital Raipur had an increase of 12% in private vehicle ownership in the past decade. Regional mobility plan for greater Raipur indicates that an average of 30 lakh passengers per month travel between Raipur and Durg. Durg and Raipur are two major urban growth centers of Chhattisgarh state. The distance between these two cities is only 40 km, which is one of the major reasons for such a heavy commute per month. For intercity public transit, bus and rail are only two available options. The frequency in the case of rail transit is lower as compared with bus transit. Whereas, long travel time is the major problem for bus commuters between these two cities. The long travel time is a result of traffic congestion [1]. One of the major disadvantages of bus services is long in-vehicle travel time resulting from frequent stops. It makes the choice unattractive for many commuters [2]. Limited-stop bus services on high-demand corridors have the advantage of shorter in-vehicle times for passengers and shorter running times that enable bus operators to serve more demand with the same number of buses and reduced operating costs [3]. Such services are successfully being operated in cities like Chicago, Montreal, New York City, and Santiago [6]. This study focuses on the improvement of bus transit, by similarly introducing accelerated operations [5]. The primary goal of the research is to minimize the total travel time of the passengers traveling in buses between Durg and Raipur.

2 Methodology

Durg and Raipur have connectivity through Durg–Raipur expressway. The public bus transportation system is privatized in Chhattisgarh. Different travel agencies are authorized by the Department of Transportation, Government of Chhattisgarh to take bus transit operations. The case of Durg–Raipur is no different.

The regional mobility plan of the Greater Raipur region indicates that there are 30 lakh commuters, who commute between Durg and Raipur monthly. Out of these, 46% of the commuters use the intercity bus service and 32% of the passengers use suburban railways. Pause times at stop points hold a significant component of travel time on transit lines. For a trip length of 20–30 km, the services sometimes become too slow due to frequent stops. The three basic accelerated operations are skip-stop, zonal, and express/local operations. If stop numbers and locations are fixed, the only way to increase speed is to introduce accelerated operations. In such operations, some of the transit units do not stop at certain stations. Accelerated operations are primarily applicable to long lines with many stations. It also requires frequent services. Thus, they apply to many rapid transit lines, particularly during peak hours. In skip-stop operations, skip-stop pairs are selected in transit lines, and alternate stops are skipped by consecutive buses. This results in reduced stop times. Therefore, the total travel time reduces. In zonal operations, a line is divided into two or more sections, referred to as zones, and each zone is served by a different set of buses. Then the first bus, leaving the city center, goes nonstop to the farthest zone and stops at all its stations. The next bus goes nonstop to the next-closer zone and serves its stations. Transit lines in large cities, mainly serving commuters from large suburban areas to city centers, such as regional rail lines in New York, London, often have a zonal operation. In local-express service, the only way to provide regular services among all stations, as well as higher-speed services stopping only at major stations, is to operate both local and express services [7]. Express/local operations are implemented in transit networks where boarding/alighting is higher between two and three major stations and the selection of station is done according to the boarding alighting data of the stations.

The objective of this study is to reduce travel time by introducing accelerated operations. It is a difficult process to identify which operation will be more beneficial to the commuters. Here comes the social cost or generalized cost function of the trip. By calculating the cost of the trip, we can decide upon which operation will be most beneficial for the passengers. A generalized cost function was created for calculating the cost of various limited-stop operations and to compare it with the standard operations. Origin Destination matrix was generated by using boarding and alighting counts for both peak hours and nonpeak hours for each trip. Based on secondary data, travel time matrices, normal fare matrices, express fare matrices, the value of time matrices were derived. Various optional models were created by using the boarding and alighting matrix. It was followed by the selection of stations for transit operations. As shown below, a social cost function was derived to compare the accelerated operations by introducing different stopping strategies. The social cost of each operation for both peak and nonpeak hours was calculated using the generalized equation and results were evaluated.

$$Social\,Cost = \sum\nolimits_{i}^{n}\left\{\left(\left(IVTT_{ij} + WT_{i}\right) \times C_{ij}\right) + Fare_{ij}\right) \times Td_{ij}\right\}$$

where

IVTT = in-vehicle travel Time between two stops (in Minutes)

WT = Waiting Time at stop j (in Minutes)

Td_{ij} = per trip Travel Demand between stop i and j.

Fare = Express or Normal Fare as Applicable

C_{ij} = Value of Time per km.

With the help of the above formula, the value of time was calculated by regression of the primary data for travel time savings. The social cost was calculated using the cost function for all the operations, i.e., skip-stop, zonal and express services, and for both Raipur–Durg, and Durg–Raipur. The social cost of accelerated operations was compared with normal operations and the result was analyzed to check the feasibility of that operation. The proposals and recommendations were made based on the results obtained through the above analysis.

3 Data Collection

Data were collected through primary surveys and secondary sources such as local authorities, mobility plans, etc. The sample size of the primary survey was 50. It was collected during the commute by intercity buses from Durg to Raipur. The secondary data consisted of bus fleet sizes, operation frequencies, fare structures, and average operational costs. The details of per capita trip rates, bus terminals infrastructure were gathered from the state urban development agencies and DPRs of buses' procurement. Public opinion survey, demand forecast, trip purpose arrival, and dispersal modes were extracted from the Comprehensive Mobility Plan of Greater Raipur Region.

The operation time of the bus services was found to be from 5:30 AM to 9:00 AM, the peak hours of the service were found to be between 8:00–11:00 AM and 4:30–6.30 PM. For peak-hours, the operational frequency was 5 min while frequency for nonpeak hours was 10 min. The buses had a seating capacity of 28 passengers and a standing capacity of 14 passengers. There were 13 stops in between Raipur and Durg while 11 between Durg and Raipur and it took an average of 110 min and 90 min, respectively, for the journey.

4 Observation and Results

The survey results along with secondary data were analyzed. A boarding alighting survey was conducted en-route for both peak and nonpeak hours. The data indicated that the maximum boarding would occur at Raipur bus station for both peak and nonpeak hours followed by Tatibandh, whereas maximum alighting would occur at Durg bus stop, followed by Powerhouse. The maximum onboard passengers were found between Tatibandh and Khursipar. For peak hour, the maximum on vehicle load was observed between Tatibandh and Bhilai-3.

In the case of Durg–Raipur, for both peak and nonpeak hours, maximum boarding would occur at Durg followed by Powerhouse. Whereas, maximum alighting would occur at Raipur followed by the powerhouse. In this case, the onboard passengers were higher between Supela and Kumhari.

Per day travel demand to and from Durg to Raipur was collected from the State Urban Development Authority. It was observed from the OD Matrix and boarding-alighting count that the major commuting station pairs were Durg–Raipur, Durg–Powerhouse, Durg Tatibandh, Supela–Raipur, Powerhouse–Durg, Raipur–Durg, Raipur–Powerhouse, Raipur–Nehru Nagar, and Tatibandh to Durg.

5 Analysis

The major problem identified during the primary survey of the passenger was a long travel time for Durg–Raipur. Most passengers traveling in between these bus stops, and the passengers commuting to intermediate sections of the route, faced congestion and overcrowding. A different set of accelerated operations was planned based on the boarding alighting count, O-D matrix, and passenger survey. The stopping strategy formulated using accelerated operation is shown in Fig. 1. This stop-strategy was considered for the calculation of generalized cost between Durg and Raipur.

The parameters to evaluate the operations and to find out the best options for intercity travel were travel time savings and economic feasibility. The evaluation was done after calculating the social cost by analyzing per trip travel demand, in-vehicle travel time, fare, and value of travel time. All the matrices were used for calculations and the results were analyzed. The generalized cost calculated is discussed in Table 1. The calculation shows that introducing skip-stop operation is beneficial as the reduction in total social cost and was found to be greater than 5% when the frequency of 5, 10, and 15 min was applied. Forzonal operation, the reduction was found to be greater than 23% as shown in Table 2 (Fig. 2).

Raipur–Durg (Nonpeak Hours)

Fig. 1 Accelerated operations for analysis

Table 1 Social cost of accelerated operations, Raipur–Durg (Nonpeak hours)

Frequency	All stop	Zonal 1	Zonal 2	Express1	Express2	Skip Stop 1	Skip stop 2	Zonal (Combine)	Skip Stop (Combine)	Reduction (Skip- Stop)	Reduction (Zonal)
5 Min	76,663	39,049	19,488	11,639	12,125	29,656	42,846	58,537	72,502	5.43	23.64
10 Min	82,184	41,515	20,842	12,377	12,860	31,697	45,671	62,357	77,368	5.86	24.13
15 Min	87,705	43,981	22,196	13,115	13,595	33,738	48,496	66,177	82,234	6.24	24.55

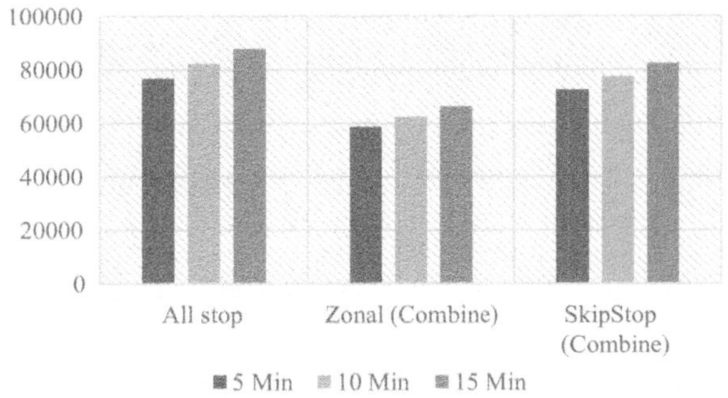

Fig. 2 Social Cost of Accelerated Operation Raipur–Durg (Nonpeak Hours)

Raipur–Durg (Peak Hours)

For peak hours, the reduction was more than 11.36% and for zonal, it was even high, i.e., 33%. The reduction in cost was greater than peak hours because travel demand was higher during peak hours (Table 3 and Fig. 3).

Durg–Raipur (Peak Hours)

In the case of Durg–Raipur for skip-stop operation, the cost reduction varied from 3 to 8%. While in zonal operation, the cost was increased by 33%. It was concluded that zonal operation would not be advisable for Durg–Raipur (Fig. 4 and Table 4).

Durg–Raipur (Nonpeak Hours)

For Durg–Raipur, the cost for accelerated operation for both zonal and skip-stop operation was higher in comparison to all stop operation (Fig. 5).

6 Discussion

The accelerated operations were evaluated and analyzed, and it was found that for skip-stop operations, travel demand should be higher, for nonpeak hours implementing skip-stop or express stop operation would lead to an increase in waiting time.

Table 2 Social cost of accelerated operations, Raipur–Durg (Peak hours)

Frequency	All stop	Zonal 1	Zonal 2	Express1	Express2	Skip- Stop 1	Skip stop 2	Zonal (Combine)	Skip Stop (Combine)	Reduction (Skip- Stop)	Reduction (Zonal)
5 Min	65,236	33,126	12,892	7379	7578	24,238	33,379	46,018	57,617	11.68	29.46
10 Min	69,851	35,188	13,795	7849	8037	25,862	35,583	48,983	61,445	12.03	29.88
15 Min	74,466	35,188	14,698	8319	8496	27,486	37,787	49,886	65,273	12.35	33.01

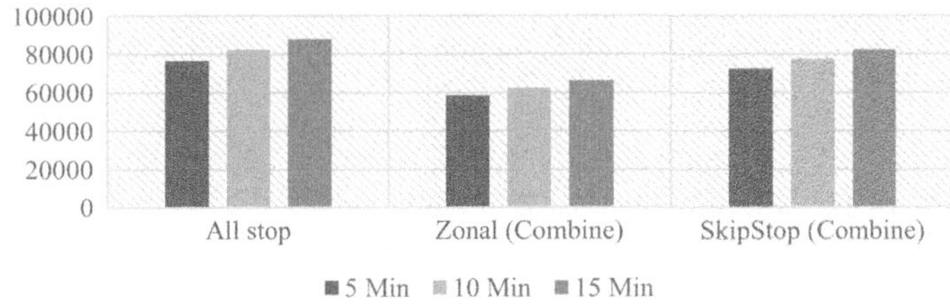

Fig. 3 Social Cost of Accelerated Operation Raipur- Durg (Peak Hours)

In the case of Raipur–Durg, both zonal and skip-stop operations, the cost was reduced for both peak and nonpeak hours. In Durg–Raipur, for nonpeak hours, both zonal and skip-stop operations had a higher cost than all stop operations. While for peak hours, skip-stop operations had a lower cost than all stop operations. Different scenarios were evaluated, and skip-stop operation was selected over the zonal operations and expresses operation. The reason for selecting the skip-stop operation is discussed herewith.

The travel pattern in the corridor is many too many. It is not advisable for zonal operation in case of many to many travel patterns. Although the difference between the generalized cost for zonal operation and all stop operation is found to be very high, the capacity cannot be utilized by the bus and it would reduce the revenue. In nonpeak hours, the Skip stop operation is proposed for the peak hours only.

Considering the O-D matrix of the transit corridor, and waiting time for the bus, following two skip-stop strategies both for Raipur–Durg, and Durg–Raipur should be proposed for the transit corridor, this would help in reducing the travel time and would be optimized for a small social cost.

Durg–Raipur (Skip-stop Operation) (Fig. 6).

The travel time between Durg–Raipur was reduced to 85 min from 97 min similarly the other details are as follows. The travel time between the stops in the above scenario is shown in Table 5.

Table 3 Social cost of accelerated operations, Durg–Raipur (Peak hours)

Frequency	All stop	Zonal 1	Zonal 2	Express1	Express2	Skip- Stop 1	Skip stop 2	Zonal (Combine)	Skip Stop (Combine)	Reduction (Skip- Stop)	Reduction (Zonal)
5 Min	53,265	30,474	40,850	16,923	15,748	26,139	34,446	71,324	48,974	8.06	−33.9
10 Min	56,409	32,471	43,873	18,027	16,750	28,051	37,014	76,344	53,988	4.29	−35.34
15 Min	61,236	34,468	46,896	19,131	17,752	29,963	39,582	81,364	59,002	3.65	−32.87

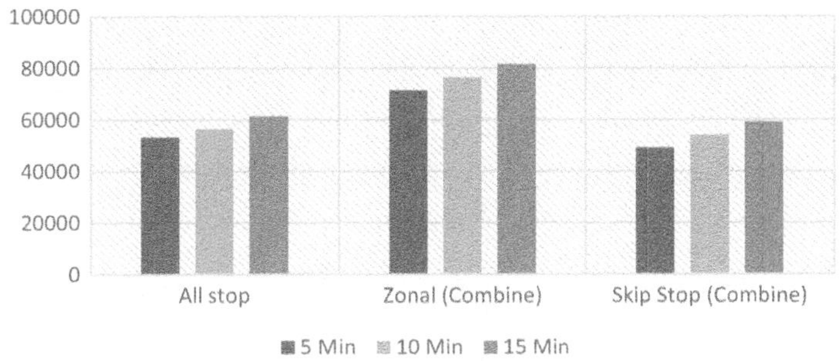

Fig. 4 Social cost of Durg-Raipur (Peak hours)

The travel time savings achieved in this case is shown in Table 6.

The study presented a method to calculate and compare the social cost of the accelerated transit operations with normal all stop operations. It should be noted that the running time of a bus was assumed to be a constant which, as a matter of fact, maybe random or fuzzy under the influence of traffic signals, road conditions, and driving psychology. Therefore, it is necessary to treat the transit scheduling problem under a random or fuzzy environment [4]. Following significant conclusions regarding the study can be drawn:

- The savings obtained by offering limited-stop services are significant, amounting in some cases to more than 5% over a normal service visiting every stop on a line at its optimal frequency.
- The demand variability among different O–D pairs, should be an important factor in the design of limited-stop services: among the scenarios studied, the greater the demand variability the higher the benefits associated with the implementation of limited-stop services.

Table 4 Social cost of accelerated operations, Durg (Nonpeak hours)

Frequency	All stop	Zonal 1	Zonal 2	Express1	Express2	Skip- Stop 1	Skip stop 2	Zonal (Combine)	Skip Stop (Combine)	Reduction (Skip- Stop)	Reduction (Zonal)
5 Min	36,893	22,483	15,578	12,270	6929	18,998	25,568	38,061	44,566	−20.8	−3.17
10 Min	40,034	23,968	16,747	13,082	12,002	20,392	27,492	40,715	47,884	−19.61	−1.7
15 Min	43,175	25,453	17,916	13,894	17,075	21,786	29,416	43,369	51,202	−18.59	−0.45

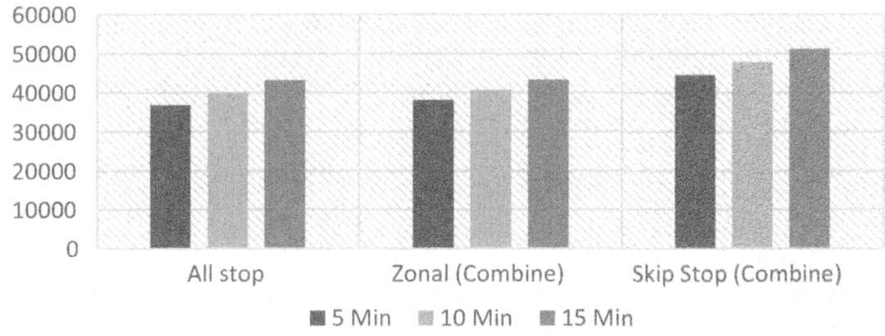

Fig. 5 Figure 24 Social cost of accelerated operation durg Raipur (Non-Peak Hours)

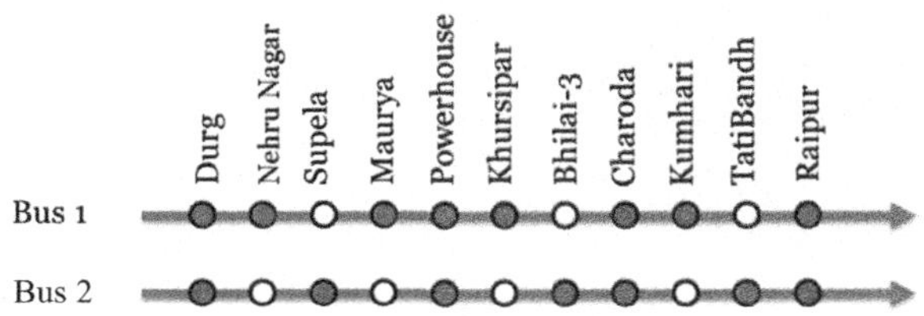

Fig. 6 Proposed skip stop operation Durg–Raipur

Table 5 Travel time after implementing accelerated operation and reduction in travel time. (Durg–Raipur)

	Durg	Nehru Nagar	Supela	Maurya Talkies	Power House	Khursipar	Bhilai 3	Charoda	Kumhari	Tatibandh	Raipur
Durg		9	13	16	19	26	30	37	52	61	85
Nehru Nagar				7	12	17		29	45		78
Supela					6	13	17	24		48	72
Maurya Talkies					5	9		22	38		69
Power House						9	15	22	35	41	63
Khursipar								13	17		58
Bhilai 3								6		30	55
Charoda									0	0	0
Kumhari											33
Tatibandh											22
Raipur											

Table 6 Travel time after implementing accelerated operation and reduction in travel time (Durg–Raipur)

	Durg	Nehru Nagar	Supela	Maurya Talkies	Power House	Khursipar	Bhilai 3	Charoda	Kumhari	Tatibandh	Raipur
Durg		3	6	6	7	5	8	8	7	9	9
Nehru Nagar				3	2	2		2	1		3
Supela					1	2	2	2		0	0
Maurya Talkies					0	3		0	3		1
Power House						0	1	1	2	2	4
Khursipar								1	4		5
Bhilai 3								0		1	1
Charoda									1	3	2
Kumhari											2
Tatibandh											2
Raipur											

References

1. Leiva, C., Muñoz, J. C., Giesen, R., & Larrain, H. (2010). Design of limited-stop services for an urban bus corridor with capacity constraints. *Transportation Research Part B: Methodological, 44*(10), 1186–1201.
2. Vuchic, V. R. (2017). Urban transit: Operations, planning, and economics. Wiley.
3. Chiraphadhanakul, V., & Barnhart, C. (2013). Incremental bus service design: Combining limited-stop and local bus services. *Public Transport, 5*(1–2), 53–78.
4. Niu, H. (2011). Determination of the skip-stop scheduling for a congested transit line by bilevel genetic algorithm. *International Journal of Computational Intelligence Systems, 4*(6), 1158–1167.
5. Cao, Z., Yuan, Z., & Zhang, S. (2016). Performance analysis of stop-skipping scheduling plans in rail transit under time-dependent demand. *International Journal of Environmental Research and Public Health, 13*(7), 707.
6. Wang, D. Z., Nayan, A., & Szeto, W. Y. (2018). Optimal bus service design with limited stop services in a travel corridor. *Transportation Research Part E: Logistics and Transportation Review, 111,* 70–86.
7. Guzman, W. M., Peszynski, K. J., & Young, L. W. (2014). Zonal operations: A method to rationalise operations. Computers in Railways XIV: Railway Engineering Design and Optimization, WIT Press, United Kingdom, pp. 353–365.

Carrying Capacity and Performance Assessment as Quantitative Tools to Assess the Sustainability of Coastal Tourism Sector

Riya Sethia and Bimal Puthuvayi

Abstract This paper attempts to perform a sustainability assessment of coastal tourism in a region through quantitative measurement tools. Coastal tourism is one of the dominant economic activities in coastal areas, often susceptible to overexploitation through unregulated anthropogenic activities. Hence, it results in an immediate need to understand the impacts and devise strategies to ensure coastal sustainability. An objective understanding of the impacts of such anthropogenic activities demands the formulation of quantitative tools to measure the benefits and the impacts. Quantitative assessment of sustainability is a complex process as it involves tangible and intangible factors. The paper proposes a quantitative tool, performance assessment, which compares economic benefits incurred from coastal tourism with environmental and social impacts. It is derived from the Green Productivity Index. It attempts to measure the implications as monetary expenditure needed to mitigate the impacts. Carrying capacity assessment is another tool adopted to determine the thresholds for various thematic areas associated with coastal tourism. It also attempts to measure the Effective Real Carrying Capacity, which implies the volume of tourism an area can sustain with available physical, social and economic capacities. Action-oriented understanding is developed through contextually identified impacts and indicators. It helps to calculate the existing economic benefits and cost of impacts per tourist and hence determines the area's sustainable loading capacity based on effective real-carrying capacity. These results can be further used to optimize the limiting factors and enhance tourism carrying capacity sustainably through planning interventions.

Keywords Anthropogenic impacts · GPI · Monetary valuation · ERCC

R. Sethia (✉) · B. Puthuvayi
National Institute of Technology Calicut, Calicut, India
e-mail: riyasethia135@gmail.com

A. Jana and P. Banerji (eds.), *Urban Science and Engineering*, Lecture Notes in Civil Engineering 121, https://doi.org/10.1007/978-981-33-4114-2_2

1 Introduction

1.1 Concept of the Project

Coastal tourism is one of the dominant contributors to the coastal economy of a region. It has been and continues to be the primary economic driver for many countries in the world. These incessant economic benefits lead to imbalance due to unregulated anthropogenic activities. Overexploitation of resources, loss of biodiversity, pollution and sociocultural impacts are the major anthropogenic activities defining this imbalance between coastal economic growth and sustainable coastal planning [1]. The impacts of coastal tourism on sustainability form a vicious cycle wherein impacts on the environment result in declining attractiveness and hence affecting economic benefits. The impacts can be classified into three broad dimensions, i.e., environmental, social and economic [1]. Thus this project attempts to identify these impacts and formulate tools to quantify the extent of impacts of the coastal tourism sector on the sustainability of the coast. Further, the assessment results are used to compare sustainability among selected destinations for the project and propose the application of the tool at various replicable locations.

1.2 Identification and Delineation of the Study Area

Identification of study area. Predominantly, the decision to choose project location in Kerala, which is better known by its tagline God's own country, was made in line with its established image as a significant coastal tourism destination in the country. To choose a study area within the state, two major factors considered were—economic benefits and high unsustainable and over-exploited tourism activities. Ernakulum district generated INR 9541.64 cr. and INR 10533.78 cr. from the tourism sector in 2017 and 2018, respectively, which corresponds to 22% and 29.05% of the total earnings from the tourism sector in the state [2, 3].

Being the most preferred destination by domestic and foreign tourists, coastal destinations in Ernakulum is highly overexploited and degraded due to the unregulated functioning of coastal tourism activities. Hence, Ernakulum district is considered as a suitable study area for the project, and significant beaches in the district are selected to conduct a detailed study. The rationale for choosing beaches within the district is based on factors like popularity of beaches among tourists (domestic and foreign), annual footfall trend (as recorded from Kerala Tourism Statistics) and status of most visited or exploited destinations (referred from articles/reviews/news). Two beaches selected for detailed study under the scope of the project are Fort Kochi and Cherai Beach to Munnambam Beach stretch, as shown in Figs. 1 and 2 and Table 1.

Delineation of study area boundary. The boundary is based on existing administrative boundaries, proximity from beach and accommodations within

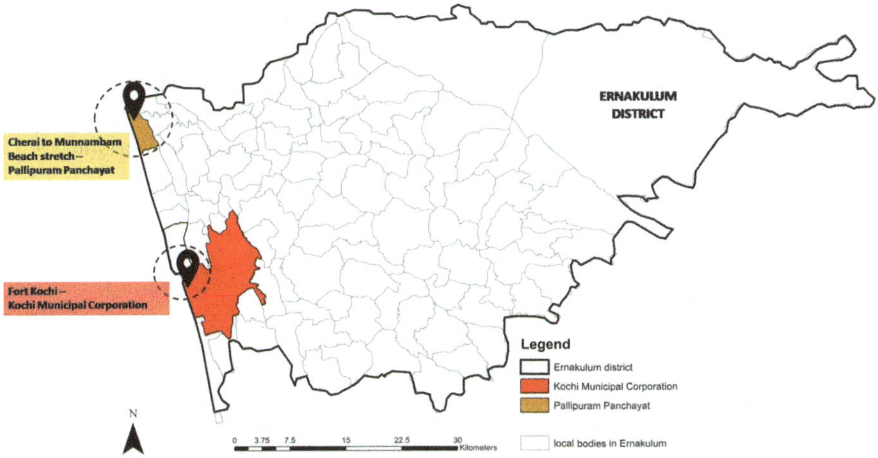

Fig. 1 Map showing coastal areas chosen for study in Ernakulum district [4, 5]

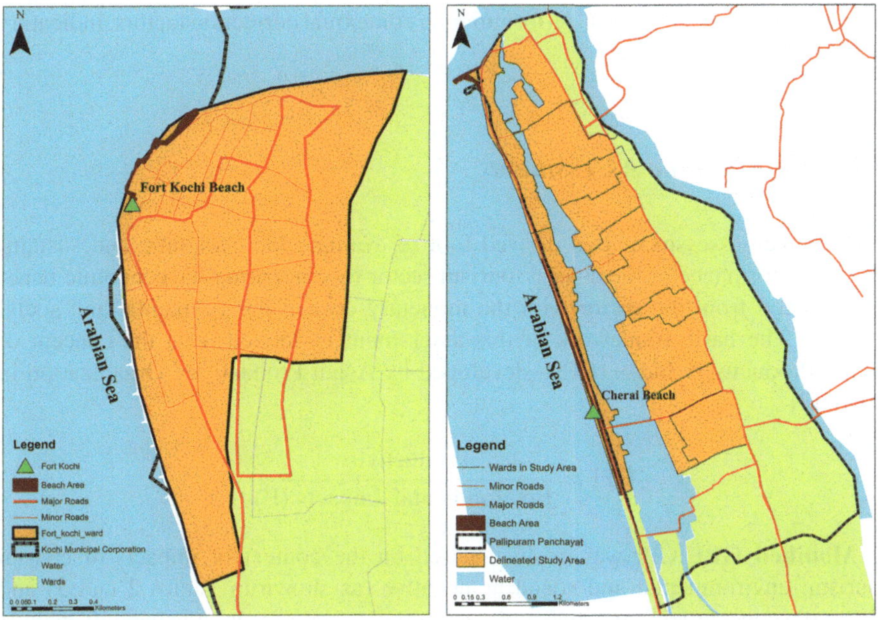

Fig. 2 Delineated area in Fort Kochi and Cherai to Munnambam stretch respectively [4, 5]

Table 1 Study area necessary details [5, 6]

Details	Fort Kochi	Cherai to Munnambam beach
Population and HH	10,279 and 2578	26,364 and 6591
Area	1.66 km^2	5.19 km^2

walking/cycling distance. The study area at Fort Kochi is defined as an administrative ward boundary, i.e., ward 1 of Kochi Municipal Corporation [4]. The boundary at Cherai beach is based on natural features like water body, existing main road, existing wards and extent of hotels/resorts within walking distance from the beach. The stretch chosen extends from Cherai to Munnambam beach, comprising of 12 wards in Pallipuram panchayat [5].

2 Methodology

Various studies on methods to measure the impacts of unregulated tourism on coastal sustainability are available. Qualitative methods adopt techniques like beach appraisal through perception surveys, local community response analysis, etc. Quantitative techniques in most studies mainly incorporate carrying capacity analysis (CCA) conducted in multiple ways. This paper attempts to propose a new method (comparing benefits with social and environmental impacts) along with modification in conventional method of CCA (formulating contextual correction factors, indicators and thresholds).

2.1 Performance Assessment

Performance assessment is a derived tool to evaluate the economic and sustainability performance of the coastal tourism sector by comparing the economic benefits incurred from the sector with the monetary cost of environmental and social impacts. The basic framework of this assessment is derived from the concept of Green Productivity Index (GPI—developed by Asian Productivity Organization in 1996) [7].

$$\text{GPI} = \frac{\text{Productivity}}{\text{Environmental Impacts (E)}} \tag{1}$$

Modified GPI. GPI was evolved based on the context of impacts of coastal tourism; environmental and social as negative (as shown in Tables 2 and 3) and economic considered as positive under the project scope. Economic benefits include revenue from local businesses and employment generated; prominent positive social impact. The idea is to compare positive versus negative impacts in the form of a ratio.

$$\text{GPI} = \frac{\text{Productivity of a sector (x)}}{\text{Environmental and Social Impacts of the sector (y1 + y2)}} \tag{2}$$

Table 2 Identified environmental impacts and indicators for monetary valuation of impacts

Environmental impacts	Indicator	Monetary measurement
Reduction in vegetation/forest cover/wetlands	Amount of loss (ea') = Initial area – Existing area	Ecosystem valuation based on cost/unit of area for ea'
Excessive water use causing shortage	Shortage of water (eb') = Amount of water in shortage	Cost of production for eb' quantity of water
Improper solid waste management	Amount of solid waste not collected and treated (ec')	Cost of collection/treatment of ec' quantity of waste
Untreated night soil	Amount of untreated liquid waste (ed')	Cost of treatment of ed' quantity of wastewater
Source [9]	*Source* [8]	
Monetary valuation (y1)		**Environmental cost**

Table 3 Identified social impacts and indicators for monetary valuation of impacts

Social impacts	Indicator	Monetary measurement
Economic inequality—local resentment	Limit to local employment (sb') = Total jobs generated – Local jobs provided	Potential total pay worth of sb'
Shortage of medical facilities	Number of required medical facilities (sc')	Cost of provision for sc'
Increase in major crimes	% of tourists unwilling to visit	Reduction in profit due to decreasing tourists
Low satisfaction and complaints by Tourists	% of tourists unwilling to visit	Reduction in profit due to decreasing tourists
Culture erosion—local resentment	% of locals feeling tourism influencing local culture	
Source [9]	*Source* [8]	
Monetary valuation (y2)		**Social cost**

The productivity of a sector is calculated as revenue generated from coastal tourism (x). Environmental and social impacts (y1 + y2) are calculated as the mitigation cost of these negative impacts. The concept is referred from PwC's TIMM methodology [8]. If GPI [x/(y1 + y2)] > 1, economic benefits are higher than the cost of impacts. If GPI [x/(y1 + y2)] < 1, cost of impacts are higher than economic benefits.

2.2 Carrying Capacity Assessment

Carrying capacity in its basic form refers to the maximum number of people who can use a place without an unacceptable alteration in the physical environment and an unacceptable decline in the quality of recreational space [10]. It is a well-established tool to assess the sustainability of tourism in an area through various methods and techniques. The methods adopted in most studies include Boullon Formula [11], Grid-based assessment using the multicriteria scheme, Indicator-wise assessment through defined thresholds [14] and computing PCC, RCC and ERCC method [12, 13]. Grid-based assessment works efficiently for continuous longer stretches and hence is not suitable for the project as in this case, two confined study areas are selected. Two methods selected with modification (determining contextual thresholds, indicators and correction factors) w.r.t. the study areas are: Indicator-wise assessment method and Computing Effective Real-Carrying Capacity [13] (ERCC) method. Method of calculating ERCC is highly practical and accurate as it considers natural, infrastructural and managerial limitations in the study area unlike Boullon formula, which calculates capacity solely based on the beach area. The second method, i.e., indicator-wise assessment, provides a more profound and detailed insight about specific factors in terms of high, medium or low carrying capacity. This enables mainstreaming of interventions contextual to limitations and impacts for any destination.

Determination of indicators and respective thresholds. The relevant indicators under each component (physical, social and economic) for various types of tourism impacts identified and measured based on the DSPIR process are referred from the EEA report [9]. Analysis of each indicator is done by comparing with the defined threshold values for each through standards/benchmarks, physical limits, etc. referred from case studies [14] as shown in Table 5.

Computing PCC, RCC and ERCC. Boullon formula (PCC) considered just beach area and space required per tourists to calculate carrying capacity. In contrast, RCC and ERCC incorporate correction factors or limiting factors, i.e., factors that restrict the tourist activities in an area [12, 13].

$$\text{Physical Carrying Capacity (PCC)} = \frac{\text{A}}{\text{Au}} * \text{Rf} \tag{3}$$

where Rf = rotation factor = daily open period/average time of visit.
A = area of beach, Au = area required per user

$$\text{Real Carrying Capacity (RCC)} = \text{PCC} * \text{Cf1} * \text{Cf2} * \ldots \text{Cfn} \tag{4}$$

$$\text{Effective Real Carrying Capacity (ERCC)} = \text{RCC} * \text{Cf1} * \text{Cf2} * \ldots \text{Cfn} \tag{5}$$

$$Cf = 1 - \frac{Lmx}{Tmx} \tag{6}$$

where Cf = correction factor, Lmx = limiting magnitude of variable x and Tmx = total magnitude of variable x.

Correction factors, indicators and their thresholds are formulated based on case studies, study area characteristics and coastal tourism, as shown in Tables 4 and 5.

Table 4 Contextual correction factors to calculate ERCC

Thematic areas	Category	Correction factors
Weather	Environment	$\mathbf{Cf_w}$ = [1 − (number of days of excessive heat/Total number of days in a year)] $\mathbf{Cf_{w'}}$ = [1 − (number of days of extreme rain/Total number of days in a year)] $\mathbf{Cf_{w''}}$ = [1 − (number of days of excessive wind/Total number of days in a year)]
Beach quality	Environment	$\mathbf{Cf_{b1}}$ = [1 − (eroded beach area/initial beach area)]
Tourism infrastructure (bed spaces)	Social	$\mathbf{Cf_t}$ = [1 − (number of unoccupied beds/Total number of available beds)]
Water	Environment	$\mathbf{Cf_{wa}}$ = [1 − (number of accommodation and vendors facing shortage/Total number of accommodation and vendors)] $\mathbf{Cf_{wa'}}$ = [1 − (number of households facing shortage/Total number of households)]
Waste	Environment	$\mathbf{Cf_{sw}}$ = [1 − (amount of solid waste not collected or treated/total amount of solid waste generated) $\mathbf{Cf_{lw}}$ = [1 − (number of households and accommodation without septic tank/total number of households and accommodation)
Overcrowding	Social	$\mathbf{Cf_o}$ = [1 − (per tourist additional space required as per existing footfall/Prescribed area per tourist as per WTO standards)]
Safety	Social	$\mathbf{Cf_s}$ = [1 – (number of crimes against tourists/total number of crimes recorded)]
Satisfaction level	Social	Tourists: $\mathbf{Cf_{st}}$ = [1 − (number of tourists not satisfied/total number of tourists) Locals: $\mathbf{Cf_{sl}}$ = [1 − (number of households not satisfied from tourism development in the area/total number of households in the area)

Source Cifuentes's methodology [12, 13]

Table 5 Effective carrying capacity indicators and defined thresholds for Indicator-wise Analysis

Thematic area	Indicators	Thresholds
Overcrowding	Overcrowding—area of accessible coast occupied by a tourist	[11] $10m^2$ (WTO), If > 10 (High cc) < 10 (Low cc) = 10 (Medium cc)
Tourism infrastructure	Shortage of beds in peak season	If > 0 (Low cc) = 0 (High cc)
	Shortage of beds in offseason	
Water	% of HH facing water shortage	If > 0% (Low cc) = 0% (High cc)
	% of tourist facilities facing water shortage	
Solid and liquid waste	Volume of solid waste collected or treated/total solid waste generated	If = 1 (High cc) < 1 (Low cc)
	Number of households and accommodation with septic tank/total number of households and accommodation	
Land	Number of beds per hectare	[15] If <= 50 beds/hectare (High cc) > 50 beds/hectare (Low cc)
	Ratio of—Tourist beds/permanent population	[15] If < 0.5 (High cc) > 0.5 (Low cc) = 0.5 (Medium cc)
Employment	Ratio of—Migrants employed/Total employed	If > 50% (Low cc) = 50% (Medium cc) < 50% (High cc)
Health	Shortage of medical facilities = Existing medical facilities—Prescribed standards as per URDPFI	If = 0 (Medium cc) >= 1 (High cc) < 1 (Low cc)
Satisfaction level—tourists	Ratio of—Number of tourists satisfied/Number of tourists not satisfied	If = 1 (Medium cc) < 1 (Low cc) > 1 (High cc)
Satisfaction level—locals	Ratio of—Number of locals satisfied/Number of locals not satisfied	If = 1 (Medium cc) < 1 (Low cc) > 1 (High cc)
Source [9]	*Source* [9, 14]	

Table 6 Sample size determined for each survey type

Survey type	Fort Kochi			Cherai to Munnambam beach		
	Population	Condition applied	Sample Size	Population	Condition applied	Sample Size
Households	2578	95% confidence level, 10% margin of error	**93**	6591	95% confidence level, 10% margin of error	**95**
Tourists	Infinite		**97**	Infinite		**97**
Accommodation providers	215		**67**	75		**43**
Vendors	300		**73**	32		**25**

2.3 Data Requirement and Collection

The calculation of indicators involves finalizing a contextual and flexible data requirement list and data collection methodology, which is followed by an analysis of results. The data requirement involves both primary and secondary data sources based on the availability of data. The process also includes the scope of the pilot survey to check the effectiveness of data collection methods, approach to analysis, determine sample size, etc. The major secondary data sources are Kochi Municipal Corporation, Pallipuram Panchayat Office, Kerala Water Authority, Tourist Police Station and NASA Meteorological data. The three primary data collection techniques proposed are: Observation survey, Questionnaire survey (local residents, tourists, vendors, accommodation providers) and Interview of ward councilors.

Sampling technique and sample size. Sampling technique for household survey is simple random sampling, for tourists and vendors is random sampling on beach area and for accommodation providers is stratified sampling, i.e., more samples in beach proximity. Sample size calculated based on condition applied is explained in Table 6.

3 Results and Discussion

3.1 Performance Assessment

The mitigation cost of identified environmental and social impacts is estimated and compared with the economic benefits to calculate GPI, as shown in Table 7.

Inference. In Fort Kochi, economic benefits and cost of impacts are almost equal, indicating a potential threat to sustainability if planning measures aren't proposed. In Cherai beach, as profits are slightly dominant over impacts, sustainability measures are necessary to reduce any future increase in cost with increase in tourists' footfall.

Table 7 Calculation of GPI for both study areas

	Fort Kochi beach	Cherai to Munnambam Beach
Revenue/cost of environmental and social impacts	$\frac{\text{Rs.8,72,95,862}}{\text{Rs.7,36,95,880}}$	$\frac{\text{Rs.2,70,53,734}}{\text{Rs.1,46,0,002}}$
GPI Value	1.18 (0.68 future potential if crimes occur)	1.85 (0.85 future potential if crimes occur)
Net Benefit % = $x - \frac{y1+y2}{x} * 100$	15.5%	46%
Source for results derived—Analysis based on Primary and Secondary data collected		

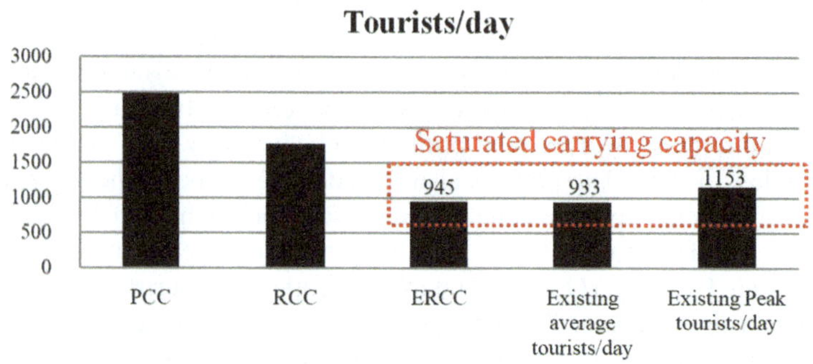

Fig. 3 Comparing PCC, RCC, ERCC and existing footfall in Fort Kochi study area

3.2 *Carrying Capacity Assessment*

Inference. In Fort Kochi, carrying capacity is saturated with overcrowding, waste management and inequality in jobs as the primary concerns. In Cherai Beach, there is the potential to increase footfall as per the existing footfall scenario. Still, water availability and waste collection/treatment have low carrying capacity, as shown in Figs. 3, 4 and Table 8. Quantitatively assessed results can be directed to policy/strategic/design proposals in the context of study area limitations using the detailed knowledge of specific indicators, its cost of impact, estimated quantity required and exact extent of augmentation to be proposed.

4 Conclusion

The paper has attempted to develop a system to compare economic versus socioenvironmental performance and determine sustainable loading capacity. The results calculated for two different locations can be utilized to frame specific proposals, optimizing limiting factors and enhancing carrying capacity. Fort Kochi has already

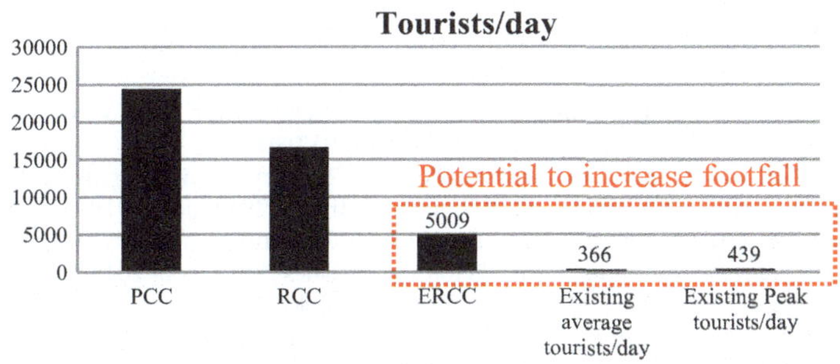

Fig. 4 Comparing PCC, RCC, ERCC and existing footfall in Cherai study area

Table 8 Effective carrying capacity—indicator wise analysis

	Cherai Beach Stretch			Fort Kochi Beach Stretch		
Thematic Area	High	Medium	Low	High	Medium	Low
Overcrowding	green				yellow	
Tourism Infrastructure	green			green		
Water			red	green		
Solid & Liquid Waste			red			red
Land availability	green			green		
Inequality in jobs	green					red
Health	green			green		
Satisfaction: Tourists	green			green		
Satisfaction: Locals	green			green		
Source – [9]	Source for results derived – Analysis based on Primary and Secondary data collected					

reached its carrying capacity with the cost of impacts equating to economic benefits. In contrast, Cherai has a high carrying capacity yet to be utilized with economic benefits almost double to the cost of impacts but holds certain limiting factors. The overall purpose is to guide tourism development, keeping sustainability intact.

Limitation. The formulated GPI considers the cost of impacts as mitigation cost; however, multiple ways of monetary valuation of impacts may be considered. Also being subjective and perception based, an attempt to convert social impacts into monetary terms can be challenged. Hence identifying other suitable ways of comparing benefits and impacts holds potential exploration. The impacts and limitations considered to calculate GPI and ERCC are based on study area context; thus may require minor revision when replicated at other destinations.

Acknowledgements We are thankful to government offices for providing data, faculties at NITC and friends to assist in the survey.

References

1. Zahedi, S. (2008). Tourism impact on coastal environment. *WIT Transactions on The Built Environment, 99,* 45–57. https://doi.org/10.2495/CENV080051.
2. D. of Tourism. (2018). *Kerala tourism statistics 2018*. Thiruvananthapuram [Online]. https://www.keralatourism.org/tourismstatistics/tourist_statistics_2018_book20191211065455.pdf.
3. D. of Tourism. (2017). *Kerala tourism statistics 2017*. Thiruvananthapuram [Online]. https://www.keralatourism.org/tourismstatistics/tourist_statistics_2017_book20181221073646.pdf.
4. K. M. Corporation. (2013). *KMC ward boundary map*. Kochi.
5. P. Panchayat. (2008). *Pallipuram panchayat map and ward details.pdf* (p. 2). Pallipuram.
6. GOI. (2011). *Primary census abstract total table for India*. Census of India. Retrieved January 25, 2020, from https://censusindia.gov.in/DigitalLibrary/Archive_home.aspx.
7. Nachimutha, M. D. G., Selladurai, V., & Santhi, P. (2006). Green productivity indexing: A practical step towards integrating environmental protection into corporate performance. *International Journal of Productivity and Performance Management, 55*(7), 594–606. https://doi.org/10.1108/17410400610702179.
8. A. Group. (2017). *Total impact measurement and management report 2016/2017*. https://doi.org/10.1017/CBO9781107415324.004.
9. E. N. EEA. (2001). *Defining, measuring and evaluating carrying capacity in European tourism destinations*. Athens
10. Coastlearn. (2018). Sustainable tourism—tools—assessing carrying capacity. Retrieved October 16, 2019, from https://www.biodiversity.ru/coastlearn/tourism-eng/tools_acc.html.
11. Rajan, B., Varghese, V. M., & Anakkathil Purushothaman, P. (2013). Beach carrying capacity analysis for sustainable tourism development in the South West Coast of India. *Environmental Research, Engineering and Management, 63*(1), 67–73. https://doi.org/10.5755/j01.erem.63.1.2648.
12. Sridhar, R., Yuvaraj, E., Sachithanandam, V., Purvaja, R. P., & Mageswaran, R. R. T. (2016). Tourism carrying capacity for beaches of South Andaman Island, India. *Intech, i*(tourism), 13. https://doi.org/10.5772/62724.
13. Dar, S. N., Shah, S. A., & Wani, M. A. (2016). Tourism carrying capacity assessment for Leh Town of Ladakh region in Jammu and Kashmir. *International Journal of Current Research, 8*(2), 26403–26410.
14. Castellani, V., & Sal, S. (2012). Carrying capacity of tourism system: Assessment of environmental and management constraints towards sustainability. In: *Visions for global tourism industry—Creating and sustaining competitive strategies*, May 2014. https://doi.org/10.5772/38750.
15. N. C. F. S. C. Management, *Carrying capacity of beaches of for providing shacks & other temporary seasonal structures in private areas Goa.*

Spatial Distribution of Socioeconomic Factors and Its Impact on Urban Land Use Dynamics: An Agent Based Modeling Approach

Vivek Kumar Singh, Vaibhav Kumar, and Arnab Jana

Abstract In spite of existing development guidelines, the current policy cannot accommodate the increasing resource demand. Therefore, leading to haphazard growth of the cities, especially in the developing nations. The unplanned advancement leading to extreme heterogeneity and complex socioeconomic profiles pose a great challenge in modeling their growth. To address these challenges, a novel Land-use Simulation and Decision-Support system (LSDS) for cities has been designed. The system comprehensively explores the interference and impact of natural environment, socioeconomic factors, spatial neighborhoods, individual choice, and national macro-policies on land use and cover change. LSDS applies the cellular automata (CA) model to examine land use land cover (LULC) changes over time, and agent-based model (ABM) to observe the role of different socioeconomic drivers in the LULC prediction process. When applied to the Dehradun city of India, quantitative analysis of spatial patterns achieved an overall built-up area accuracy of 10.62% (relative error) and Kappa accuracy value 79.73%. It was observed that most of the urban agriculture areas converted to medium compact residential areas occupied by higher and middle-income groups. An expansion of low-density residential in the southern and northeastern parts of the city was observed. Most of the land in the northeast part of Dehradun city was predicted to be owned by high-income group. The study can benefit the decision-makers in generating various redevelopment scenarios and efficient resource planning.

Keywords Urban dynamics · Spatial cognition · Sustainable development · Agent-based modeling · LULC

V. K. Singh
Centre for Atmospheric Studies, Indian Institute of Technology Delhi, New Delhi 110016, India
e-mail: december.keviv@gmail.com

V. Kumar (✉) · A. Jana
Centre for Urban Science and Engineering, Indian Institute of Technology Bombay, Mumbai 400076, India
e-mail: vaibhav.kumar.cs10@gmail.com

A. Jana
e-mail: arnab.jana@iitb.ac.in

A. Jana and P. Banerji (eds.), *Urban Science and Engineering*, Lecture Notes in Civil Engineering 121, https://doi.org/10.1007/978-981-33-4114-2_3

1 Introduction

More than half of the population is predicted to live in cities by the year 2030. Better planned cities with well managed urban spaces can be a potent tool to cater the demand [1]. Land use and land cover (LULC) modeling system using analytical frameworks is one of such tools. However, such a system requires modeling of complex interaction between physical and socio-economical drivers [2, 3]. The challenge is even bigger due to the gap in the literature concerning modeling at finer scales, especially in the cities of developing nations [4].

The microscopic land-use growth is extremely sensitive by local constraints [5, 6]. These constraints have various constituents such as population density, number of households, literacy rates, construction of roads, schools and hospitals, distance from CBD (Central Business District), police stations, administrative policies, etc. Further, socioeconomic factors, as local constraints also directly correlate to the patterns of urban land use [7]. The existing research for developing nations is less focused on modeling these constraints, thus leading to unrealistic solutions [8]. The main objective of the study is to address this gap.

In recent decades, Remote Sensing (RS) and Geographic Information Systems (GIS) have given answers to these problems [9]. Researchers have coupling GIS tools with geostatistical models to capture the city growth dynamics [10]. Cellular Automata (CA), first developed in the 1940s, is one such modeling technique [11]. The lattice-based approach has been widely adapted in land use prediction [6, 11–13]. Despite several advantages, CA has limitations regarding (a) reaction to an external event (b) modeling neighbor character (c) absence of variance in transition rule which defines the success of the model [4].

Further, a significant disadvantage of cellular automata modeling is its inability to incorporate socioeconomic factors in the simulation, which play an essential role in the urban growth prediction. As a result, a new era of generic algorithms was developed to incorporate causal factors (such as socioeconomic), which give better understanding of urban growth. One such generic algorithm is Agent Based Modeling (ABM) and simulation [14].

This technique captures the complex network of interactions and connections that make up real systems and makes it possible to model complex systems involving emergent patterns and unexpected changes and events [15]. It also allows capturing the emergent phenomena that result from the interactions of individual entities. In the past, many studies have applied ABM in urban land use planning [15–17]. In a recent study, Liu et al. [12] coupled CA and ABM models to deduce better results in predicting future urban growth. However, the exclusion of socioeconomic details influencing neighborhoods remains a drawback of the research. In this study, we aim to overcome this limitation by modeling granular socioeconomic datasets and physical parameters such as circle rate of land parcels in predicting future household type.

1.1 Contribution of the Study

This study proposes a Land-use Simulation and Decision-Support system (LSDS), which integrates an ABM, and a CA model, over a GIS-based system with the following contributions:

- Effectiveness of LSDS is demonstrated while predicting LULC, while modeling socioeconomic factors.
- ABM is implemented to identify and predict the granular land use attributes such as suitable areas for different residential classes based on socioeconomic factors [17], heterogeneous nature of the urban environment, and their response to various socioeconomic variables such as literacy, financial status, circle rate, neighborhood, etc.
- The system can be extended to multiple scales and use-cases for informed decision-making.

2 Case Study: Dehradun City

To apply the methods Dehradun, the capital of the Indian state of Uttarakhand, was selected as the case. Field survey and digitalization techniques were applied to map urban features like roads, settlements, urban agriculture, industries, institutions, urban forest, public and semi-public institutions, commercial and recreational areas (parks and gardens). Urban settlement areas were classified into three groups: low-density residential areas, medium-density residential areas, and high-density residential areas, based on footprint compactness. Spatiotemporal distribution of urban land use pattern of Dehradun city was prepared for the years 2007, 2013, and 2019 (see Fig. 1). These maps are prepared using Landsat 7 satellite data {March month (cloud free)}. Table 1 lists urban services and socioeconomic factors that were considered in the study. The urban service datasets such as distances were derived using spatial methodologies and census data socioeconomic datasets were procured from the decision-making agency Dehradun Nagar Nigam. Figure 2 shows the circle rate map for the study area.

3 Simulation Framework

This aim to simulate the spatiotemporal urban growth scenario is achieved through various sequential steps, as explained in Fig. 3. In the first major step (Step 1), the CA model is developed to predict LULC for 2019. In the second major step (Step 2), the CA-based model is combined with the developed AMB model for more accurate results for the year 2019. The IDRISI GIS Analysis tools in the TerrSet software was used to implement the CA-Markov model. The Recursive Porous Agent Simulation

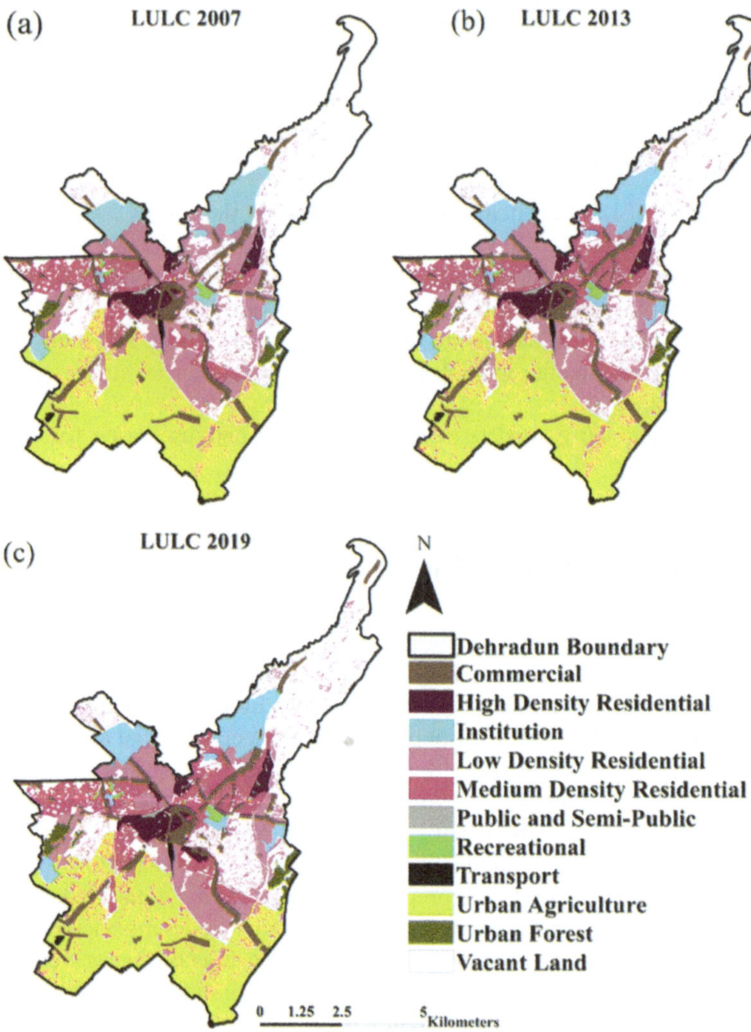

Fig. 1 LULC maps for the year (**a**) 2007, (**b**) 2013, and (**c**) 2019

Table 1 List of urban services and socioeconomic factors used in the study

Urban services	Socioeconomic factors
Circle rate, distance to roads, schools, and hospitals	Population density, household density, literacy rate, income (census data)

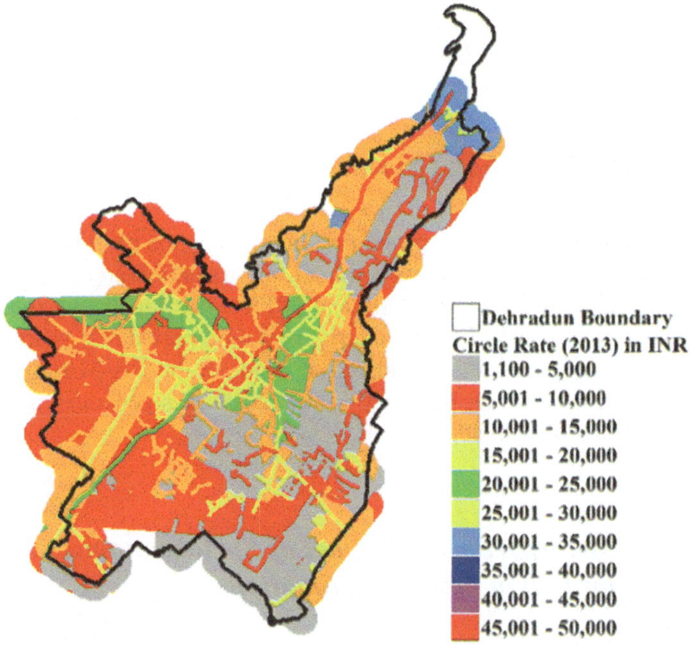

Fig. 2 Circle rate map

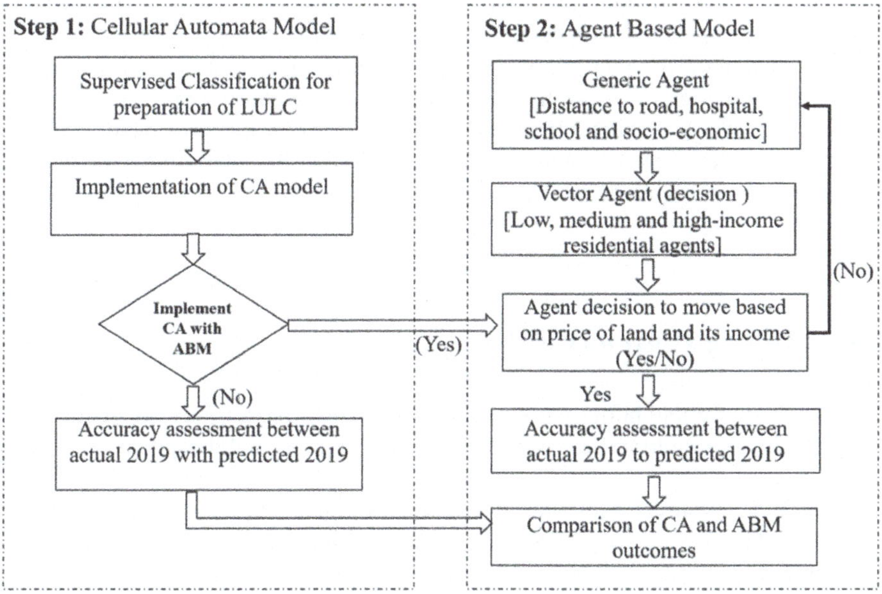

Fig. 3 Steps involved in developing the LSDS

Toolkit (Repast) a free and cross-platform, simulation toolkit was used to implement the ABM model.

3.1 CA Model for Urban Growth Estimation

Growth expansion depends on the properties inherited by the neighboring cell of the grid [6, 13]. In CA-Markov prediction model, the transition area matrix is calculated for all the classes for LULC 2007 and 2013. After calculating the transition area matrix and transition probability matrix for a time-period of 6 years for LULC 2007–2013, these matrices are considered as input to the CA-Markov model for the prediction of LULC 2019. Predicted LULC 2019, was generated and accuracy assessment was carried with actual 2019.

3.2 ABM to Incorporate the Impact of Socioeconomic Factors on Urban Growth Estimation

Agents are classified into two groups: vector group and generic group (see Table 2). Here low-income residential means residential areas that house people having less income. Similarly, medium and high income corresponds to areas having citizens from high-income class. Such details were derived using the census datasets. Vector group consist of those agents who take decision based on the outcome of the generic agents. The generic group agents are defined each for socioeconomic factors. Their interaction produces outcomes that are considered by vector agents to make decisions. Interactions of an agent with multiple agents and with the environment are derived from proximity rules which are the conditions that influence any land use land cover change.

3.3 Defining Proximity Rules

Rule-1: Household builds a house where the size of the house is adequate.
Rule-2: Household moves to areas where schools and hospitals are easily accessible.
Rule-3: Neighborhood qualities play a role in influencing household choice.
Rule-4: The socioeconomic status of the area influences the type of house chosen.
Rule-5: Households will move to areas where transport routes are accessible.

Table 2 List of vector agents and generic agents developed in the study

Vector agents	Generic agents
Low-income residential, medium-income residential and high-income medium residential	Distance to roads, schools, and hospitals, population density, circle rate, household density, literacy rate, and built-up
Based on the income group:	
Low-income residential:	Circle rate (10,000)
	Distance to road (3–4 km)
	Distance to school and hospital (3–4 km)
	Distance to built-up and number of household density
	And literacy rate
Medium-income residential:	Circle rate (10,000–25,000)
	Distance to road (1–3 km)
	Distance to school and hospital (1–3 km)
	Distance to built-up and number of household density
	And literacy rate
High-income residential:	Circle rate (25,000–40,000)
	Distance to road (1–3 km)
	Distance to school and hospital (1–3 km)
	Distance to built-up and number of household density
	And literacy rate
Commercial:	Circle rate (25,000–40,000)
	Distance to road (1–3 km)

3.4 Simulation Algorithm

Step 1 and step 2 are considered for CA and step 3 to 5 are used to incorporate CA with ABM.

Step 1: A matrix size of 5 * 5 grids was iterated over the grid to obtain changes in the urban LULC pattern at intra city level. Classified map of 2007 and 2013 in Fig. 1a, b were also resampled to the same grid size. Using the Markov chain process, transition rules were obtained. These rules describe the probability of each cell to retain or change its suitable state.

Step 2: CA was run for a classified map of 2007 and 2013 (see Fig. 1b) using transition probability matrix and area matrix. The classified predicted map of 2019 was deduced, and accuracy assessment was carried out.

Step 3: Proximity rules mention in Sect. 3.1 are implemented on each of low-density, medium-density, and high-density residential built-ups.

Step 4: Agents decide the effect of the neighborhood (Eq. 1) is based on socioeconomic status, existing agents, and physical drivers such as distance to road, school, and hospital.

$$\mathrm{NE}(d_t) = K \cdot d(m)^{i,j} \tag{1}$$

where, d(m) is the Euclidean distance of each driver, m is the socioeconomic driver, K is the proportional constant which is directly proportional to the distance to various socioeconomic drivers (D_{ij}) from a grid (i, j), for NE (d_m) $\in [-1, 1]$. Selection of new grids for agent movement from current grid ($Lulc_{ij}$) to new grid is based on the maximum number of available neighborhood grids. Equations (2)–(6) describe the procedure to identify the neighbors and perform the transition of one class to another. Suitability weightage (S_{ij}) of each LULC class k for grid (i, j) ($Lulc_{ij}$) for time t (2007 in our case) is calculated in Eq. (2) and for (t + 1) which is 2013, in Eq. (3).

$$S_{ij}(k)_t = P\big(Lulc_{ij}(k) \times W_t(k)\big) \tag{2}$$

$$S_{ij}(k)_{t+1} = P\big(Lulc_{ij}(k) \times W_t(k)\big) \tag{3}$$

Here, W_t (k) denotes the weight of each class k calculated from transition probability matrix obtained in Step 2. Equation (4) corresponds that LULC class k is equal to proportional constant K.

$$Lulc_{ij}(k) = K \sim D_{ij}(i = 0 \ldots n, j = 0 \ldots n) \tag{4}$$

Equation 5 calculates the suitability weightage ratio used for the predicted time-period (pt). In our case, pt is considered as 2019, for which we have predicted the LULC.

$$S(kpt) = S_{ij}(k)_{t+1}/S_{ij}(k)_t \tag{5}$$

The maximum number of neighborhood grid ${N^k}_{pt}$ is deduced as follows:

$${N^k}_{pt} = {N^k}_t + \frac{d(N^k)}{d(t)} * S\big(k_{pt}\big) \tag{6}$$

${N^k}_{pt}$ = maximum of the grid for class k, $\frac{d(N^k)}{d(t)}$ = rate of change of cell, and ${N^k}_t$ = Number of cells allocated for each class.

Step 5: Each rule defined in Sect. 3.1 was validated for each grid (i, j). After every rule is satisfied, the agent takes move to a new grid.

Step 6: The process continues until all the rules are satisfied for each cell. Once all grids are visited the algorithms stops.

3.5 *Accuracy Assessment*

In this study, Kappa statistics was selected as the metric for accuracy assessment. The value of Kappa statistics lies between 0 and 1. The closer the value is to one more accurate is the outcome.

4 Results and Discussions

It can be noted that low-density residential and medium-density residential has shown a higher tendency to occupy more area by converting urban agriculture or vacant land into the built-up area as shown in Table 3. It was found that low-density residential classes converted to medium-density residential classes at a transition probability rate of 0.3% from the year 2013 to 2019, in the CA model, whereas in ABM, the rate was observed to be 1.7%. These areas were mostly toward the western and southern parts of the city. Few low-density residential areas were identified in the outskirts of Dehradun city . These residential areas are on the southern and eastern side of the city.

It can be noted that due to less availability of land in the center of the city, high-density residential classes spread in the southeast part of the city. The northeastern part of the city witnessed a similar growth pattern. From 2007 to 2013, a trend was observed in the northeastern part of the city, where vacant land converted into low-density residential areas. Similarly, the southern and southeastern parts of the city witnessed similar conversion trend, in which agricultural land converted into low-density residential areas. The trend was also observed in the medium-density residential areas. Accuracy assessment was done for each class by comparing the predicted 2019 with actual 2019 LULC. Overall, kappa accuracy was deduced to be 79.73%. It was found that the relative error in the area occupied by households was 10.62%. Figure 4 shows the final predicted map.

5 Conclusion

The study proposes a Land-use Simulation and Decision-Support system (LSDS), which is an artificial intelligence simulation model. LSDS integrates microscopic ABM and CA model on the GIS platform. The system can solve the microscopic influence of socioeconomic and service utility drivers in quantifying spatiotemporal urban growth at the micro level. Taking the case of Dehradun city, in India, the

Table 3 Transition area matrix used to describe the number of cells is classified into the same LULC categories from 2007 to 2013

LULC categories	1	2	3	4	5	6	7	8	9	10	11
1	196,780	0	0	0	0	0	0	0	0	0	0
2	0	75,591	0	0	0	0	0	0	0	0	0
3	0	0	266,613	0	0	0	84	0	0	0	0
4	0	0	61,789	347,223	0	0	0	0	0	0	0
5	0	0	0	0	186,720	0	0	0	0	0	0
6	0	0	0	0	0	2823	0	0	0	0	0
7	0	0	0	0	0	0	11,092	0	0	0	0
8	0	0	0	0	0	0	0	6471	0	0	0
9	0	0	0	19,902	0	0	0	0	689,756	0	0
10	0	0	0	0	0	0	0	0	0	40,279	0
11	0	0	24,442	9345	0	0	0	0	0	2303	738,342

Note 1—Commercial, 2—High Density Residential, 3—Medium Density Residential, 4—Low Density Residential, 5—Institution, 6—Public and Semi-Public, 7—Recreational, 8—Transportation, 9—Urban Agriculture, 10—Urban Forest and 11—Vacant Land

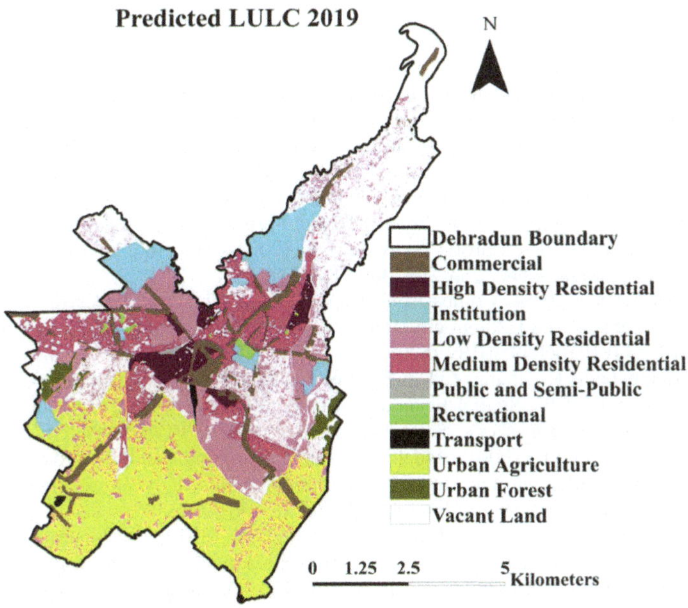

Fig. 4 Predicted classified map of Dehradun for the year 2019

system successfully demonstrates how socioeconomic factors affect the urban land use pattern, what the different drivers affect the growth at intra city level, and how these were integrated with geo-computational models for future prediction. Various transition of LULC classes were successfully captured. With limited land resources and dynamic property of urban land-use patterns, major changes were observed in the southeast and southwest parts of the city. The northeastern part of the city had a large area under vacant land class, which converted into settlement classes. A similar trend was observed in the southern part of the city. The results were obtained at very high accuracy which highlights the significance of applying agent based modeling in capturing the interaction between socioeconomic drivers and LULC dynamics.

The set of rules considered in the study has some limitations due to heterogeneous nature of agents such as the dynamics of the market price to buy land. Limiting the group of certain classes in the LULC to maintain certain green areas can also be implemented. Further, establishing the relationship between vector agent and generic agent and adding more socioeconomic variables remains as future work. Addressing the limitation also remains our future work. The development could predict the accurate growth profile of the city; thus it can help decision-makers foresee the demand for informed decision-making the diverse urban setup in a country like India.

References

1. Kumar, V., Jana, A., & Ramamritham, K. (2020). Simulating fire-safe cities using a machine learning-based algorithm for the complex urban forms of developing nations: A case of Mumbai India. *Geocarto International*, 1–16. https://doi.org/10.1080/10106049.2020.1756463.
2. Makido, Y., Shandas, V., & Ferwati, S. (2020). Predicting urban growth. In V. Shandas, C. Skelhorn, & S. Ferwati (Eds.), *Urban adaptation to climate change: The role of urban form in mediating rising temperatures* (pp. 75–92). Cham: Springer International Publishing.
3. Brown, D.G., Walker, R., Manson, S., & Seto, K. (2004). Modeling land use and land cover change. *Land change science* (pp. 395–409). Springer.
4. Mondal, B., et al. (2020). Comparison of spatial modelling approaches to simulate urban growth: A case study on Udaipur city, India. *Geocarto International, 35*(4), 411–433. https://doi.org/10.1080/10106049.2018v.1520922.
5. Shi, L., Taubenböck, H., Zhang, Z., Liu, F., & Wurm, M. (2019). Urbanization in China from the end of 1980s until 2010—Spatial dynamics and patterns of growth using EO-data. *International Journal of Digital Earth, 12*(1), 78–94. https://doi.org/10.1080/17538947.2017.1400599.
6. Li, X., & Yeh, A.G.-O. (2000). Modelling sustainable urban development by the integration of constrained cellular automata and GIS. *International Journal of Geographical Information Science, 14*(2), 131–152. https://doi.org/10.1080/136588100240886.
7. Pan, H., Zhang, L., Cong, C., Deal, B., & Wang, Y. (2019). A dynamic and spatially explicit modeling approach to identify the ecosystem service implications of complex urban systems interactions. *Ecological Indicators, 102*, 426–436. https://doi.org/10.1016/j.ecolind.2019.02.059.
8. Herold, M., Couclelis, H., & Clarke, K. C. (2005). The role of spatial metrics in the analysis and modeling of urban land use change. *Computers, Environment and Urban Systems, 29*(4), 369–399. https://doi.org/10.1016/j.compenvurbsys.2003.12.001.
9. Sudhira, H. S., Ramachandra, T. V., & Jagadish, K. S. (2004). Urban sprawl: Metrics, dynamics and modelling using GIS. *International Journal of Applied Earth Observation and Geoinformation, 5*(1), 29–39. https://doi.org/10.1016/j.jag.2003.08.002.
10. Chen, M., Arribas-Bel, D., & Singleton, A. (2019). Understanding the dynamics of urban areas of interest through volunteered geographic information. *Journal of Geographical Systems, 21*(1), 89–109. https://doi.org/10.1007/s10109-018-0284-3.
11. Wolfram, S. (1983). Statistical mechanics of cellular automata. *Reviews of Modern Physics, 55*(3), 601–644. https://doi.org/10.1103/RevModPhys.55.601.
12. Liu, D., Zheng, X., & Wang, H. (2020). Land-use simulation and decision-support system (LandSDS): Seamlessly integrating system dynamics, agent-based model, and cellular automata. *Ecological Modelling, 417*, 108924. https://doi.org/10.1016/j.ecolmodel.2019.108924.
13. Li, X., & Yeh, A. G.-O. (2001). Zoning land for agricultural protection by the integration of remote sensing, GIS, and cellular automata. *Photogrammetric Engineering & Remote Sensing, 67*(4), 471–478. Retrieved February 07, 2015 [Online].
14. Thill, J.-C., & Dragicevic, S. (2017). *GeoComputational analysis and modeling of regional systems*. Springer.
15. Wahyudi, A., Liu, Y., & Corcoran, J. (2019). Generating different urban land configurations based on heterogeneous decisions of private land developers: An agent-based approach in a developing country context. *ISPRS International Journal of Geo-Information, 8*(5), 229. https://doi.org/10.3390/ijgi8050229.
16. Roy, D., & Lees, M. (2020). Understanding resilience in slums using an agent-based model. *Computers, Environment and Urban Systems, 80*, 101458. https://doi.org/10.1016/j.compenvurbsys.2019.101458.
17. Maretto, R. V., Assis, T. O., & Gavlak, A. A. (2015). Simulating urban growth and residential segregation through agent-based modeling. In *2010 Second Brazilian Workshop on Social Simulation (BWSS)* (pp. 52–57). Retrieved February 07, 2015, from https://ieeexplore.ieee.org/xpls/abs_all.jsp?arnumber=6030014 [Online].

Deciphering the Indore Story: Assessing Municipal Solid Waste Management in the Last Mile of Operations

Subodh Wagle and Arpit Arora

Abstract The municipal solid waste management (MSWM) sector has remained quite neglected in India. As a result, most rapidly growing cities are eclipsed by heaps of garbage left in the open. Bulk of literature on the MSWM sector discusses a certain range of technoeconomic and managerial issues as major challenges. This paper, however, takes the policy perspective with the focus on operations in the last mile of the sector. It investigates the role of front-line workers involved in the last mile functions of collection and transportation of the solid waste in the city of Indore, which rose from the 149th rank in 2014 to 1st rank in 2017 in the national-level cleanliness survey and then remained at the top in 2018 and 2019. The paper concludes by underscoring critical importance of various reforms in the human resource management (HRM) in the city of Indore, such as stringent attendance system and systematization of working procedures in the last mile of operations. These reforms were supported by innovative tactical measures for ensuring accountability of citizens and front-line workers and by accommodating crucial actors from the old system. Together, these measures have helped Indore to rise and remain at the top-position in Cleanliness Ranking in the country.

Keywords Municipal solid waste management · Indore · Front-line worker

1 Background

1.1 Introduction

Rapid urbanization and swift economic development have not only resulted in the increase in population and areas of cities but have also exerted substantial pressure on the existing infrastructure of cities across the country [9]. In India, the level of

S. Wagle · A. Arora (✉)
CTARA, IIT Bombay, Mumbai, India
e-mail: arpit.arora@iitb.ac.in

S. Wagle
e-mail: subodh.wagle@iitb.ac.in

A. Jana and P. Banerji (eds.), *Urban Science and Engineering*, Lecture Notes in Civil Engineering 121, https://doi.org/10.1007/978-981-33-4114-2_4

urbanization has risen from 27.8% in 2001 to 31.6% in 2011 and it is projected that nearly 50% of country's population will reside in urban settings by the year 2021 [13].

With urban local bodies (ULBs) striving for improvements in services like water, electricity, and transportation, a crucial public service—municipal solid waste (MSW) management—has remained neglected in India [14]. This led to a situation in which rapidly growing cities are eclipsed by heaps of garbage left in the open [11]. Effective MSW management in the country experiencing exponential population growth and significant improvements in living standards has now become a burning issue, though; the concerned ULBs have no solution to effectively address it [8].

1.2 Current Scenario of MSW Management in India

MSW, being one of the most neglected sectors across cities, typically has a limited budget allocation. Further, around 95% of the allocated budget is spent on activities of waste collection and transportation. Less than 5% of the budget is left for disposal and treatment of waste [1]. The collection system relying on bins located across communities is the predominant mode of waste collection in the country with no standardization of capacity, size, or locations of these community bins [15]. Furthermore, across the country, segregation of waste at source is practiced at a marginal level.

Another highly problematic practice involves the use of same vehicles for dumping the waste at dumping sites as well as for collecting waste from community bins. These vehicles are open from top and uncovered. This leads to spilling of waste on roads, leading to spread of unhygienic conditions across the city. In most urban regions, waste is disposed by directly dumping it in low-lying areas outside the city boundaries. No specific plans are prepared or implemented for management of the dump sites and the MSW is not weighed or compacted while dumping [7]. Additionally, the practice of unauthorized burning of MSW at landfills in order to reduce the volume of waste is prevalent across the country. This is often attributed to the rag pickers in most cities who operate without any regulation [12].

Thus, the existing MSW management system in most cities is overwhelmed by multiple challenges and issues. Some of the major issues underscored in the literature include: lack of awareness among policymakers, administrators and citizens about waste segregation, lack of financial resources in the MSW management sector, ineffective implementation of rules on ground, lack of planning and coordination among multiple authorities involved, and use of dilapidated equipment and old technology [2, 6, 8, 15].

1.3 Structure of the Paper

While the bulk of literature on the MSW sector thus highlights the above-mentioned issues, this paper takes a policy perspective with focus on operations in the last mile of the MSW sector. With the country witnessing a dismal scenario in the MSW sector that is discussed in the previous section, performance of the city of Indore in managing the activities related to MSW provides hope of transformation. The city jumped from 149th position in 2014 in the national-level cleanliness rankings to the top position in 2017 and remained at the top for 3 consecutive years. The existing literature credits this miraculous transformation of the city of Indore to the efforts of the Indore Municipal Corporation (IMC). This research focuses on the changes in human resource management, especially in the last-mile operations, and effects of these changes on front-line workers—i.e., workers who work at the lower end of the organizational bureaucracy and are in direct contact with citizens or clients—and their performance.

Coming to the structure of the paper, in the second section after this introduction, the paper discusses the research methodology adopted for conducting the study and provides a brief introduction to the field of study: Indore. The third section highlights the last-mile operations in the activities of collection and transportation of MSW in the city of Indore. The section introduces multiple actors involved in the activities of MSW collection and transportation and various stages at which these actors operate and interact with each other. The fourth section delineates major changes in human resource management carried out by the IMC and discusses the effects of these administrative changes on the performance of front-line workers. The paper concludes by outlining key factors responsible for transformation of the city of Indore, emphasizing on the role of front-line workers and their efforts in making Indore the cleanest city in India.

2 Literature Review and Research Methodology

As mentioned before, majority of literature discussing issues related to MSW management emphasizes on importance of awareness among citizens regarding the segregation of waste, role of dilapidated equipment and technology, and lack of financial resources in the sector of SWM. The existing literature neglects the importance of role played by human resources which are responsible for "getting the job done" at the front lines. This research highlights the critical importance of the role played especially by front-line workers. It focuses on the last-mile operations and delves on the work and performance of front-line workers involved in carrying out functions of collection and transportation of MSW in the city of Indore.

The city of Indore, unlike its less commercial cousin and administrative headquarter of the state, Bhopal, is considered as a commercial hub of the state of Madhya Pradesh with little to offer in terms of charm and visual beauty [3]. The city houses approximately 3 million people [4] with a daily floating population of 3–5 hundred thousand people [10]. The city is spread in an area of approximately 275 km^2 with a population density of about 10,000 people/km^2. According to IMC reports, the city generated approximately 1,115 MT of waste per day in the year 2017.

The MSW management sector in Indore, like many cities, was plagued by several issues till 2016. The services of waste collection and treatment, provided by private contractors, were of poor quality. The dustbins in the city were in dilapidated condition, creating an ugly look for the city with very low levels of waste handling capacity. The IMC staff was demoralized and citizens had a very poor image of them. Public participation in MSW management activities was negligible and insufficient infrastructure and machinery added to the misery [5].

As mentioned before, due to the initiatives taken by the municipal corporation and local residents, the city reached the top spot in Cleanliness Survey rankings in 2017, and it continued to be at top for the consecutive 3 years of 2017, 2018, and 2019. The city received ODF (open defecation free) status in December 2016. Efficiency of collection and transportation was found to have increased with introduction of GPS tracking systems along with stringent monitoring of tasks carried out by front-line workers by IMC. The workers were found to be working more diligently and actively and residents were found to become more responsible and receptive toward the strategies introduced by the IMC.

In this background, this research investigates a general research question: How changes in operating systems and human resource management helped improve performance of front-line workers in MSW management sector in the Indore city? This exploratory and descriptive research takes a qualitative approach, using semistructured interviews and nonparticipant observation as the main data collection methods.

The data are collected by interviewing stakeholders from mixed land-use zones in the city, which have both, commercial and residential activities. The study conducted in mixed zones provides an opportunity to simultaneously understand different strategies employed by the administration for commercial and residential generators of garbage. The research primarily relies on interviews of 26 respondents from different stakeholder groups such as: front-line workers, their immediate supervisors, higher level municipal officials, and NGO representatives. The method for the selection of respondents involved purposive sampling across the categories followed by snowball sampling within the category. The respondents were selected from different levels of the bureaucracy of the municipal corporation in order to ensure a holistic and detailed understanding of the transformation process and the impact this transformation had on the performance of front-line workers according to officials posted at different levels of the Indore Municipal Corporation. For analysis of qualitative data, the study used the technique of thematic analysis.

3 Last-Mile Operations in Indore: Collection and Transportation of MSW

This section provides an overview of the organization structure of Municipal Corporation in the Municipal Solid Waste Management sector (Please refer to Fig. 1). The section also presents the description of various stages in the functions of collection and transportation of MSW as organized in the city of Indore. It also identifies various factors involved in each stage of collection and transportation (please refer to Fig. 2).

The Municipal Solid Waste Management plan comes under the Indore Smart City Mission headed by the Municipal Commissioner. One out of the six Additional Municipal Commissioners is responsible for the MSWM in Indore. The Additional Municipal Commissioner is assisted by the City Sanitary Inspector along with two Assistant City Sanitary Inspectors. The Project incharge is responsible for ensuring the entire system of MSWM (explained in Fig. 2) works smoothly. The project incharge is assisted by the assistant project incharge followed by computer operators for each of the 19 zones in the city.

Multiple front-line workers are involved at different levels and stages of the waste management system. The workers who directly interact with residents during

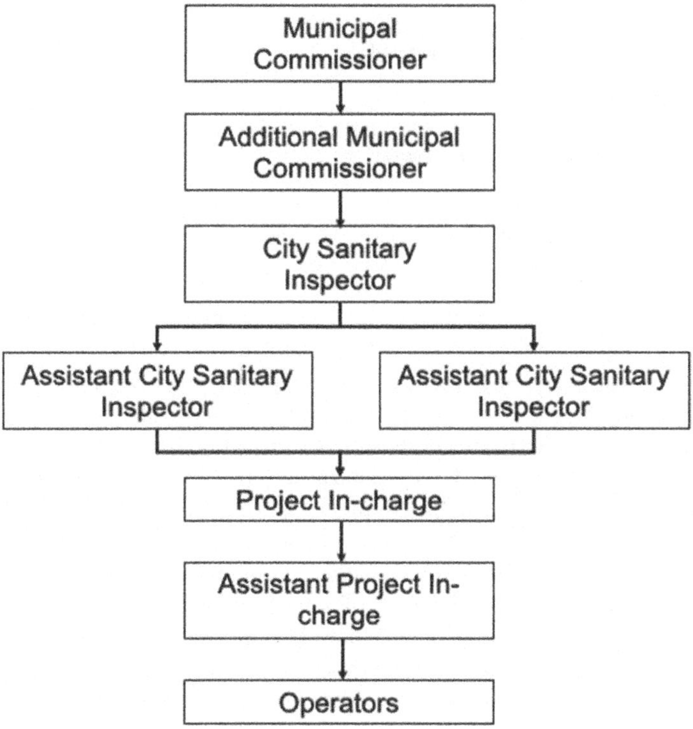

Fig. 1 Organization structure of municipal corporation for MSW management

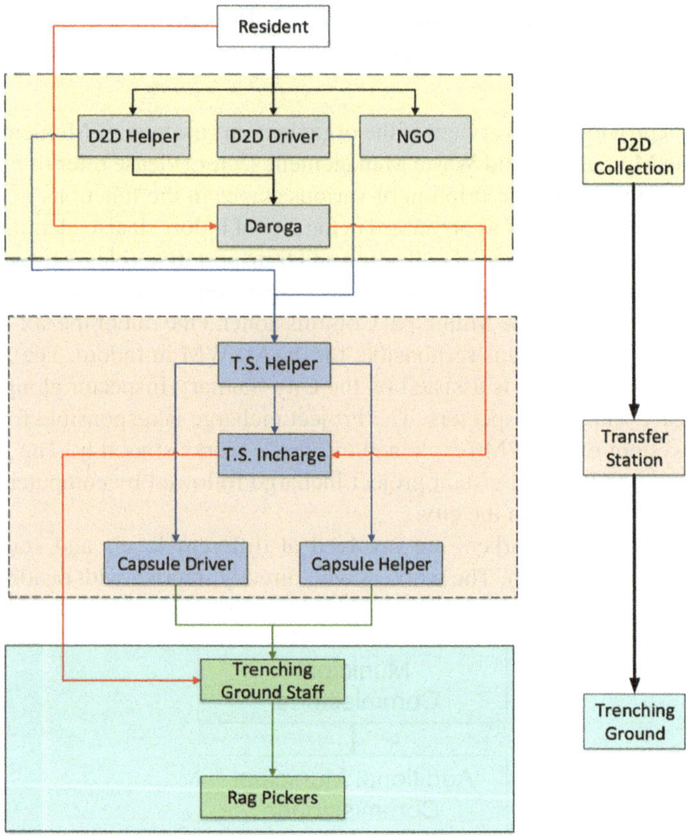

Fig. 2 Actors involved in MSW management

waste collection process are: D2D (i.e., door-to-door) helper, D2D vehicle driver and NGO representative accompanying the helper and driver. The NGO representative is responsible for carrying out awareness campaigns and collecting feedback about the services provided by the corporation from residents. With the onus of segregation on the residents, the helpers are responsible for ensuring that segregation of waste in three categories (wet, dry, and domestic hazardous) is done properly by the resident. The vehicle driver assists the helper in this task. The daroga, who is the immediate supervisor of the front-line workers, is accountable to the citizens for the services provided by the workers. The daroga handles complaints from the citizens regarding issues related to cleanliness and quality of work provided by the front-line workers. According to the officials and residents, daroga is the person who "takes heat" from all the directions. S/he has to ensure that workers complete their designated amount of work and with desired quality. In case of absenteeism of workers or any other issue faced by the workers, the daroga is expected to resolve these issues and ensure the completion of work. On the client side, the daroga is responsible for acknowledging and acting upon the grievances of clients. In addition to handling the issues of workers and clients, the daroga is also supposed to report the progress of the ward-level MSW activities to her/his superior daily.

The collected segregated waste from households in D2D vehicles is sent to a transfer station for unloading. At the transfer station, all three types of waste (dry, wet and domestic hazardous) are weighed using a three-way weigh bridge for each D2D vehicle. The transfer station helper is responsible for keeping a record of the quantity of different types of waste unloaded from each vehicle. The station helper is also responsible for ensuring that the waste being dumped at the station is properly segregated. Once the waste is received at the transfer station, it is then compressed, filled separately (dry and wet) and transported to the trenching ground in the capsules (special type of vehicle). Each capsule has a capsule helper along with the driver.

The capsule helper assists in dumping of waste at the trenching ground (T.G.). The capsule helper, along with the T.G. staff, ensures that the waste is dumped at a designated spot. The disposal spots for dry waste and wet waste are different at the T.G. facility. The dry waste is dumped at the Material Recovery Facility (MRF) where rag pickers segregate the dry waste further in 13 different categories. The T.G. staff is responsible for ensuring that segregation is carried out as effectively as possible and that each rag picker gets equal share of waste to segregate.

4 Major Changes in Human Resource Management and Their Effects

This section of the paper presents major administrative changes related to human resource management carried out in Indore. It also presents major improvements in the performance of front-line workers due to these administrative changes, and the effect of improved performance of front-line workers on the state of the MSW sector in the city.

4.1 Major Changes in Human Resource Management

One of the key administrative changes in the human resource management (HRM) in Indore was introduction of a well-defined, strict, multimodal attendance system. The new HRM system involved taking attendance twice a day, both, at the start and end of a shift. The system of biometric attendance was introduced as per the guidelines of National Cleanliness Survey. Attendance registers were still used to cross-check availability of a particular worker on the day and avoid problems with network, affecting the biometric system.

The previous system saw workers completing their designated work in two to three hours, with no fixed shift timings. In the new system, mandatory 8 h long shift was introduced for front-line workers. This measure combined with introduction of biometric attendance at the start and end of the shift guaranteed availability of workers throughout their shift. This ensured well-defined work timings for the workers and availability of workers whenever need arise.

The old HRM system lacked suitable mechanisms to monitor and ensure quality of work performed by front-line workers. This resulted in managerial and on-field problems because it was easier for workers to escape the responsibility of delivering adequately good quality of work output. Along with the multimodal attendance system, VTS (vehicle tracking system) was incorporated in the new HRM system to ensure strict monitoring at the lowest level of operations. Frequent visits by darogas were made mandatory to check the quality of work output delivered by front-line workers.

In the previous system, definite work locations were not assigned to worker. Their work areas had to be changed frequently, depending on the availability of other workers. Instead, the tool of deployment chart was introduced in the new system for systematic and well-defined work allocation. This ensured that each worker had a designated working spot and allocated work area. The darogas were given some discretion to alter the allocation of work but only in certain special cases.

Along with these administrative changes discussed above, regular acknowledgment and appreciation of work done by front-line workers were made as part of the new HRM system. This helped to boost the morale of front-line workers. The workers gave a positive response to such appreciation and it encouraged them to keep working diligently.

4.2 *Major Improvements in Performance of Front-Line Workers*

Introduction of strict, multimodal system of attendance resulted in significant reduction in rate of absenteeism of front-line workers. In addition, mandatory attendance twice in a shift compelled workers to remain available during designated number of hours. This step, however, was opposed by certain sections of front-line workers, which mainly included workers having a permanent position. Nonetheless, the absenteeism rate was found to have reduced from 60 to 70% in the previous system to 20 to 30% in the new system.

In addition to increase in the quantum of work output due to well-defined shift timings, the quality of work also improved. The workers were compelled to stay at their spots throughout their shift. Further due to reduction in absenteeism rate, more front-line workers were available for work and hence the work pressure was not significant on an individual worker. This had a positive impact on the quality of service provided by workers. Some front-line workers, especially young females working in the morning shift, however, expressed concerns over such fixed timings for entire shifts. This was because such timings made it difficult for them to deal with their family responsibilities such as looking after young children and performing other household chores.

As a result of reduction in work pressure, sufficient time became available to each worker to complete her/his allocated work effectively. Thus, work environment significantly improved and instances of sudden increase in work pressure for worker have reduced drastically. The workload has become relatively stable as compared with the previous system. Despite extension of work hours, in order to achieve cleanliness targets, the number of front-line workers was increased. Significant increase in staffing led to significant increase in the wage bill, leading to increased financial burden. However, the salary of front-line workers remained the same and a smaller number of people were offered permanent posts by IMC. This issue was considered as a severe threat to the sustainability of the system.

Along with improvement in work environment and health conditions, introduction of segregation of waste at source, other measures led to improvement in compliance of citizens. This resulted in significant reduction in drudgery, filth handing, and health risk unlike in the previous system wherein workers were forced to collect mixed waste from large community bins using hand-held ploughs. Mandatory provision of PPE (personal protective equipment), according to the Cleanliness Guidelines, and provisioning of additional health benefits to workers increased the occupational safety of front-line workers.

Accountability of front-line workers was found to have increased with introduction of the deployment chart. All these measures resulted in effective cleaning of each area. This was because, with this measure, it became easier to identify the defaulters who delivered work outputs of below-par quality.

Furthermore, motivated front-line workers were able to shift the onus of segregation of waste on their clients (citizens). Through sustained pressure and through tactics involving both appeal and coercion, the workers succeeded in enforcing a behavioral change in the residents of the city. Citizens also responded positively to efforts of front-line workers by accepting the responsibility of segregation of waste at the source and by adhering to regulations of IMC.

4.3 Effect of Improved Performance of Front-Line Workers on MSW Sector

Due to the sustained improvement in performance of front-line workers, Indore witnessed a remarkable transformation in its MSW management sector. Door-to-door collection of MSW is now prevalent throughout the city. By providing satisfactory services, the front-line workers are successful in eradicating the entrenched vested interests of middlemen called jagirdaars, which were active in the previous system. These actors possessed significant nuisance-making ability because they possessed intimate knowledge of ground-level functioning of MSW management system in the city. Additionally, they were aware of vulnerability of the system and thus, could pose a serious threat of sabotaging the new system. Elimination of these actors significantly reduced challenges faced by IMC in implementing any new initiative.

Thus, as a result of efforts of front-line workers and with citizens responding positively to initiatives of IMC, efficiency of collection and transportation of MSW increased significantly. The positive feedback from citizens served as an impetus for IMC, which was once a demotivated and untrustworthy organization.

Finally, the improved performance of the overall MSW management system of the city helped Indore win the title of cleanest city in the national-level cleanliness survey of cities for the first time in 2017. The city retained the top spot in 2018 and 2019 as well.

5 Conclusion

With introduction of technology-based interventions like biometric attendance, one of the major issues of absenteeism of front-line workers in MSW management department was successfully dealt with. As a result, absenteeism rate reduced to 30% from 60% in the previous system.

In the case of front-line workers, due to changes in the attendance system, effective monitoring and stringent attitude adopted by local authorities toward absenteeism and quality of performance, there has been significant reduction in workload on a particular front-line worker in comparison to the previous system.

Serious reduction in drudgery was affected and due to provisioning of facilities like PPE and medical services, there was a significant improvement in working conditions of front-line workers. In addition, regular appreciation of work was found to keep workers motivated. Another key reason for the diligent and effective working of workers was transfer of onus of waste segregation on clients/citizens. This led to significant improvement in quality of work output from workers and acted as an incentive for them to continue their work diligently.

Thus, the study concludes by pointing at major factors underlying transformation of the city of Indore: (a) appropriately designed and effectively implemented initiatives to change the human resource management system fundamentally, (b) successful strategies and mechanisms to ensure diligent and effective performance of front-line workers, and (c) successful strategies and mechanisms to ensure citizens' compliance.

Though this study is entirely focused on the city of Indore, the findings of the study are certainly relevant for a wide range of Class 1 cities in the country. It is very much possible for these cities to incorporate administrative changes affected by the Indore Municipal Corporation as most of these cities face problems and barriers that are very similar to those faced by the municipal administration of the city of Indore.

References

1. Annepu, R. K. (2012). *Sustainable solid waste management in India* (Vol. 2, No. 01). Columbia University, New York.
2. Asnani, P. U. (2006). Solid waste management. *India infrastructure report*, 570.
3. Bhargava, A. (2017). *How Indore became garbage-free and beat every other city to it.* Retrieved February 11, 2019, from https://www.thebetterindia.com/114040/indore-madhya-pradesh-clean-garbage-free-india/.
4. Census of India Website. (2011). *Office of the registrar general & census commissioner, India.* Retrieved April 9, 2019, from https://censusindia.gov.in/pca/pcadata/Houselisting-housing-MP.html.
5. Indore Municipal Corporation. (2017). *India's cleanest city–Indore.* Retrieved February 13, 2019, from https://cdn.cseindia.org/docs/photogallery/slideshows/06_20171212_IMC-SWM-Final-Indore.pdf.
6. Joshi, R., & Ahmed, S. (2016a). Status and challenges of municipal solid waste management in India: A review. *Cogent Environmental Science, 2*(1). https://doi.org/10.1080/23311843.2016.1139434.
7. Kumar, S., Bhattacharyya, J., Vaidya, A., Chakrabarti, T., Devotta, S., & Akolkar, A. (2009). Assessment of the status of municipal solid waste management in metro cities, state capitals, class I cities, and class II towns in India: An insight. *Waste Management, 29*(2), 883–895. https://doi.org/10.1016/j.wasman.2008.04.011.
8. Kumar, S., Smith, S., Fowler, G., Velis, C., Kumar, S., Arya, S., et al. (2017). Challenges and opportunities associated with waste management in India. *Royal Society Open Science, 4*(3), 160764. https://doi.org/10.1098/rsos.160764.
9. Mani, S., & Singh, S. (2016). Sustainable municipal solid waste management in India: A policy agenda. *Procedia Environmental Sciences, 35,* 150–157. https://doi.org/10.1016/j.proenv.2016.07.064.
10. Nidugala, G., & Pant, A. (2017). Cleanest cities in India 2017: Indore's meteoric rise to no. 1 spot. *Journal of Public Affairs, 17*(4), e1674. https://doi.org/10.1002/pa.1674.
11. PWC. (2017). *Waste management in India—shifting gears.* Retrieved January 12, 2019, from http://www.indiaenvironmentportal.org.in/files/file/waste-management-in-india-shifting-gears.pdf.
12. Sharholy, M., Ahmad, K., Mahmood, G., & Trivedi, R. (2008). Municipal solid waste management in Indian cities—A review. *Waste Management, 28*(2), 459–467. https://doi.org/10.1016/j.wasman.2007.02.008.
13. Sharma, K., & Jain, S. (2019). Overview of municipal solid waste generation, composition, and management in India. *Journal of Environmental Engineering, 145*(3), 04018143. https://doi.org/10.1061/(asce)ee.1943-7870.0001490.
14. Swaminathan, M. (2018). How can India's waste problem see a systemic change? *Economic & Political Weekly, 16.* Retrieved January 19, 2019, from https://www.epw.in/engage/article/institutional-framework-implementing-solid-waste-management-india-macro-analysis.
15. Vij, D. (2012). Urbanization and solid waste management in india: Present practices and future challenges. *Procedia—Social and Behavioral Sciences, 37,* 437–447. https://doi.org/10.1016/j.sbspro.2012.03.309.

ICT and Its Use in Solid Waste Collection Systems for Route Optimization

Nikhil Kumar, Mayank Gupta, and Arnab Jana

Abstract Solid Waste Management has been a significant challenge for rapidly urbanizing cities of developing nations, consequently increasing the stress on environmental sustainability. The existing system of "door to door" waste collection is mostly random, resulting in high operating costs. To address the limitations, the transportation of waste from an urban area to the landfill needs adequate infrastructure and optimal resource allocation strategies. In this study, we proposed a methodology for waste collection in a city using a capacitated vehicle routing problem. Optimized routes were assigned to collection vehicles based on 1 day prior information from citizens regarding waste collection. We assumed the use of Information and Communication Technology to collect information from citizens regarding collection from their bins on a day. For this study, a dataset for Bandra (Mumbai) was created by geotagging of houses and marking the road network. The objective function was set to minimize the distance traveled by the garbage pickup vehicles. We simulated 40 scenarios for a particular day to analyze the effectiveness of methodology by extracting estimation of distance traveled to the finish collection process. The results showed that when the number of pickup locations was reduced, there was a significant reduction in the distance traveled, and time was taken by vehicles. For 70% coverage of bins, an average distance saving of 22.8 km was achieved. The methodology showed significant potential for efficient decision-making for SWM.

Keywords Information and communication technology (ICT) · Solid waste management (SWM) · Geographical information system (GIS) · Route optimization · Resource allocation

N. Kumar (✉) · M. Gupta · A. Jana
Centre for Urban Science and Engineering, Indian Institute of Technology Bombay, Mumbai 400076, India
e-mail: nikhil.89.kumar@gmail.com

M. Gupta
e-mail: mayankgupta1990@gmail.com

A. Jana
e-mail: arnab.jana@iitb.ac.in

A. Jana and P. Banerji (eds.), *Urban Science and Engineering*, Lecture Notes in Civil Engineering 121, https://doi.org/10.1007/978-981-33-4114-2_5

1 Introduction

Solid waste management (SWM) is a significant concern for all the countries, irrespective of the development scenario in the country. With global population growth, development, and urbanization, there is an increase in solid waste generation [1]. The upsurge in the amount of Municipal solid waste (MSW) has created a need for proper waste management policies for the collection and disposal of it. Among all the steps for waste management, solid waste collection (SWC) is the most demanding. Collection and transportation of waste is the major cost contributor and can reach up to 70% of the total waste management budget [2]. Due to which there is a concern regarding the efficiency of SWC methods and route designs [3]. There have been numerous studies that have tried to reduce the expenditure for SWM and increase collection efficiency [4] using optimization models. Various optimization models for routing and transportation of waste have been tried to reduce the cost of the municipal solid waste management system (MSWMS) [5].

The most commonly applied approach is the vehicle routing problem (VRP), which leads to an effective waste collection route. There are possibilities of constraints in a VRP problem. The most common is the capacity constraint. When a capacity constraint is added to a VRP, it is named the capacitated vehicle routing problem (CVRP). Studies use CVRP approach with different algorithms and software. The metaheuristic approach is popular for the optimization of CVRP [6]. A large number of software packages are available for waste collection optimization.

Despite having better optimization techniques and software, there is a gap in data for analyzing demand. Due to the lack of a proper monitoring system, the SWC process becomes inefficient. In the case of bins which are not full yet, it will be a waste of resource to travel to an empty bin [7]. The data gap can be covered by the use of Information and communication technology (ICT). Multiple studies have used geographic information systems (GIS) in combination with geographic position system (GPS), radio frequency identification (RFID), general packet radio service (GPRS), and Wi-Fi technologies as parts of ICT to overcome the data gap. However, most of the studies have not considered bin capacity or waste level. Thus this study aims to fill the research gap concerning the use of ICT and CVRP for SWC route optimization.

This study focuses on proposing a framework for an ICT interface that involves people's participation and transparency of data to all the bodies involved in MSWMS. The study is executed in a simulation environment where assumptions have been made for analyzing demand. A CVRP model with a meta-heuristics algorithm has been used. The objective of the study is to check the feasibility of the proposed methodology concerning reduction is the distance traveled by SWC vehicles.

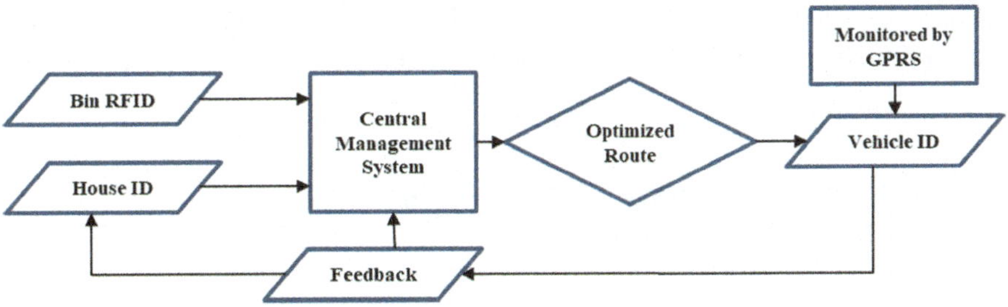

Fig. 1 Information and communication technology framework to be used for waste collection

2 ICT-Based SWM Framework

The ICT-based SWM framework uses ICT tools to monitor and manage the waste from its point of creation to the point of its disposal with a focus on the SWC. The principal objective of this framework is to create real-time data and information modules for all the stakeholders involved in MSWMS of a city. For ICT use in SWC systems, there is a need for data that include, geotagging of waste bins, geotagging of houses, collection of road network data, size of the bins, the volume of waste, type of waste, number of vehicles, types of vehicles, number of waste collection workers and other SWM related data. All the information is stored on a GIS platform under the central management system (CMS) in the municipal authority. In Fig. 1, the ICT framework is presented. The first step is to get a request for waste removal from community bins (RFID bins if they are full) and household/apartment bins with the help of a mobile app-based system. The CMS collects details on the location, amount and type of waste will be known. Based on the information, CMS will carry out the CVRP optimization to provide routes and assign to vehicles to pick up the waste. The vehicles will be monitored with the help of GPRS, and the collection details and locational feedbacks will be sent to the CMS and the app user. The app will also help in creating a support directory, which will look after the user's complaints and suggestions.

3 Methodology

To apply the proposed framework and test its viability in increasing waste collection efficiency, a CVRP model was used with a metaheuristics algorithm. A dataset was prepared for evaluating the possible travel distance savings after route optimization. A total of 40 scenarios were created for a particular day to analyze the effectiveness of methodology by extracting estimation of distance traveled, fuel usage, and CO_2 emission to the finish collection process. Thirty scenarios were simulated to cover collection from bins where it was considered that only 65–95% of them were full. Ten scenarios were considered when 70% of the bins were full. The full bins were randomly selected using a random function. The base case has been considered to be an optimized route for 100% of the bins to be full.

3.1 Mathematical Formulation, Assumptions, and Simulation Setup

The mathematical formulation of the CVRP problem where the decision variable depends on the capacity of the vehicle C, decision variables are modeled according to Eqs. (1) and (2).

$$X_{ijk} = \begin{cases} 1, & ifvehiclektraversesarc(i,j) \\ 0, & otherwise \end{cases} \tag{1}$$

$$Y_{ik} = \begin{cases} 1, & ifbinibelongstorouteofvehiclek \\ 0, & otherwise \end{cases} \tag{2}$$

The objective function is to minimize the travel distance of the garbage pickup trucks as in Eq. (3).

$$\min \sum_{i=0}^{n} \sum_{j=0}^{n} \sum_{k=1}^{k} d_{ij} X_{ijk} \tag{3}$$

Subjected to assumptions: All the vehicles start from the depot Eq. (4).

$$\sum_{j=1}^{n} \sum_{k=1}^{k} X_{0jk} = 1 \tag{4}$$

The vehicles were empty Eq. (5).

$$\sum_{j=1}^{n} q_{0jk} = 0, \forall k = 1, 2 \ldots K \tag{5}$$

Each of the bin locations was visited and by only one vehicle Eq. (6).

$$\sum_{i=0}^{n} \sum_{k=1}^{K} X_{ijk} = 1, \forall j = 1, 2 \ldots n \tag{6}$$

Vehicles that enter a vertex also leave the node Eq. (7).

$$\sum_{j=1}^{n} X_{ijk} = \sum_{j=1}^{n} X_{jik} = Y_{ik}, \forall i = 0, 1, 2 \ldots n; k = 1, 2 \ldots K \tag{7}$$

Any bin location visited by a vehicle was emptied Eq. (8).

$$\sum_{i=0}^{n} \sum_{k=1}^{K} q_{ijk} - \sum_{i=0}^{n} \sum_{k=1}^{K} q_{ijk} = C_j, \forall j = 1, 2 \ldots n \tag{8}$$

The waste collected in the vehicle from the bins cannot exceed the vehicle carrying capacity Eq. (9).

$$\sum_{i=1}^{n} c_i X_{ijk} \leq C, \forall j = 0, 1 \ldots n; k = 1, 2 \ldots K \tag{9}$$

After collection, all the vehicles return to the depot Eq. (10).

$$\sum_{i=1}^{n} \sum_{k=1}^{K} X_{i0k} = 1 \tag{10}$$

The distance between the two nodes was the same in both directions Eq. (11).

$$dist_{ij} = dist_{ji}, \forall i = 0, 1, 2 \ldots n; \; j = 0, 1, 2 \ldots n \tag{11}$$

A tabu search metaheuristic optimization method was applied to determine the origin–destination matrix for the shortest path and time between the waste pickup locations along with the network. The algorithm constructs an initial solution by inserting pickup location one at a time onto the most appropriate route, which is improved upon by resequencing pickup locations of each route, moving pickup location from one route to another, and exchanging pickup location between routes.

3.2 Measuring Parameters

It is noteworthy that CO_2 emission is not modeled in optimization in finding the best path. Instead we have found distance-based path and deduced carbon emission of them to analyze their emission efficiency. The results were then compared depending on the distance traveled by a vehicle, the fuel consumption, and the CO_2 emission. Fuel consumption was calculated using Eq. 12 [8].

$$f_{consumption} = f_{total} - (f_{t_{empty}} + f_{t_{full}})/W \tag{12}$$

Here, $f_{consumption}$ was the fuel consumption for collection (L/tonne), f_{total} was the total fuel in the vehicle to travel the given distance as well as the wait time, f_{t_empty} was the amount of fuel it took the empty vehicle to reach the first bin location, f_{t_full} is the fuel consumed from the last bin location to the depot and W is the amount of waste collected. In the case of this study, since we considered the location of depot just in the middle outside the site, f_{t_empty} and f_{t_full} have been considered zero. CO_2 emission rates were calculated using Eq. 13 [9].

$$E_{CO_2} = W \times D \times EF_{fuel}/(F \times W_{ave}) \tag{13}$$

Here, W is the waste transported by one vehicle, D is the distance traveled by the vehicle, EF_{fuel} is the CO_2 emission factor of the fuel, F is the fuel consumption rate Eq. 14 and W_{ave} is the average waste collected by all the vehicles Eq. 15.

$$F = D/(f_{collection} \times W) \tag{14}$$

$$W_{ave} = \frac{\sum W}{No.of vehicles} \tag{15}$$

4 Results and Discussions

4.1 Study Area

A dataset was prepared in GIS using the existing scenario of Bandra, Mumbai, the total area of 10-km^2. Figure 2a shows the location of Bandra and the building footprint of the area. A total of 3478 buildings were considered for the study. In Fig. 2b, the location of bins has been shown. The bin locations were assumed near the buildings depending on the type of the building and its connectivity to the major roads. A total of 2434 bin locations were made. A random volume of waste was allotted to every bin from 1, 2, to 5 units, it was done using a random function. In Fig. 2c, the road network has been shown with the nodes and vertices. The nodes represent the places

a b c

Fig. 2 Study area. **a** Building footprint and road network. **b** Location of bins. **c** Network where a 360° turn is possible for a vehicle. The location of the depot was assumed just outside of the selected site for easier comparison of results.

To validate the efficacy of the proposed ICT framework and route optimization, a total of 40 simulation scenarios were created and compared against each other. The assumptions made for the vehicle carrying capacity were 500 volume units. The maximum number of orders a vehicle can go to 170. To pick up waste from a bin takes a maximum of 2 min. A total of 15 vehicles were considered for the study looking into the total volume of waste at full area coverage SWC. Here we have considered that not all bins are full on a particular day. Hence a percentage of the bins were full and need to be emptied.

4.2 All Bins Are Full Condition

Based on the methodology described thus far, a CVRP model is applied to simulate the results considering that all bins need to be emptied. In Fig. 3, the optimized route for waste pickup was shown according to the vehicle numbers. The result of the optimized route can also be used to decide which vehicle and crew will leave to pick up waste on what route. All 15 vehicles were needed to go out for pickup and return to the depot within the assumed maximum time of 480 min. All the vehicles were full before reaching the assumed maximum visited bin location count of 170 per vehicle. The total distance traveled to collect all the waste was 172.7 km. The average distance traveled by a garbage vehicle is 11.5 km, and the average stop time for a vehicle to pick up waste is 335 min. Engine of the vehicle was kept running during the transfer of waste from the bin to the garbage vehicle. The cumulative time for which the vehicles are running was 5577 min, and the amount of CO_2 emission is 1.2 metric tons. Using the emission factor of the moving vehicle and that of a waiting vehicle were used to calculate the total carbon emission. The results from simulation

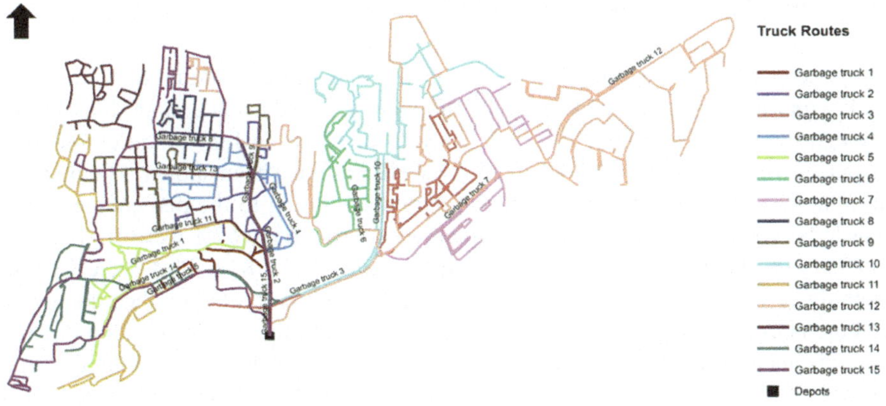

Fig. 3 Simulated route of the trucks at full coverage scenario

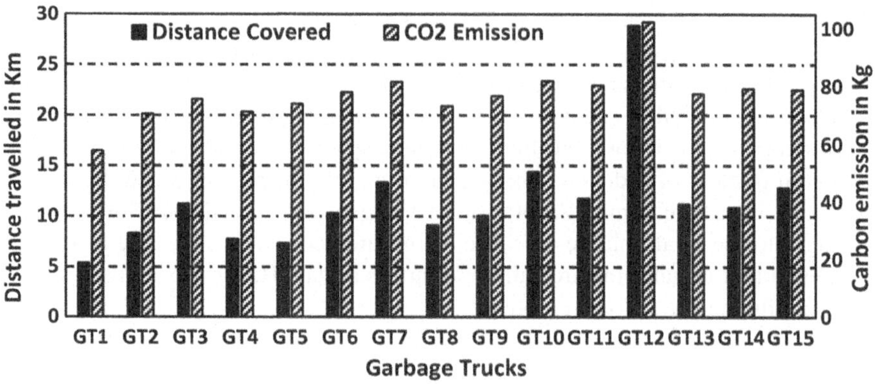

Fig. 4 Distance traveled and carbon emission by each garbage truck at full coverage scenario

have been considered as the base case condition, and all the later results have been compared to it to see the amount of savings possible when the ICT framework is used.

In Fig. 4, the breakup of the distance traveled by each vehicle to pick up the waste and the amount of carbon emission by them. Depending on the location of the bins, garbage truck 1 (GT-1) traveled the least distance of 5.4 km and was full at 132 order count. Whereas GT-12 traveled 28.8 km and was full at 166 order count. Depending on the bin count, each vehicle visits the wait time was calculated. The significant change in carbon emission is seen cause of the waiting time as the difference between the distances traveled by the vehicles was quite minimal at a city scale of things. GT-14 and GT-11 are full have visited a total of 170 bin locations, but the distances traveled by them are small to be in the time constraint of 480 min.

4.3 Varying Coverage from 65 to 95%

Thirty scenarios were assumed, with 65–95% of the bins considered to be full. For each scenario, the bins considered to be full were selected using a random function. Though the waste volumes in the bins stayed the same. In Fig. 5, the cumulative distances traveled by the vehicles in all the scenarios have been shown. The number of vehicles required to collect the waste varied according to the percentage of the bins to be visited. Vehicles required to pick waste from 65 to 67% were 10, from 68 to 75% were 11, from 76 to 81, 83% were 12, from 81, 82, 84 to 88% were 13, and from 89 to 95% required 14 vehicles. The number of bins visited by the added new vehicle is quite less in the next following scenarios. In the scenario with 89%, the total distance traveled to visit all the bins is 173.2 km, which is higher than the total distance for a 100% bin visit scenario. Though the total distance was more, but it uses one less vehicle to complete the collection process. In the scenario with 84% bin

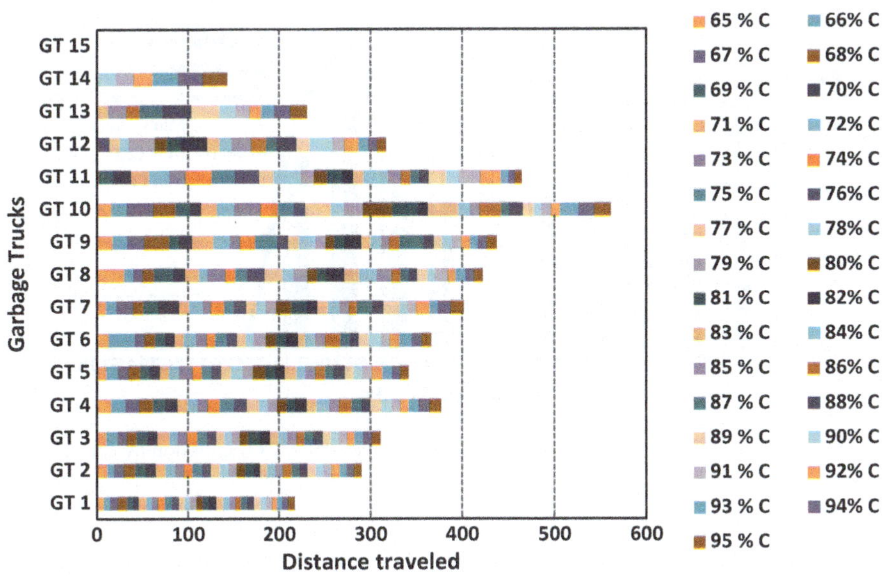

Fig. 5 Cumulative distance traveled by each garbage vehicle for coverage percentage scenarios from 65 to 95%

visited, the number of vehicles used is 12, which is one less than that of scenario 82–83% bin visit scenario. However, a gradual increase in the total cumulative time taken by the vehicles to collect waste for the increasing percentage of bins can be seen. The cumulative time taken to collect waste from 65% of bins visited was 3730 min, which gradually increase to 5297 min at 95% of bins visited. Thus, the time of vehicle operation allows us to estimate the carbon emissions in each scenario. Increase in number of simulations for each scenario with random selection of bins would allow to estimate potential carbon emission. However, due to high computational demand, we have limited our simulations.

4.4 Ten Simulations at 70% Coverage with Randomly Selected Bin Locations

In order to assess the variation in allocation of resources and carbon emission depending on the spatial variation of bin orders, we did cross-sectional simulations at demand of 70% of total bins for Bandra, Mumbai. A total of 10 simulations were conducted where 70% of the bins were considered to be full for waste collection. The locations of the bins were selected using a random function in all the scenarios. The volume of the waste was kept unchanged for all the bins. In Fig. 6, the mean of the distance traveled, and carbon emission by each vehicle was represented with its maximum and minimum values in different scenarios. GT-10 had the most distance

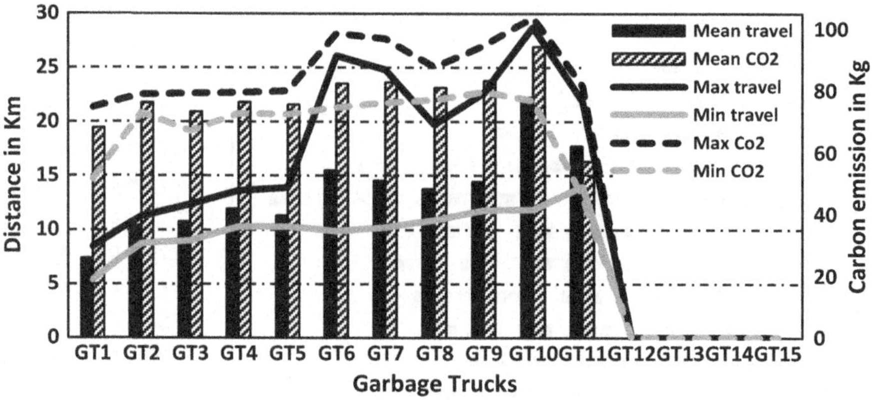

Fig. 6 The mean, minimum, and maximum distance traveled and CO_2 emission by each garbage truck at 70% coverage

covered at 22.22 km, with a maximum of 28.7 km and a minimum of 11.84 km. GT-1 travels the least distance with an average of 7.3 km, a maximum of 8.4 km, and a minimum of 5.3 km. In the case of carbon emission vehicles, which visited more number of bins had more stoppage time, which leads to an increase in the collection time, which was directly proportional to the carbon emission from the vehicle. GT-10 average carbon emission was 0.094 metric tons, GT-11 average carbon emission was 0.057 metric tons. To collect waste, the total average distance traveled by all the vehicles combined came to be 150 km, with the maximum at 151.24 km and a minimum of 142.47 km. The cumulative average time taken for the waste collection was 4005 min, with minor variations in the minimum and maximum values. The average carbon emission for the collection process came to be 0.85 metric tons.

5 Conclusion

The study mainly emphasized on improvement of the SWC system using the CVRP model integrated with ICT for the collection of data. A GIS environment was created to test run the assumptions for collection and transportation schemes. Forty scenarios were created by considering the number of bins full and ready for vacating in percentages (65–95%), the bin locations were selected randomly. The results confirm the importance of considering only full bins for vacating as there was a considerable amount of savings in fuel and less carbon emission happened during the collection process. In case where all the bins were considered full the total distance travelled by the vehicles came to be 172.7 km and total carbon emission was 1.2 metric tons.

In the case of full bins, percentage being 65–95%. The results showed less usage of vehicles, and the cumulative carbon emission was less than the base case scenario.

Though in some of the scenarios, the total distance traveled crossed 172.7 km due to the less number of vehicles being used.

At 70% of the bins to be full and ready for pick up, the results showed an average of 150 km, which is 13% less than the full coverage scenario. The average cumulative time for collection is 4005 min, which is 28% less than the time taken to collect waste from all the bins. The low carbon emission came to be 0.85 metric tons, which was 29% less than the emission produced by visiting all the bins.

The results of the study demonstrated the importance of route optimization technique for MSW collection for minimal fuel consumption and carbon emission. The work also recommends the use of ICT for a transparent, user-friendly system for the data collection on full bins. In future works, the very same methodology can be used for other delivery and pickup services too.

References

1. Phillis, Y. A., Grigoroudis, E., & Kouikoglou, V. S. (2011). Sustainability ranking and improvement of countries. *Ecological Economics, 70*(3), 542–553.
2. Ghose, M. K., Dikshit, A. K., & Sharma, S. K. (2006). A GIS based transportation model for solid waste disposal—A case study on Asansol municipality. *Waste Management, 26*(11), 1287–1293.
3. Nuortio, T., Kytöjoki, J., Niska, H., & Bräysy, O. (2006). Improved route planning and scheduling of waste collection and transport. *Expert Systems with Applications, 30*(2), 223–232.
4. Shekdar, A. V. (2009). Sustainable solid waste management: An integrated approach for Asian countries. *Waste Management, 29*(4), 1438–1448.
5. de Oliveira Simonetto, E., & Borenstein, D. (2007). A decision support system for the operational planning of solid waste collection. *Waste Management, 27*(10), 1286–1297.
6. Kuo, R. J., Zulvia, F. E., & Suryadi, K. (2012). Hybrid particle swarm optimization with genetic algorithm for solving capacitated vehicle routing problem with fuzzy demand—A case study on garbage collection system. *Applied Mathematics and Computation, 219*(5), 2574–2588.
7. Faccio, M., Persona, A., & Zanin, G. (2011). Waste collection multi objective model with real time traceability data. *Waste Management, 31*(12), 2391–2405.
8. Larsen, A. W., Vrgoc, M., Christensen, T. H., & Lieberknecht, P. (2009). Diesel consumption in waste collection and transport and its environmental significance. *Waste Management & Research, 27*(7), 652–659.
9. Lin, T. P. (2010). Carbon dioxide emissions from transport in Taiwan's national parks. *Tourism Management, 31*(2), 285–290.

Association Between Built-Environment and Livability: Case of Mumbai Slum Rehabs

Ahana Sarkar, Nikhil Kumar, Arnab Jana, and Ronita Bardhan

Abstract Daylight and ventilation effective design guidelines remain under-researched in the current habitat design policies of India. This study is of its first kind where environmental metrics and occupant health status were investigated together in the recently developed two hyperdense multirise slum rehabilitation housing (SRH) of Mumbai. A mixed-mode research methodology involving in-situ environmental sensor deployment, computational simulation coupled with household survey and statistical analysis was adopted here to investigate the interrelationship between built-environment geometrics, environmental characteristics and occupant health. Computational Fluid Dynamics (CFD) simulations predicted indoor airflow elucidated that the apartments lacked adequate airflow due to dearth of cross-ventilation. Sky View Factor (SVF) and Daylight Autonomy (DA) levels showcased that the tenements lacked adequate daylight. As a validation, regression analysis elucidated that the households with higher household size, staying at lower floor levels, devoid of exhaust fans, with shut windows, were found to have higher probability of incidence of Tuberculosis. These results conclude a strong association between built-environment and health status. This study would further aid in formulating forthcoming habitat design guidelines and rehabilitation designing policies.

Keywords Natural ventilation · Computational fluid dynamics · Daylight · Sky view factor · Low-income · Liveability · Built-environment · Health

A. Sarkar (✉) · N. Kumar · A. Jana · R. Bardhan
Centre for Urban Science and Engineering, Indian Institute of Technology Bombay, Powai 700046, India
e-mail: ahana.sarkar@iitb.ac.in

R. Bardhan
Department of Architecture, University of Cambridge, Cambridge, UK

A. Jana and P. Banerji (eds.), *Urban Science and Engineering*, Lecture Notes in Civil Engineering 121, https://doi.org/10.1007/978-981-33-4114-2_6

1 Introduction

The multidimensional concept of liveability embraces cognate notions of sustainability, quality of life, and more importantly health of communities. Yet, detrimental living condition leading to degraded liveability ultimately exposes the low-income population to greater health risks. Poor built-environment design often contributes to degraded liveability in these communities. This phenomenon gets aggravated especially in cities of developing nations where inevitable urban growth couples with extreme land and housing shortage.

The metro cities have recently resettled the slum dwellers to hyperdense low-income high-rise rehabilitation apartments with secured tenure, free housing and individual-level basic infrastructure [1]. Yet, these slum rehabilitation policies focusing only on occupancy maximization principle barely consider the liveability parameters like indoor environmental quality, thermal comfort etc. which indirectly affect the occupant health [2]. Therefore, a systematic process-oriented evaluation of these liveability parameters becomes urgent.

The effect of built-environment on the indicators of physical liveability like daylight levels, indoor air quality, ventilation rate and thermal comfort, which indirectly impact occupant health turns integrally crucial. Furthermore, distinguishing the roles played by built-environment design in controlling the significant indicators of natural daylight and ventilation has the prospective to add to advanced design methods to formulate forthcoming climate-sensitive slum rehabilitation housing (SRH) design policies. This assessment becomes exigent especially for the affordable housing, where space and financial complexities restrain the population from utilizing active lighting and cooling techniques to attain desirable indoor comfort levels.

Literature on the association of architectural design and environment has majorly focused on parameters like building disposition, orientation, canyon ratio, courtyard, balcony etc. While the effect of neighborhood, building-envelope and interior level built-environment design parameters on natural ventilation potential has been explored widely, speck of researches on the same parameters on indoor daylight levels elucidates the blind-spot in literature [3, 4]. There also lies a distinct lack of comprehensive research that deals with aggregated impact of same built-environment design parameter on both natural ventilation and daylight quotient. Only a few studies have further linked this association to the occupant health conditions.

Therefore, the novelty of this study lies in embracing an environmental planning analysis from aggregated daylight and natural ventilation perspective on housing design taking Mumbai slum rehabilitation housing (SRH) as case study. The evaluation tool employed here highlights *how built-environment design parameters affect the liveability and indirectly health conditions of the low-income population?* While natural ventilation and daylight are well-established preconditions of livability, poor health conditions are fallouts of degraded livability. The research aims to (a) understand the built-environment design in different SRHs, (b) identify the particular built-environment parameters that significantly impact the daylight and natural ventilation levels in these SRHs, and (c) to understand the impact of built-environment variables on present occupant health-status within these SRHs. As the built-environment design is governed by city-level Development Control Regulations (DCRs), the only alternative remains is to modify these guidelines so as to warrant that the health of people residing in these tenements is not compromised due to mere design faults. This study, therefore by observing the health and liveability facet of slum rehab policies, investigates the architecture-environment interlinkage, thus pointing the current blind-spot of SR schemes. The inferences from this study would aid in formulating the low-income habitat design guidelines of developing nations especially in the global south in future.

2 Background and Study Area

These recently developed hyperdense multirise slum rehabilitation housing are mostly located at disadvantageous sites often sharing adjacency with dumping grounds, biomedical incinerators, refinery as well as ecologically sensitive zones like mangroves, coastal lines etc. This study has considered two SRHs of the M-east ward of Mumbai as case study areas (see Fig. 1).

Fig. 1 Study area (two slum rehabilitation colonies in Mumbai)

2.1 *Natwar Parekh Compound*

The Natwar Parekh SRH compound, constructed in 2008 under in-situ slum rehabilitation program consists of 4800 dwelling units allocated in 59 housing blocks. These tenements are accessed via single-loaded corridor around the periphery of the block, with the toilet and kitchen abutting through an internal ventilation duct. The tenements are single multipurpose rooms of approximately 21 m^2 area consisting of an unsegregated kitchen and individual level toilets. The net housing density is 1103 dwelling units/ha.

2.2 *Lallubhai Compound*

The Lallubhai Compound, constructed in 2003 in Markund, Mumbai consists of seven-floor high 65 towers and around 6318 dwelling units, thus raising the density to 1099 DU/Ha. The one-room tenements of 21.47 m^2 with individual toilets and unsegregated kitchen are placed against a two-meter wide double-loaded corridor and are connected to a common stair-well.

3 Data and Methods

A mixed-mode methodology involving computational simulations coupled with statistical analysis was adopted here to investigate the relationship between built-environment, environmental characteristics and health indicators (see Fig. 2). The study was conducted in two broad stages: (i) simulating the indoor daylight and ventilation conditions of the SRHs followed by (ii) testing the association of the occupants' health conditions with built-environment variables.

The study initiated with built-environment detail measurements of the building blocks as well as tenements, so as to understand the physical conditions of the housing. This was coupled with in-situ environmental sensor deployment, where lux meter and Testo 480® anemometer were utilized to retrieve the daylight and natural ventilation data within the tenements. These experimental data were utilized as boundary conditions for the simulations. While computational fluid dynamics (CFD) simulations were performed to interpret the indoor airflow characteristics, Sky View Factor (SVF) and Daylight Autonomy (DA) were estimated to elucidate the indoor and outdoor daylight levels. The second stage of the study design consisted of a paper-based household survey within these two SRHs to collect information concerning current built-environment conditions and illness symptoms of the residents.

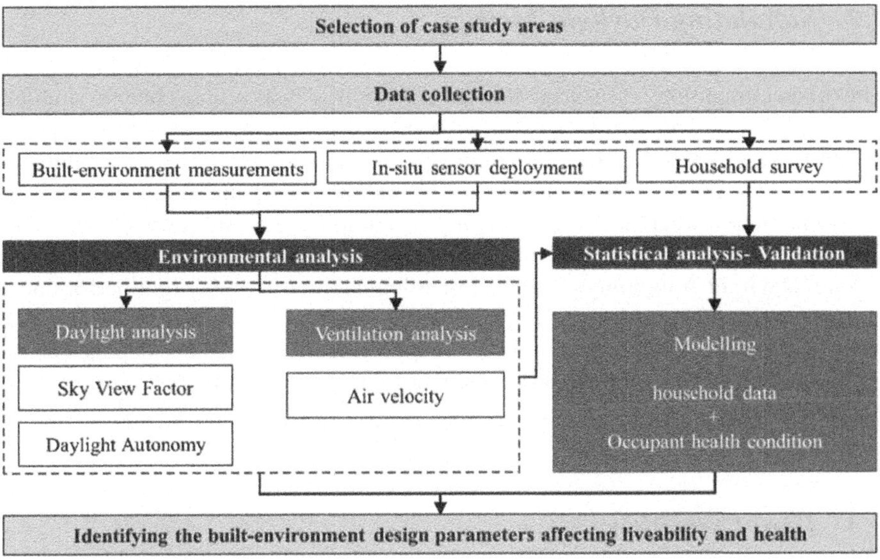

Fig. 2 Methodology adapted in this study

3.1 Simulating Airflow Distribution

This study used CFD simulations in the commercially available CFD tool of ANSYS FLUENT v16.2 to investigate the indoor airflow distribution. However, numerical estimation of natural ventilation becomes challenging owing to unreliability like stochastic behavior of wind flow [5]. Hence this study adopted a deterministic simulation approach using double precision, three dimensional, parallel, and finite volume pressure-based solver. Fine tetrahedral mesh with three times refined meshing was applied in the window (here, air-inlet) and door (here, air-outlet) to resolve the high gradient regions of the flow field with higher accuracy and precision level. The Reynolds-Averaged Naiver–Stokes (RANS) standard k-ϵ turbulence model was employed to calculate the flow fields. Second-order UPWIND scheme and SIMPLE (Semi-Implicit Method for Pressure-Linked Equations) algorithm were utilized to discretize the convection term in the governing equation and couple the pressure and velocity components, respectively. Hybrid initialization with a minimum of 10,000 iterations were used to attain the convergence. Apart from the grid independence test, the solution was considered to converge till 10^{-6} (RMS) of the residuals of the air velocity profile in all the simulated models. The CFD simulations were supported by in-situ environmental sensors. The air was assumed to enter the unit through the window (velocity-inlet) and escape through the door (pressure-outlet). The middle of the window was maintained at a constant atmospheric temperature of 300 K (26.85 °C) with 5% turbulent intensity. The pressure was constant across inside and outside boundaries of the tenement unit, while the ventilation was solely wind driven.

3.2 Simulating Daylight Metrics

The urban climatology measuring tool of "Sky-Helios" was utilized here to simulate the environmental indices of sky view factor (SVF). SVF, being a geometrical concept is a dimensionless parameterization of the quantity of visible sky at a given point and is expressed as a value ranging between zero and one. While the value zero refers to the phenomenon where taller flanking objects completely obstruct the sky visibility, the value of one refers to the situation where the entire hemisphere of sky is clearly visible. Daylight Autonomy (DA), a climatological metric of daylight performance is expressed as the percentage daytime hours received by a given point in space above a specified illumination level (here, 300 lx). The DA level also depends on the building orientation, contextual weather conditions and geographical location of the point in space.

3.3 Data Collection

In October 2018, a preliminary paper-based random survey was conducted in the SRHs to understand their current health status and built-environment conditions. Any household member staying in the house, and at least one from each floor was selected as the respondent. The questionnaire was segregated into three major sections starting with the socioeconomic details including age, gender, occupation, household size, duration of residency, and household income. While the health status questions retrieved information on incidence of upper respiratory diseases including Tuberculosis, the housing conditions related information included occupant behavior data of window operating schedule, floor level, room layout, and active cooling technique usage etc.

A total of 3548 households were surveyed, from Lallubhai and Natwar Parekh compound. Binary logistic regression models were estimated to investigate how built-environment and user behavior impact incidence of occupant ill-health conditions, where reported cases of Tuberculosis (an upper respiratory disease) was considered as dependent variable whereas the built-environment conditions were taken as independent variables. While natural ventilation and daylight are well-established preconditions of livability, poor health conditions are fallouts of degraded livability variables. Around 167 households in Natwar Parekh compound and 187 households in Lallubhai compound used to shut more than 50% of the windows to use them as storage spaces. Additionally, around 1123 households in Natwar Parekh and 976 households in Lallubhai compound had household size of equal to or more than 5, indicating overcrowding in these colonies.

4 Results and Discussion

4.1 Analyzing Indoor Airflow Distribution

The CFD predicted indoor airflow distribution in the tenement units is illustrated in Fig. 3. The "dark blue" bands represent low-air velocity zones (0–0.14 m/sec), inferring that natural ventilation would not be sufficient to provide adequate thermal comfort and desirable indoor air quality within the living spaces. Whereas the "green to red" bands represent high indoor air velocity (0.60–0.96 m/sec), where thermal comfort can be attained without the aid of any electromechanical ventilation techniques. Subjective interpretation of Fig. 3 reveals that the indoor air velocity of the tenements ranged between 0.15 and 0.85 m/sec when outdoor velocity was measured 0.98 m/sec, indicating that built-environment design affects indoor air movement pattern.

Natwar Parekh tenement performed poor in terms of airflow performance with average velocity of 0.15 m/sec in the living space. This can be attributed to the window and door locations, where the air-inlet and outlet (here, door) was located on the same side, thereby, reducing cross-ventilation effectiveness. Since the air entering the living space, moved out through the door without getting distributed in the active zones, the air exchange rate in the living space turned extremely low. Similar airflow characteristic was observed in Lallubhai compound, where despite the window and door were placed opposite to each other, the cross-ventilation was

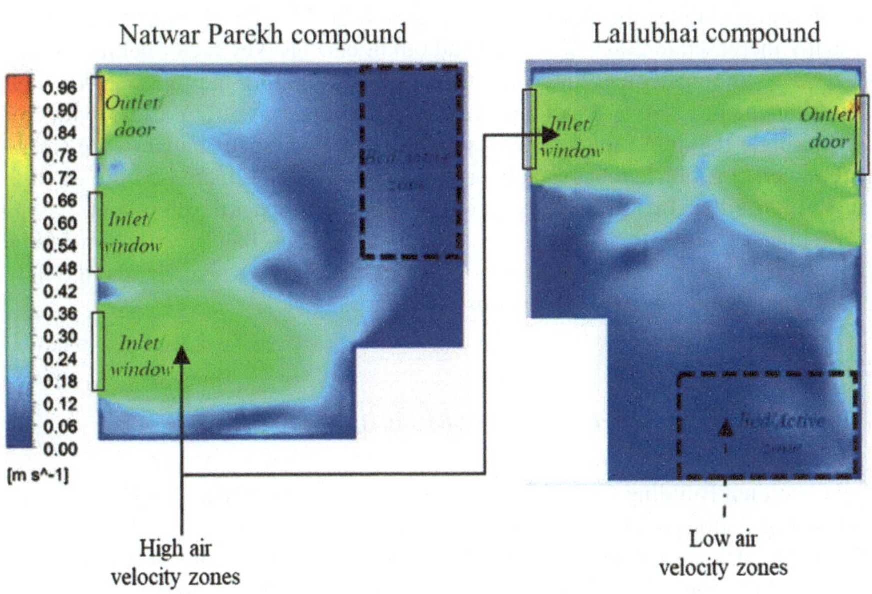

Fig. 3 CFD predicted indoor air velocity in Natwar Parekh and Lallubhai Compound

enabled only at a particular section, while prevailing low air velocity over the living zones. The average air velocity near the air plume measured 0.66 m/sec, however, the indoor airflow was ranged as low as 0.06–0.15 m/sec in the rest of the living space. This phenomenon is majorly due to the location of the outlet with respect to inlet.

The evidence-based thermally comfortable air velocity in tropical countries ranges between 0.2 and 1.5 m/sec [6]. Natwar Parekh and Lallubhai compound with average air velocity of 0.06 m/sec is much lower than the reported comfortable air velocity.

4.2 Analyzing Daylight Levels

SVF analysis, (see Fig. 4) demonstrated the importance of building disposition in impacting urban microclimate. Natwar Parekh and Lallubhai Compound performed had SVF values of 0.1 and 0.2, respectively. The hyperdense building disposition with tall and bulky building blocks surrounding narrow alleys deteriorated the view of sky from the ground. These low SVF values indicated that these colonies also lack access to sunshine and adequate daylight [7].

DIVA-for-Rhino was used for simulating the environmental metric of Daylight Autonomy (DA) [8]. The daylight availability data concerning annual occupancy schedule of 8 a.m.–6 p.m. for winter months and 7 a.m.–5 p.m. for rest of the year were utilized as boundary conditions. The simulated DA levels in Lallubhai compound measured up to 30% only, when a single building was considered. On accounting the surrounding buildings, the DA levels decreased to 5–10% for 1st and 2nd floor and gradually increased to 20–30% for 5th and 6th floors. The key factor behind the low DA levels is the inadequate inter-building gaps, eliminating the inflow of daylight into the tenements. Natwar Parekh performed better in terms of DA levels, with 65% DA for single building scenario. This phenomenon can be attributed to the single-loaded open corridor layout of Natwar Parekh SRH, which increased the daylight availability on the corridor. However, due to highly compact development, the DA level reduced to 10% for lower floors, when surrounding blocks were considered. For Natwar Parekh, Fig. 4ic illustrates that only the corridor received the daylight, keeping the lower-floored tenements in dark.

4.3 Analyzing Built-Environment Design

While National Building Code of India (NBCI) and Greater Mumbai Development Control Regulations (GDCR) had recommended a housing density of 500 DU/Ha and 200 DU/Ha, it was relaxed for special SRA guidelines, i.e., DCR 33(10), where no maximum threshold for density has been mentioned. This has led to the development of SRHs with housing density as high as 1100 DU/Ha. Despite, the household size of Mumbai is 4.6, but for SRH colonies, but for SRHs it is 5.3, thereby further

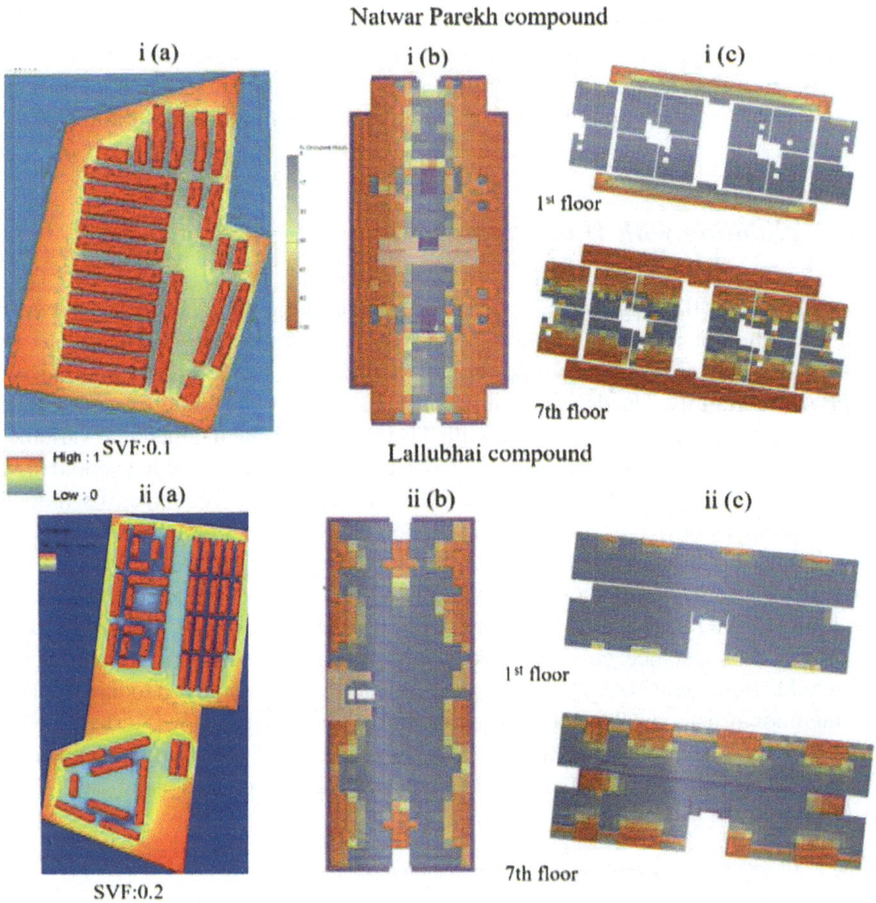

Fig. 4 Daylight simulations of Natwar Parekh and Lallubhai Compound, i(a) and ii(a) Sky View Factor of SRHs, i(b) and ii(b) Daylight Autonomy levels while considering single, i(c) and ii(c) Daylight Autonomy levels of 1st floor and 2nd floor while considering neighboring buildings

increasing the population density. While the prescribed and desirable inter-building gap should be one-third of the adjacent building height, the canyon ratio (Height of building: width of adjacent alley) for the SRHs was found to be 8.33. Another major built-environment design parameter is the window number, size and location. While the prescribed window-to-wall ratio (WWR) is 17%, the WWR values were 10 and 16% for Lallubhai and Natwar Parkeh compound, respectively. Additionally, the windows in Natwar Parekh compound are located on the same side and adjacent to the open common corridor, owing to which the occupant tend to keep them shut in lieu of maintaining privacy, safety and security. This has led to lack of indoor airflow thereby lowering the air exchange levels. Additionally, due to extreme space constraint the occupants used to close windows and use them as storage spaces, further deteriorating the indoor ventilation rate. The higher household size already resulting in over crowdedness, coupled with extreme housing density, ill-designed building arrangement, lower inter-building gaps, inadequate integrated open spaces, and low WWR values resulted in poor daylight and indoor natural ventilation levels.

4.4 Validating with Household Data

Binary Logistic (BL) regression was estimated to explore the relationship between built-environment parameters and reported cases of Tuberculosis (see Table 1). Around 10% of the households reported cases of at least one member being infected by TB in their family. The dependent variable considered in the models was "whether the household had at least one person with TB infection." The independent variables consisted of built-environment were the floor-level of the tenement, household size, the number of exhaust fans available in the apartment, and percentage of windows that was shut to utilize as storage space. The BL estimates demonstrated that the households staying in higher floor levels had lower reported cases in Natwar Parekh compound. Another significant factor was the crowding effect or the household size. For both the SRHs, the factor of household size turned positive and statistically significant. For instance, in Natwar Parekh compound, out of 40 eight-membered households (tenement size: 18.53 m^2), 10 households were reckoned to have at least one member affected by TB. Next, when the impact of ventilation was tested, the BL models estimate demonstrated that the number of exhaust fans in a household had a negative relationship with the reported cases of TB. However, this parameter turned

Table 1 Binary logistic regression model estimation results

	Natwar Parekh Compound						Lallubhai compound	
	Duration less than 5 years (N = 858)		Duration more than 5 years (N = 920)		Duration less than 5 years (N = 626)		Duration more than 5 years (N = 1144)	
Variables	Beta	Std Er	Beta	Std Er	Beta	Std Er	Beta	Std Er
Constant	−2.389***	0.376	−2.754***	0.362	−3.095***	0.451	−2.762***	0.259
Floor level	−0.152***	0.059	−0.048	0.052	–	–	–	–
Household size	0.159***	0.041	0.137***	0.035	0.057	0.080	0.147***	0.039
Exhaust fan	−0.101	0.273	−0.343	0.231	0.181	0.374	−0.459*	0.251
Window_shut	0.004	0.004	0.011**	0.005	0.017**	0.008	0.002	0.006
Model summary								
2loglikelihood	550.067		658.356		333.897		779.170	

Note Significance level—*** at 99% CI, ** at 95% CI, * at 90% CI significant only for the households residing more than 5 years in Lallubhai compound. Also, the households with higher percentage of windows closed reported higher cases of TB. These results indicate that ventilation either mechanical or naturally wind driven, turns integrally crucial for maintaining healthy indoor environment.

5 Conclusion

This study adopts a pluri-dimensional methodological approach to explore the relationship between built-environment design and liveability especially health conditions in low-income resettlements. This study identified that the architectural design parameters of integrated open space, inadequate inter-building gaps at neighborhood scale, internal corridor design, floor-level, window number, size and location at building envelope level has a strong relation with site-based and indoor ventilation potential and daylight availability levels. Moreover, the window opening behavior might impact the indoor health conditions. This study through statistical analysis indicated that the built-environment has a strong association with occupant health status (especially, upper respiratory diseases like Tuberculosis).

Use of exhaust fans, openable windows and high-level air-outlet or ventilator should be promoted in the SRH tenements for efficient ventilation rate. Buildings in a site should be laid and oriented in a way to enable air penetration and daylight availability. In this case, rational building disposition is necessary. The urban canyon ratio, a crucial morpho-metric should be designed in a way to enable airflow in lower floors of the buildings. The regulation of height of building to width of adjacent road in the current DCR should be modified. The parameter of building footprint or ground coverage ratio should also be taken into consideration while designing the mass scale housing. Adequate integrated courts or open green spaces should be provided for enabling adequate ventilation and daylight. The other crucial environmental metrics such as humidity, CO_2, and distinct indicators of TB incidence like proximity to patients, personal hygiene etc. have not been reflected in this study and hence needs further investigation. These afore-mentioned design parameters need to be incorporated in the bye-laws for formulation of forthcoming low-income housing in cities of developing nations especially in global south.

References

1. Lueker, J., Bardhan, R., Sarkar, A., & Norford, L. K. (2020). Indoor air quality among Mumbai's resettled populations: Comparing Dharavi slum to nearby rehabilitation sites. *Building and Environment*. https://doi.org/10.1016/j.buildenv.2019.106419.
2. Bardhan, R., Debnath, R., Malik, J., & Sarkar, A. (2018). Low-income housing layouts under socio-architectural complexities: A parametric study for sustainable slum rehabilitation. *Sustainable Cities and Society, 41*:126–138. https://doi.org/10.1016/j.scs.2018.04.038.

3. Montazeri, H., Blocken, B., Janssen, W., & Hooff, T. CFD analysis of wind comfort on high-rise building balconies: validation and application. *Iawe. Org*, pp 1674–1681. Retrieved from https://www.iawe.org/Proceedings/BBAA7/H.Montazeri.pdf
4. Sarkar, A., & Bardhan, R. (2019). Optimal interior design for naturally ventilated low-income housing: a design-route for environmental quality and cooling energy saving. *Advances in Building Energy Research,* 1–33.
5. Indraganti, M. (2010). Adaptive use of natural ventilation for thermal comfort in Indian apartments. *Building and Environment, 45*(6), 1490–1507.
6. Tantasavasdi, C., Srebric, J., & Chen, Q. (2001). Natural ventilation design for houses in Thailand. *Energy and Buildings, 33*(8), 815–824. https://doi.org/10.1016/S0378-7788(01)00073-1.
7. Lin, T., Tsai, K., Hwang, R., Matzarakis, A. (2012). Quantification of the effect of thermal indices and sky view factor on park attendance. *Landscape and Urban Planning, 107*(2).
8. Kazanasmaza, T., Grobe, L. O., Bauer, C., Krehel, M., & Wittkopf, S. (2016). Three approaches to optimize optical properties and size of a South-facing window for spatial Daylight Autonomy. *Building and Environment*. https://doi.org/10.1016/j.buildenv.2016.03.018.

NO ROAD CITIES; Study on the Effects of Road Infrastructure on Land Use, Environmental Quality and Human life—A Case of Sarovaram Bypass, Calicut, Kerala

Arun Baby M. Wilson and M. A. Naseer

Abstract Roads are the backbone of transportation network. As described in Wegener's Land use—Transportation cycle, transportation brings in changes in land use, which accelerates the growth. In India, the car ownership is skyrocketing which posing a high threat on roads to bear a much higher volume of vehicular traffic than their capacity. A very important fact we ignore is that all these growths are happening at the cost of our environment. Environmental degradation is happening as road infrastructure is getting further developed. This study attempts to analyze the impact of road transportation on the environment and human life. On identifying the adverse impacts of road infrastructure, a proposal is being put forward highlighting its advantages and effects on the environment and human life. The main objective of the study is to analyze the changes that can occur if the already laid road infrastructure is scrapped off from the ground surface and provide with other means where sustainable alternative transportation modes are available.

Keywords Sustainability · Environmental engineering · Road transportation · Wegener's land use—transportation cycle · Surface paving

1 Introduction

The cities are now being filled with vehicles and services related to it. The change in land use pattern is so evident that the open and recreational spaces are decreasing as the land uses related to the transportation sector is increasing drastically. The impact that these changes causing upon our ecosystem is alarming [1]. While discussing all these issues, consider a situation where there is no road present. There is a concept being discussed worldwide among the ecocentric communities which is being addressed as the "PARKING TO PARKS" concept. The concept tells the idea of converting the parking plots into park spaces which eventually benefits the quality of our life.

A. B. M. Wilson (✉) · M. A. Naseer
National Institute of Technology Calicut, Calicut 673601, India
e-mail: ar.arunbaby@gmail.com

A. Jana and P. Banerji (eds.), *Urban Science and Engineering*, Lecture Notes in Civil Engineering 121, https://doi.org/10.1007/978-981-33-4114-2_7

Transportation is the key component of development. Every development is been always served by a transportation network, which leads to progress in economy and development. With the variety of modes of transportation, the need for transportation networks is increasing day by day as everyone is depending upon the motorized movement for various purposes. The car ownership statistics is drastically increasing in an alarming fashion as the transportation infrastructure would face a deficiency in infrastructure in the near future. As the transportation is helping in economic growth of the humanity, it impacts the environment and the human health very adversely.

While thinking about a life without roads, there is always a question on how to survive without road transportation. For a city to function without roads is not easy as it would be the most used mode of transportation. Keep that in mind, this study is trying to understand how the environment and human life is affected if the landcover used for road transportation is cut down to the minimum. To evaluate this, the concept of "No Road Cities" is put forward as an idea in which the city is free of congested roads and polluting traffic.

This paper is intended to study the environmental and land use impacts the road transportation is possessing upon a region and the changes that could be observed if the road is taken off. To put the analysis and results into real world, a study had conducted upon the impact of road transport and related activities on the land cover and the environmental conditions in an area of interest which is at the city of Calicut in Kerala, a portion of Sarovaram bypass road stretching from Eranhippalam to Arayidathupalam.

1.1 Objectives and Relevance of the Study

The major area of focus is the environmental impacts and the change in land use pattern that is happening due to the presence of the transportation infrastructure. The objectives are framed considering the key areas that need to be addressed through the study. The objectives are as follows:

1. Study the land use pattern of the area related to road transportation
2. Study on environmental conditions
3. Identification of alternative transportation methods
4. Assessment of conditions after implementing alternative methods (Fig. 1).

A study on the impact of road transportation is important because of the adverse effect it is creating upon the environment and the human life. When considering the area selected, the unique ecosystem that present is suffering from various negative impacts because of the road transportation and development happening due to the presence of the road [2].

The relevance of the study is understood based on the observations made on the following criteria as the road transportation is affecting the daily life and the environmental quality in different ways.

Fig. 1 Illustration showing the cost differences in different modes of transportation

1. Increased transportation infrastructure
   a. Attracts more traffic
   b. Addition of more roads
2. Change in land use pattern
   a. Environmental pollution
   b. Change in environmental quality
   c. Fuel combustion emissions
   d. Noise
   e. Permeability of surface
3. Increasing temperature—Heat island
   a. Change in tree cover and ground cover
   b. Heat reflectance of the paving materials
4. Importance of the eco-system present
   a. The mangrove forest present
   b. Nonrenewable unique ecosystem.

There always rise a question on this concept, “Why No Roads”? Consider all the costs associated with road transportation and it is understood that the road transportation is the costliest method of transportation when taking the account of line-haul cost. When looking into the overall cost associated with the road transportation, taking all the parameters associated, the most important cost is the cost of environmental degradation.

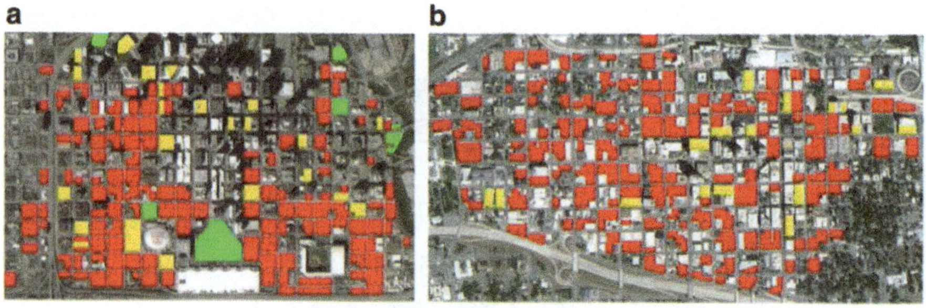

Fig. 2 Land use pattern related to road transportation; **a** Houston, TX, **b** Little Rock, AR

Table 1 Land use pattern related to road transportation of Houston, Tx and Little Rock, AR

Land use	Color code	Houston (%)	Little Rock (%)
Surface parking		21	26.5
Garage parking		3.7	2.7
Park space		2.6	0.0
Street area (including sidewalks)		39.7	32.0
Total area for rights-of-way plus off-street parking		*64.7*	61.2

1.2 *Road Transportation and Land Use*

A study had conducted in different places in the USA and found that the land area occupied for road transportation and related activities is much higher compared with other land uses and it literally killing the open spaces and atmospheric quality. Similar methodology is opted for studying various cities and the change in land use pattern is observed allover is same (Fig. 2 and Table 1).

1.3 *Methodology*

The study starts with secondary data collection on the environmental impacts of the transportation infrastructure. Then the analysis on the site identified is performed based on the data collected and on field visits. The change in climatic parameters and environmental quality are assessed according to the land use pattern of transportation infrastructure.

A baseline data analysis was performed to study the adverse effects of road infrastructure upon the environment and human life. The area selected is a stretch of Calicut city in Kerala having the most environmental importance in the city and witnessed significant changes in land use and other related changes. To establish the advantages for not having a road infrastructure along the selected stretch, a documentation on change in land use pattern with road infrastructure had done. The environmental characteristics were listed out with measured and observed changes in atmospheric conditions, pollution level, geographic properties, hydrology and social aspects. Alternative transportation modes (Pedestrian pathways, cycling tracks and Waterway transportation) have proposed using the available stretch of land and water canal. A post implementation analysis had done in terms of economic and environmental effects of the proposal with the help of data collected.

2 Literature Reviews

2.1 Geithoorn, Netherlands

Geithoorn is being called as the "Little Venice of Holland". As the nickname suggests, the village is very much similar to the Venice in its characteristics. The major attraction of the village is that there are no roads present within the village. The entire village is connected to each location through water ways and canals. The village is located on the northern part of the Netherlands and in close proximity to the Mediterranean Sea.

The waterways are laid across the village connecting all portions and every residences and buildings are accessible through water. The use of an alternative transportation system had benefitted the entire village in various aspects (Fig. 3).

Along with the extensive presence of canals and bridges, there are also numerous cycling and walking paths crisscrossing the town. This is how they promoting the zero-emission mobility to all through non-motorized transport.

Fig. 3 Images of waterway system in Geithoorn and the only ground vehicle is bicycle

The major advantages that the community is getting for not having roads in the village are satisfying the quality of life and environment in the area. The advantages are:

- Land used for transportation is 1.2% of the total land
- No heat generation due to vehicular movement—1.5° Celsius lesser temperature.
- Road accidents are not there
- No Parking spaces—improves park spaces
- Pedestrian friendly—child friendly village
- Very less air contamination—no vehicular exhaust
- No hard ground covers—not opposing the water permeability.

Geithoorn is a tourist village right now. But the use of alternative transportation system to avoid surface road transportation is making the area a very good example of preserving the nature as it is. The village is an example in explaining the fact that life is possible even without roads.

2.2 *Freeway to Greenway, Seoul*

There are regions where the society and government had realized the threat that the transportation infrastructure is posing upon the environment and people and decided to go choose environment over economy. The freeway to Green way project in Seoul is a classic example of reducing the transportation infrastructure and re-inventing the richness of nature.

The project was to remove a freeway, which was an elevated road of 50 m wide, to revitalize the urban stream lying beneath. The two sides of the freeway were totally separated with the road structure and through the project the central area became a beautiful urban recreational space and it acts as a bridge between the two sides of the street.

Key elements

The government was decided to go with the proposal of removing the 50 m wide freeway of 5.8 km length even though it will cause a huge financial loss, but ultimately it could result in better atmospheric conditions and quality spaces. The procedure as follows;

- Removal of elevated highway concrete structure.
- Daylighting of a previously covered urban stream.
- Creation of an extensive new open space along the daylighted stream.
- Creation of pedestrian amenities and recreational spaces (two plazas, eight thematic places).
- Construction of 21 new bridges, reconnecting the urban fabric (Fig. 4).

As a result of the project, they experienced and documented a huge difference in so many factors related to urban life. The below listed are the major results of

Fig. 4 **a** The 50 m wide freeway bridge in Seoul, **b** The stream turned into urban park after removal of the bridge

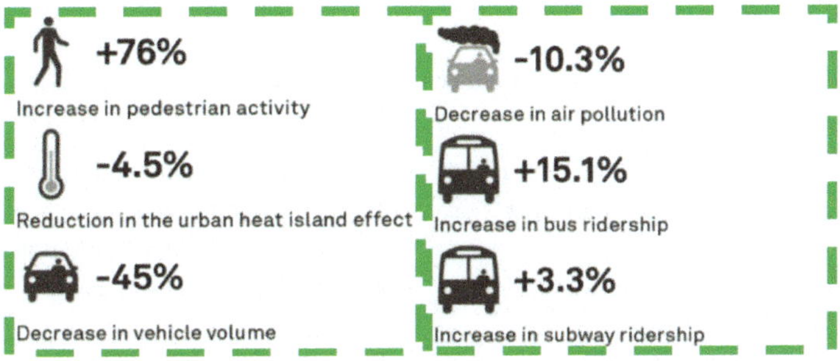

Fig. 5 The changes happened after the introduction of urban park in Seoul

the project after implementation. It shows how well the environmental quality is enhanced with the introduction of the urban park instead of transportation network. All these factors collectively enhance the quality of life and healthy lifestyle for the people (Figs. 5, 6 and Table 2).

The environment is preferred over the economy in this project where reinventing the urban stream would benefit the entire region and community. The revived urban park served as the best recreational place in the region, which attracts so many visitors both domestic and foreigners. This shows the possibility of renewing the environmental quality through the best interventions and practices such as restricting actions that degrade the environment and natural habitats.

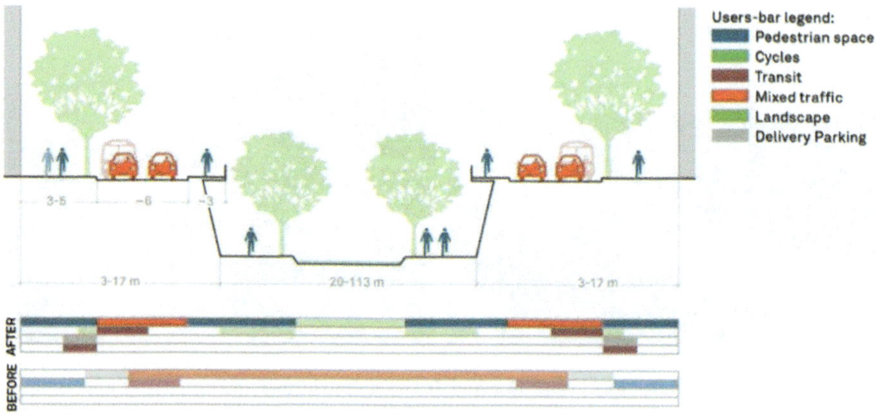

Fig. 6 Section showing the profile of the area before after the urban park

Table 2 Priorities of the project on environment over economy

Economy v/s Environment	
• Economic growth at the cost of Environment	• Environmental quality will improve
• Loss of money spent or Environmental benefits	• The character of the area would change from just a road to the hub of various activities
• Scrapping of the roads will result in a loss of huge amount of money	• Improves health and well being
• The movement becomes slower	• The emphasis is on Quality
• Emphasis on the Quantity	

3 Impact of Road Infrastructure on Human Life and Environment

The different cases discussed showed the extend of the relation between road infrastructure and the human life and environmental quality of an area. To analyze this in a real case, a stretch of Sarovaram bypass in Calicut, Kerala had taken in consideration and studied different aspects to forecast the environmental and life quality if the road is not present.

3.1 Sectors of Canoli Canal

The Canoli Canal has been divided into eight sectors for the ease of management of the cleaning project. Each sector been assigned to the residents and institutions nearby to maintain and undertake the cleaning process. The sectors are demarcated with proper signages and Sector boards are installed along the banks of the canal.

Fig. 7 Sector 6 of canoli canal—sarovaram biopark section

The sector division had done based on major landmarks and traffic intersections. The quality and cleanliness are differing from sector to sector based on the activities that have been taken place along.

The most important sector is sector 6 when considering the potential and activity along the area. Sector 6 is stretching from Eranhipalam junction to Arayidathupalam junction. Sarovaram Bio Park is along this stretch with the presence of large mangrove forest and marshy land, which is having a very unique ecosystem. The Sarovaram bio park is utilizing the biodiversity and ecosystem present along the canal. But the invasion of commercial activity had taken out the region and the area are no more suitable as a recreational place. The most pristine water that been coming from the Elanthur river is now becomes as dirtier as the wastewater from an industry on reaching sector 6 (Fig. 7).

3.2 *Sarovaram Bypass*

Sarovaram bypass is along the canoli canal stretch sector 6. The road is connecting two major nodes; Arayidathupalam junction and Eranjipalam junction. These two junctions are the nodes, which are connecting all the major roads from outside Calicut to the core city. These junctions are supposed to experience the highest peak traffic volume in the city. During office and school timing, the bypass is overcrowded and the traffic is slow moving.

The presence of unique ecosystem is giving this area more importance than any other region along the bypass. The mangrove forest is a must preserved ecosystem and hence the canal also. The land use pattern along the bypass is mostly of commercial land use, which are related to road transportation. Fuel stations and parking lots are also there along this stretch. There is not much of other activities happening along the road as it is just a connecting road between two major junctions (Fig. 8).

The area is being denoted as dream city in the proposed master plan of Calicut city for the year 2035. High density commercial land use and mixed-use land use are proposed for the region along the bypass. For reviving the Cannoli canal, there should be proposed a green corridor by scrapping of the road that is present. The nearby roads got developed because of the existing Sarovaram bypass. This constantly pulling higher volume of traffic into this region (Table 3).

With an urban park being introduced the land use pattern of the entire region will change as there will not be any commercial activity related to transportation would be happening [3].

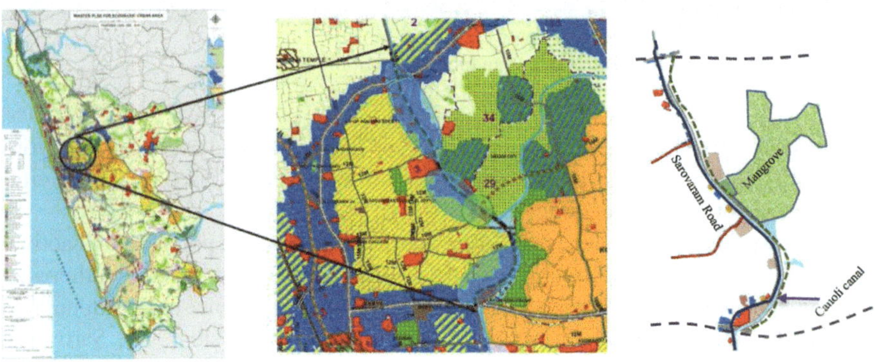

Fig. 8 Sarovaram Bio Park, Calicut, Kerala

Table 3 Land use distribution along Sarovaram Bypass

Land use	Area with road (sq.km)	Area without road (sq.km)
Water body	0.11	0.11
Mangrove	0.46	0.46
Recreational	0.01	0.20
Vacant land	0.06	0
Road	0.1	0
Fuel stations	0.006	0
Commercial—TR	0.05	0
Commercial	0.02	0.1
Public	0.01	0.01
Parking and footpath	0.06	0.02

3.3 Urban Infrastructure to Urban Park, Sarovaram Bypass

The land use pattern sows that the area is being developed based on the transportation network present and bigger future proposals had also put up in making this stretch of the road into a commercial hub. This will impact the environmental quality of the area to a very greater extent. With the help of collected data from various organizations and departments on the environmental quality, a post situation analysis is done for the same stretch of road as what would be happening if the road running through is scrapped off (Fig. 9).

Ground Cover Change

There is a change in the ground cover due to the black topping of the roads. This creates a mask over the natural ground cover of the area preventing sunlight from directly reaching the surface.

- Ground cover change—216,000 sq.m
  - Paved area—asphalt concrete, cement concrete and paving blocks
  - Heat reflection of surfacing materials
  - Asphalt—0.05–0.1 sq.km
  - Concrete and pavers—0.35–0.116 sq.km
  - Increases the temperature by 3–4 degree Celsius.

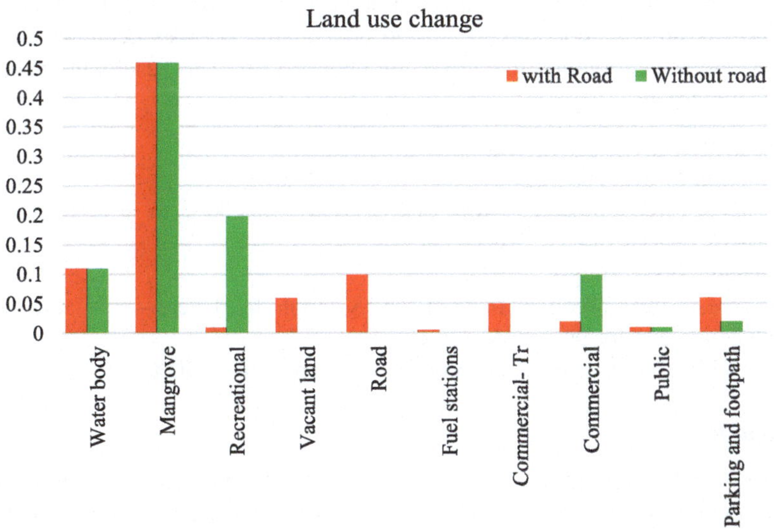

Fig. 9 Land use area distribution along the Sarovaram bypass before and after the urban park introduction

Water Permeability

As the surface is being covered, the surface paver is restricting the water penetration into the earth. This will result in reduction in groundwater table and reduced fertility of the ground.

- Water permeability reduction
  - Soil—0.1–1 cm/s
  - Asphalt Concrete—0.00001 cm/s
  - Cement Concrete—0.00000001 cm/s
  - Rate of decrease in groundwater
  - Canoli Canal—40–60 cm decrease in water level over 6 years
- Increased surface runoff
- Impact on the environment and Ecosystem
  - Earth became more dry
  - Not suitable for growing plants—drying out of plants
  - Mangrove forest depletion

Change in Tree Cover and Vegetation

The entire region is just asphalt covered and concrete paved and thus there is no vegetation is present. For making a road and related services occupied the land surfaces, the vegetation and tree cover are chopped off. This results in changes in climatic conditions.

- Change in tree cover and vegetation
- Reduction in CO_2 absorption
- One acre of tree removes 2.6 tons of CO^2/year—21,600 sq.m = 5.34 acre, could remove 13.9 tons of CO^2 every year
- Reduced oxygen supply
- One acre of trees produces oxygen for 18 people for 1 year—5.34 acre—could produce oxygen for 96 people a year
- Loss in UV shield—tree shade reduces UV exposure by 50%

Pollution levels

Because of the increased tree cover and reduced vehicular traffic, the atmosphere and the environment will see a greater increase in the environmental quality as pollution levels will be getting lowered. The level of SPM content in the air can be brought down by 30–45% within 1 year and thus enhancing the environmental quality a lot [4].

Fig. 10 Present condition of Sarovaram bypass

3.4 Post Situation of Sarovaram Urban Park

To understand the impact on human life and environmental quality, along with the forecasted data, a post situation is visualized to give it a clearer idea. Illustrations prepared to bring in the post situation are as follows (Fig. 10).

High-density commercial land use and mixed-use land use are proposed for the region along the bypass. This will attract more intensive development into the area and thus will destroy the unique ecosystem present. Considering all these factors along with the renewal of the Cannoli canal, there should be proposed with a green corridor by scrapping of the road that is present. There are nearby roads which got developed because of the Sarovaram bypass. This constantly pulling higher volume of traffic into this region. There will always be a question on removing the road as it is economically not viable, but taking the environmental benefits that are going to be achieved into account, the solution seems very much feasible [3] (Fig. 11).

The urban green corridor will have complete tree cover and vegetation with waiting benches and areas. Bicycle paths would be demarcated to encourage NMT. A complete stretch of commercial shops can be provided as the pedestrian shopping area will receive more activity than a shopping area along a road with moving vehicular traffic [5].

The entire canal has to be cleaned and renewed. But bypassing the traffic that the roads along the canal receives are much difficult in all other regions compared with

Fig. 11 Sarovaram bypass after urban park implementation

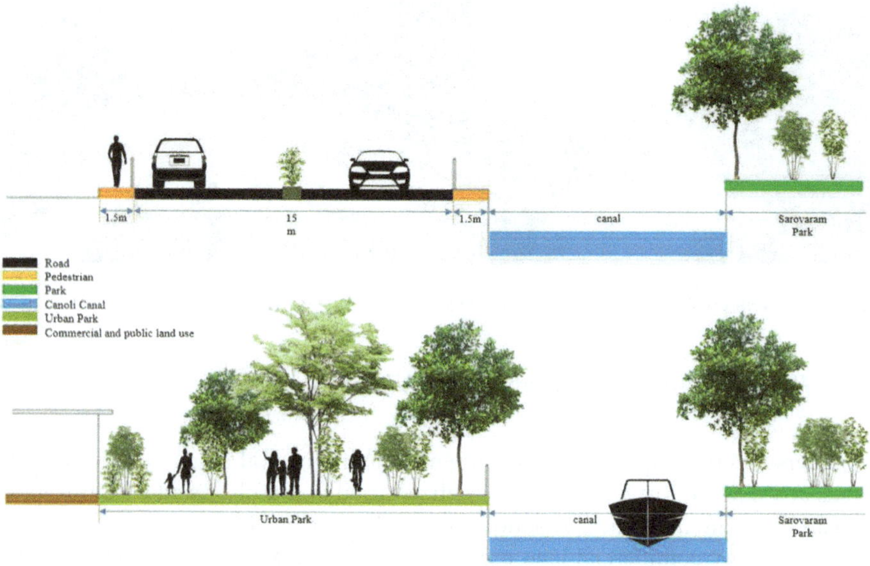

Fig. 12 Section showing Sarovaram bypass before and after urban park implementation

Sarovaram bypass. There are roads already developed near the bypass, which are connected to the main road and can be used to steer the traffic.

As in the case of freeway to greenway project in Seoul, the vehicular traffic will get reduced after some time of implementation as no one would willing to come by their own vehicles if road and parking spaces are not available. The activity being offered along the urban green corridor would attract more people as it is completely along a natural ecosystem and the presence of waterway transportation can benefit the tourism sector in future (Fig. 12).

4 Conclusion

Through this study, it is found that the road infrastructure is having a very prominent role in defining the land use pattern and the development of an area. The land being used for different activities related to road transportation is taking up the major part in the land use distribution.

Analyzing the environmental changes that could bring in by the introduction of urban parks and reducing the road infrastructure, the inferences drawn out say how the road transportation land use is killing the environment and human life. On the other hand, we have to realize that the introduction of urban park in the area of study could bring in oxygen supply for only 96 people every year. So, the amount of green cover needed to produce oxygen for the entire population lives in urban area is really huge. The fact that we still eating up our environment in the name of development without understanding that we are literally risking our own life is alarming.

The "No Road Cities" concept is an extension to the "parking to park" concept as it can bring in a very healthy environment utilizing the alternative transportation methods available and strategically distributing the transportation network so that it will not sprawl into the ecosystems present and thus not harming the environmental quality.

References

1. Aly, S. S. A., & Amer, M. S. E. *Architectural engineering and environmental design department, Arab academy for science, technology and maritime transport, Egypt*: Green Corridors as a response for nature: greening Alexandria city by creating a green infrastructure network.WIT Transactions on Ecology and the Environment, Vol. 138.
2. Louiza, H., Zeroual, A., & Haddad, D. (2015). Impact of the transport on the urban heat Island. *International Journal for Traffic and Transport Engineering., 5,* 252–263. https://doi.org/10.7708/ijtte.2015.5(3).03.
3. Eraghi, S. G., Meschi, M., & Gholampour, S. (2015). Studying the relationship between urban green corridors and sustainable urban landscape. *International Journal of Science, Technology and Society*. Special Issue: *Research and Practice in Architecture and Urban Studies in Developing Countries, 3*(2-1), 36–40. https://doi.org/10.11648/j.ijsts.s.2015030201.18.
4. Neeta, S., & Gautam, G. (2013). Air pollution due to road transportation in India: A review of assessment and reduction strategies. 8.
5. Goyal, V. S., & Advisor, U. T. (2014). Integrating and planning for non-motorized transport in urban areas, (August), 12–13.

Informal Settlement as a Self-sufficient Responsive Neighborhood: A Case of Hollywood Basti—Gulbai Tekra, Ahmedabad

Digisha Mehta, Pratyoosh Madhavi, and Yasin Kabaria

Abstract Slums in India are characterized by unhygienic living conditions like filthy streets, cramped houses, insufficient light and ventilation and poor services. Apart from this their lies an organic solution of housing brought by democratic will of the dwellers within. Though not created by architects or planners, they are housing by the people responding to their specific needs. The concern of this paper is to bring out those specific features, which should be retained while working on betterment of slums. Government's answer to this issue has always been addressed in various programs under housing needs. Since early 70s till today, there have been schemes like Environmental Improvement of Urban Slum scheme (1972), Sites and services scheme(1980), Urban basic services for the poor(1985), The night Shelter Scheme for Pavement dwellers (1990), National Slum Development Programme (1997), Valmiki Ambedkar Awas Yojana (2001), National Slum Development Programme (2006), Rajiv Awas Yojana (2013), Pradhan Mantri Awas Yojana (2015) etc. Whether to relocate and rehabilitate them, or redevelop and give them better-living conditions in the same location, government has constantly been struggling with various appropriate alternatives for finding solutions to these issues. But still one struggles to find a successful slum improvement program or project that can be studied and its approaches or principles can be adopted while finding solutions to similar issues. It is obvious that there cannot be a one size fits all solution to deal with such an issue. The specific needs and links have to be addressed in solutions provided for betterment of these slums. The intent of the paper is to understand the characteristics of the built fabric of an informal settlement by adopting the framework of Ian Bentley in the book, Responsive Environments, which is an outcome of concerns over the loss of human considerations in the modern urban environment. The methodology adopted for this paper relies on primary survey of the slum through visual observations, personal interviews and detailed documentation of one "Vaas." The case taken in this paper is of an informal settlement, which lies in a central location of Ahmedabad named

D. Mehta (✉) · P. Madhavi · Y. Kabaria
Nirma University, Ahmedabad, India
e-mail: digisha.mehta@nirmauni.ac.in

P. Madhavi
e-mail: pratyoosh.madhavi@nirmauni.ac.in

A. Jana and P. Banerji (eds.), *Urban Science and Engineering*, Lecture Notes in Civil Engineering 121, https://doi.org/10.1007/978-981-33-4114-2_8

Hollywood Basti in Gulbai Tekra. It is well known for sculpting using Plaster of Paris and has a good hold on market at local and national level. This settlement lies in a high development potential zone as it is surrounded by mixed use landuse and falls under Transit oriented Development zone as per Ahmedabad Development Plan. The settlement has gradually evolved from a small group of people native of Rajasthan, who migrated and settled here since 1856 and have blended well with the Gujarati culture over time. This paper brings forth various aspects under permeability, variety, legibility, robustness, richness and personalization that need to be considered and retained while undertaking any initiatives for upgradation or rehabilitation of these informal settlements.

Keywords Informal settlements · Spatial fabric · Social needs · Organic growth · Responsive features

1 Introduction

Informal settlements are solutions found by people themselves while struggling to cope with the challenge of financial feasibility, housing shortage, growing market prices, etc. along with the need to survive in an urban setting for economic reasons. They have found a way of surviving and living with the limited resources available to them. Although their living conditions are identified as unhealthy, filthy, densely packed, unorganized, etc. but within it, there also lies an underlying order which is a direct outcome of their social, cultural and all the basic needs. These settlements are alive with activities throughout the day owing to the manner in which they are designed and used by its residents. They are a highly democratic, personalized setting created by the users themselves, a direct reflection of them and their lifestyle embedded into their built form. Although it is extremely necessary to give them better hygienic and safer living conditions but it is also important to understand the underlying order existing within their built fabric to better understand them and their needs before any solution could be suggested.

Informal settlements are one of the prime concerns of the government and various approaches have also been considered for dealing with the same. One feels that there is a gap between what is offered through rehabilitation or redevelopment programs and the actual needs of the people whether in terms of lifestyle, culture, work-life connectivity, etc. One understands that there cannot be a one size fits all solution to deal with such an issue. The specific needs and links have to be addressed in solutions provided for betterment of these slums. Some such attempts where efforts have been made to retain the social dynamics of the slum without disturbing the work-place link along with reinforcing needed infrastructure and services are Aranya housing at Indore and IFFCO Housing, Gandhinagar by B. V. Doshi, Netajinagar Pune by Prasanna Desai, etc.

Informal settlements have been a popular topic of research for a number of aspects. But one rarely finds a study on the built fabric and spatial character of the settlement. Therefore, various approaches for study and analysis of built form by various researchers and planners have been studied for developing the framework for this study. Since the study aimed at understanding the users, their lifestyle, choices, etc., by understanding their spatial fabric which they have designed for themselves, one of the major criteria for the framework was the understanding of the degree to which the built fabric responds to its users. Lewis Mumford rightly said at a time when the planners were dealing with various urban issues and concerns owing to the Industrial Revolution and the sudden outburst of urban migration and population growth in his paper titled "What is a City?" that in a city its "….social facts are primary and the physical organization of a city, its industries and its markets, its lines of communication and traffic, must be subservient to its social needs" [7]. Aristotle in his book "Politics" defines a city as one in which each citizen should be able to know all the others by sight [4]. Besides this authors like Kevin Lynch have also focused on the imageability of a city (here built environment) giving five parameters that can be considered as primary elements that create the image of a city among its users [5]. Jane Jacobs talks about the importance of users and their built environment that makes the spaces alive and bustling or dead and unsafe [2].

This paper has taken Ian Bentley's book—"Responsive Environment" as a framework for studying different characteristics of the built environment of the slum since this book begins with the idea that "… there are important relationships between social life and the arrangement of the built environment." This book is an outcome of concerns over the loss of human considerations in the modern urban environment. This book is intended to be a handbook for designers to provide users with an essentially democratic setting enriching their opportunities by maximizing a degree of choice available to them, an idea he terms as "Responsive."

This paper is an attempt to take this framework and analyze an organic setting, Hollywood Basti—Gulbai Tekra, Ahmedabad, which is democratically made and has emerged out of the specific needs of those people. The authors have conducted a detailed survey of the built environment of Hollywood Basti. This paper tries to extract the underlying responsive features deeply rooted within the fabric which need to be retained while adopting any slum development measures.

1.1 Hollywood Basti, Gulbaitekra, Ahmedabad

In around 1856, a group of about 15 sculptor families migrated from Rajasthan and settled in Ahmedabad. Over the period of years, this group has grown to a squatter settlement of a group of about 800 families, all settled in an area of about 5.2 ha. With the growth of the city and expansion, the area occupies one of the prime locations in the center of the city of Ahmedabad. The slum is known as Hollywood Basti. The real reason behind the name is unknown. As per the anecdotal stories, the name is probably due to its carefree lifestyle [9]. A photographer Kannagi Khanna had also documented women of the basti comparing them with Hollywood actresses (Fig. 1). The slum is popular for the sale of plaster of Paris idols, specially Ganesh idols. People from all over Ahmedabad and from other parts of India come here to buy these idols (Fig. 2).

Due to the high development pressure on the area and Government's consistent efforts toward its rehabilitation and redevelopment, one senses a feeling of insecurity among the residents of the settlement. The government had attempted to relocate and

Fig. 1 Photo documentation of women of Hollywood Basti by Kannagi Khanna

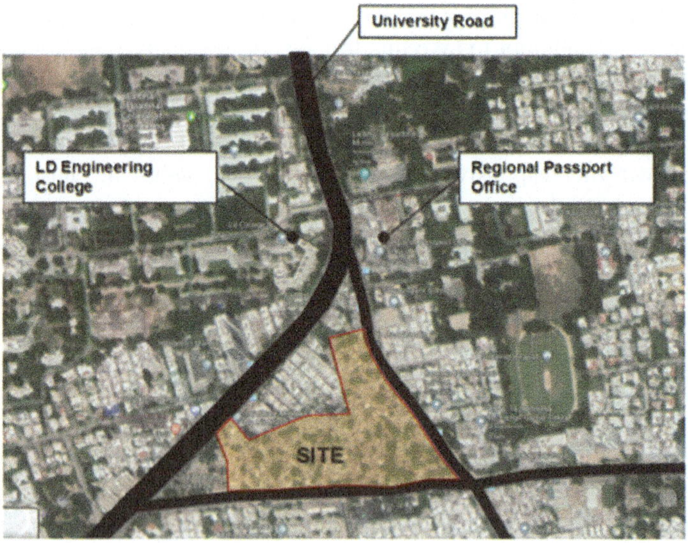

Fig. 2 Hollywood Basti and its context

resettle a few families from Gulbaitekra that had to be removed during expansion of the main road to the outskirts of Ahmedabad in a government Rehabilitation scheme in Odhav area. The scheme offering apartment to the residents completely failed as it could not achieve live-work balance and showed no consideration to their ground oriented lifestyle which existed in Hollywood Basti.

The settlement is divided into four different parts also locally termed as “Vaas.” The name of the Vaas was given after the leaders who were accepted as heads by the residents of each Vaas. With time the boundaries of these Vaas have merged. This paper has taken “Anda Mukhi no Vaas” as a sample for analyzing each feature in detail.

2 Hollywood Basti Through Lenses of Ian Bentley: Finding Order in Chaos

Ian Bentley, an architect, urban designer and an academician has done extensive research and study on designing strategies for regeneration of run-down inner-city areas and effects of property development process on urban form, building imagery, architectural theory etc. The book “Responsive Environments” is an outcome of the involvement of the authors in a series of projects, lectures and seminars held in Department of Architecture and Joint center for Urban Design at Oxford Polytechnique. The book defines responsive as that quality of built environment that provides its users with “a democratic setting, enriching their opportunities by maximizing the degree of choice available to them [1].” The book is a research, which focuses on the users for whom the built fabric is designed. Therefore, this paper uses the framework provided by the book to analyze one of the most democratic settings created by the users themselves for a better understanding of the users, their lifestyle and their preferences.

This book gives seven parameters to the designers that act as tools or devices that can be used to create and provide built environment that is responsive to its users. The seven parameters include—Permeability, Variety, Legibility, Robustness, Visual Appropriateness, Richness and Personalization [1]. Thus the existing built fabric of the settlement is viewed through the lens of each of these parameters.

2.1 Permeability

Permeability is the property of being permeable or passable, penetrable, pervious or porous [6]. Ian Bentley defines permeability in a built environment as the extent to which it allows people a choice of access through it, from place to place and is a key measure of its responsiveness. He discusses the importance of both physical and visual permeability through a built environment and how variations in the public

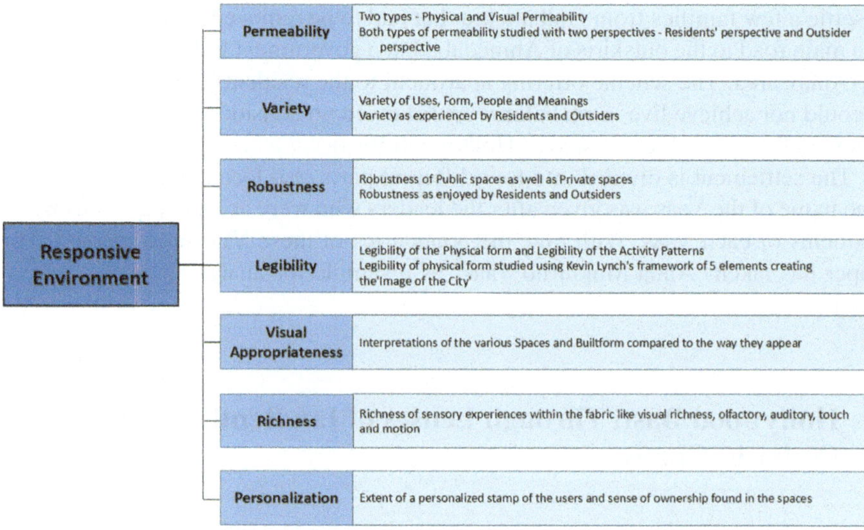

Fig. 3 Parameters for assessing responsiveness of Hollywood Basti

spaces, private spaces and their interface within the built environment have different implications for permeability [1].

Hollywood Basti, as can be seen in Fig. 4, is extremely porous and consists of numerous entry points to enter the settlement, but the permeability is observed to be unidirectional and the settlement is self-contained and introvert with no specific entry point for outsiders. The outsider contact is limited to only the periphery of the site abutting the road, i.e., the public–private interface can be seen only at the periphery of the site as seen in Fig. 4. Figure 5 shows how the narrow streets of the studied Vaas open up midway creating interesting chowk and nodes. These chowks were mostly found near existing trees of the area.

One of the major reasons for lack of outsider permeability is due to the privatization and personalization of spaces within and the manner in which the internal spaces have been organized and designed. The narrow winding streets lead to lack of visible permeability for outsiders. The way in which the internal public spaces have been personalized through various spillover of the household activities, it gives an outsider passing through the lanes, the feeling of a trespasser in a private territory. This leads to the alienation of the settlement from the surrounding context. Besides this, the settlement has also gained attention due to a number of antisocial activities existing within [9]. This can be attributed to the intense introvert nature of the fabric and lack of permeability for the outsiders (Fig. 4).

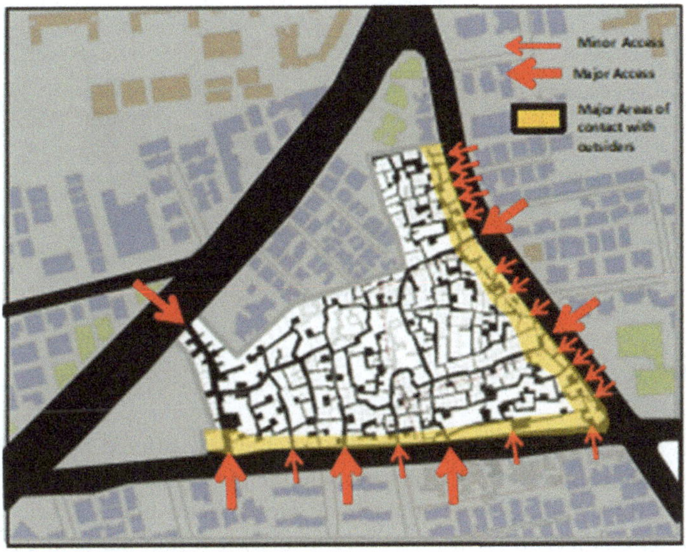

Fig. 4 Permeability of the settlement

Fig. 5 Internal permeability in 'Andamukhi Vaas'

Fig. 6 Sketch showing various activities within the settlement

2.2 *Variety*

Ian Bentley defines "variety" as the experiential choice offered by the built environment. It is a quality extremely important to the place since accessible places are valuable only if they offer experiential choice. He further explains a variety of experiences through variety in forms, uses and meanings within the built environment [1]. In Hollywood Basti, one finds a wide variety of forms, functions and people but limited to the residents of the settlement. Therefore as a resident one finds a variety of activities, functions, choice available within the settlement but as an outsider the form seems monotonous with limited choice of activities found within. One finds a resident using the settlement for almost all his daily activities, the spaces within being used to their full potential at different times of the day for varied activities and therefore constantly changing its meaning. For an outsider the streets are filthy, leftover spaces between the massing of their houses but for the residents those leftover spaces hold multiple meanings. They turn into a spill over space from their living spaces where they can sit with neighbors and have a discussion, do kitchen work like cutting vegetables, extend their idol making work, may convert to a small tea stall where they can have a chat over a cup of tea, or a space for children to play or a space to sleep for the family at night, etc. (Fig. 6)

2.3 *Robustness*

Similarly, Ian Bentley defines Robustness in a built environment as the quality of an environment or place, which offers their users to use them for many different purposes and therefore offer their users more choice than places whose design limits them to a single use. He expresses the concern of specialized spaces for specific activities designed in public spaces [1].

According to Bentley compartmentalization of activities reduces the intensity of robustness of that place. For a public space, the robustness is important because people do not generally come to enjoy one particular activity but they want to enjoy the variety the place offers every time it is visited. Public spaces that are devoid of such flexibility are not much appreciated or acclaimed by users [1].

Hollywood Basti is a very good example in terms of how beautifully it encompasses so many varied activities within the same limited space and how it constantly changes its character at various times of the day and year to accommodate the same. Robustness in Hollywood Basti is not just limited to its spaces but is a part of their lifestyle. People have various occupations throughout the year for income within their family and therefore people here are involved in occupations like animal husbandry, labor, salesman, sculptor, etc. But during the time of Ganesh Chaturthi, almost every family will be involved in the process of idol making. Thus the character of the entire settlement changes during that time. The spaces within their houses that were used for sleeping and day to day activities convert to storage spaces for idols and the family shifts to outdoor spaces for sleeping at night. The houses unlike modern urban houses do not have specific labeled rooms like bedrooms, living rooms, etc. but they have spaces where various activities can occur flexibly. Unlike urban environment, there is no clear demarcation between the private and the public realm, their household activities spill over and extend in the public spaces outside (Figs. 7 and 8).

When one looks at their public spaces including their streets and chowks, one can understand how such a minimum space can have so many multiple meanings attached to it. The street is not just a movement space but also acts like a gathering space, a space where women can sit outside their houses and do their household chores along with chatting, a space for children to play, a space where a resident can set up a tea stall in an offset and people can gather over a cup of tea near it, or a workspace where they can work on preparing the idols or a space for sleeping at night. The chowks turn into a space for a small shrine, or for gathering, or for keeping the cattle, etc. Thus each space here holds multiple meanings and it keeps changing constantly.

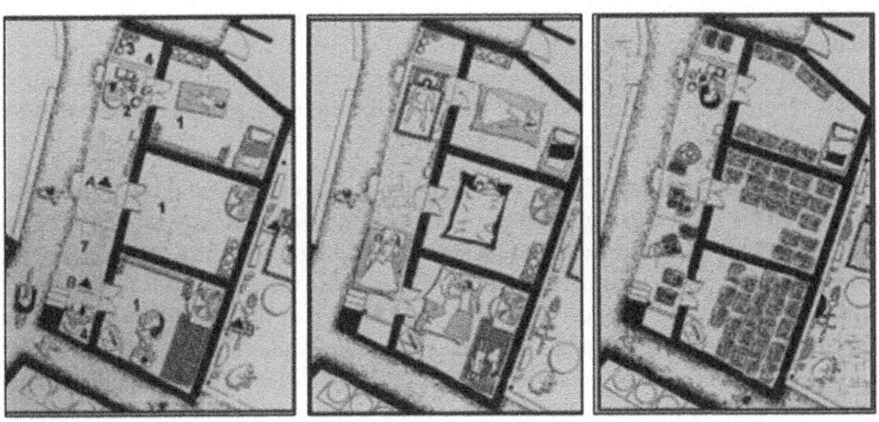

Fig. 7 Transformation of space as per the need of the users during day time (1), during night (2) and during the time of Ganesh festival (3) [8]

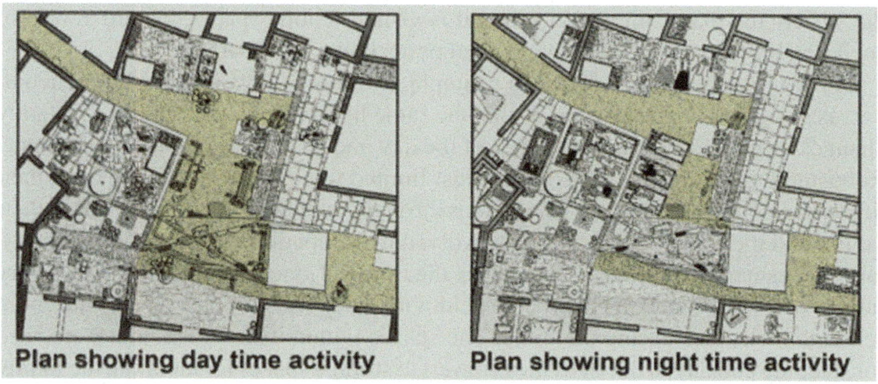

Fig. 8 Plan of part of a street amid few houses used as a spill over space for cooking, playing, drying clothes and other daily activities during day time (1) and its transformation into a sleeping space night time (2) [8]

2.4 *Legibility*

As per Ian Bentley, Legibility is the quality that makes a place graspable. He feels that people can take advantage of the qualities of Permeability and Variety that a built environment offers only when they can grasp a place's layout and the activities that happen there and therefore the quality of Legibility of a built environment is equally important. He expresses the importance of Legibility at two levels for a built environment—Physical form and Activity patterns. Ian Bentley also talks about the key physical elements that help make the physical form of the built environment more legible [1]. He quotes Kevin Lynch and suggests the consideration of the five key elements from his book "Image of the city" for Legibility, i.e., Path, Nodes, Landmarks, Edge and District [5].

Considering Hollywood Basti as one district, one needs to understand this district with two perspectives—Legibility of Hollywood Basti to an outsider and second for the residents or people within the settlement.

Legibility to an outsider—This district is surrounded on three sides by prime roads. The two roads shown in Fig. 9 form the edge of the settlement, which acts as the interface where outsiders meet the residents. This edge is lined with shops by the residents and act like a landmark of the city too owing to the activity of selling Ganesh idols. Out of the three corners formed the one shown in the Fig. 9 acts as a major node of the settlement. There lies a temple named Bodiyadev shrine at the node. This temple is considered to be nearly 200 years old and acts like an important landmark for the dwellers as well as outsiders. The space around is also used for celebrations and festivals throughout the year.

Legibility within the district—All the streets within the built fabric are narrow winding pedestrian streets formed as the resultant spaces between the houses. They

Fig. 9 Streets acting as edge and temple acting as landmark for outsiders

vary in size, scale, enclosure and are full of different activities at all times of the day. The scale is very intimidating, making it uncomfortable for an outsider to enter. But as one enters these streets open up to larger open spaces acting as major nodes for people to gather and communicate. These nodes are formed generally near existing trees many times also consisting of a small shrine or temple or spaces for the cattle, thus reinforcing its character strongly. There are many shrines within the fabric (Fig. 10), which act as a landmark at a very small scale but the scale of these shrines is too small to make it legible and appear as a surprise suddenly while walking along the paths.

2.5 Visual Appropriateness

Ian Bentley feels that the interpretations people put on a place should not conflict with its activities and appearance. This quality he terms as visually appropriate [1]. Hollywood Basti is a self-contained and introvert development, grown over a period of time responding to self-need. The purpose of house building by residents has never been to get an approval by outsiders over aesthetics or its appropriateness. The entire cluster is like an uninviting private space meant only for them. The only public space or an interface where they come in contact with the outsiders is the edge along the main road. This is a commercial edge, which gives idol making a market for the

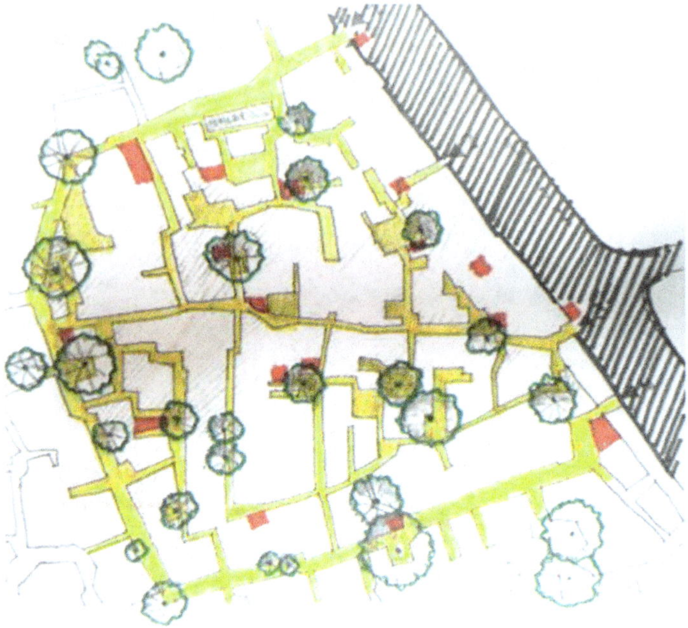

Fig. 10 Plan showing location of shrines acting as landmarks within 'Andamukhi Vaas'

residents. Hollywood Basti has fetched maximum benefit of its core location and the prime roads it is surrounded by. The edge is visually appropriate for the purpose it intends to serve. It attracts more buyers and acts as an active interface between the outsiders and the residents.

2.6 Richness

According to Bentley, Richness is the quality of the varied sensory experiences a place offers to its users [1]. With respect to sight the settlement is a kind of unpredictable labyrinth. It is full of variations in terms of scale of spaces, light, activities, etc. as one moves through them. For an outsider, its winding streets generate curiosity in terms of what the space offers next. As one moves through these lanes one finds a variety of various personalized spaces each unique in their own manner. In this respect, it is important to mention the spillover spaces of the houses, which accommodates various household activities, a rare sight in an urban context.

The place provides a varied kind of sensory experience related to hearing and smell. The neighborhood feels highly alive and active with the sounds of varied activities like the sound of children playing in the lanes, people chatting at tea joints, women chatting with each other sitting at the entrance while doing household chores,

Fig. 11 Sketches showing residences with personalized facades and spaces created by the dwellers

sound of animals like cow, goat and hen reared by the people, etc. Along with that one also constantly experiences various smells of Plaster of Paris, food stalls, the cattle, the drains and the sewage not properly managed, etc. while moving through its streets.

The fine grain of this fabric makes this experience unavoidable by any person who visits this place.

2.7 *Personalization*

Ian Bentley describes Personalization as the degree to which people can put their own stamp on a place [1]. Since Hollywood Basti is completely designed by its residents and does not follow any rules, one finds an extremely personalized approach within the fabric. Each house, cluster or even the public spaces within the fabric have been designed for specific need of a collective group of residents or an individual with their personal approach. Unlike modern urban housing facades, where without a house number it is difficult to reach, here everyone has a personalized mark of identity since each house has been designed by the resident himself and it is difficult to find two similar houses (Fig. 11).

3 Conclusion

As Ian Bentley says that design should be humane, we find through the case of Hollywood Basti that the humane features which are part of democratic design get lost in modern design. A lot of learning come from this study under different parameters. The extreme internal permeability offered by Hollywood Basti for their own dwellers leads to high social bonding. But the lack of outsider permeability has also led to issues of various antisocial activities and therefore it is felt that there is a need to increase the outsider permeability up to a certain extent. Similarly, the variations in private and public spaces are reflection of their lifestyle, culture and tradition. Hollywood Basti provides high degree of robustness. Multiple use of space at its optimum can be learned from this settlement. The transformation of spaces throughout the day and then through the year can be taken as learning for space design providing flexibility. Personalization of space is also something getting lost in modern modular design. In Hollywood Basti, we see that though facing same situation the design of each house is different. The play with outside and inside space is also different. People are not bound by labeled spaces, they use the space the way they want and what suits them best.

Any betterment plan for the slums should retain the essence of the place. The solution should be specific to the settlement addressing their need and lifestyle.

References

1. Bentley, I. (1985). *Responsive environments*. Burlington: Elsevier Ltd.
2. Jacobs, J. (1961). *The death and life of Great American Cities*.
3. Khanna, K. (2012, May). *Going Hollywood*. Retrieved from www.betterphotography.in: http://www.betterphotography.in/features/hollywood/12661/.
4. Kitto, H. (1996). The Polis. *The City Reader*.
5. Lynch, K. (1960). *Image of the city*. MIT Press.
6. Merriam Webster. (n.d.). *Definition of permeability*. Retrieved from www.merriam-webster.com: https://www.merriam-webster.com/dictionary/permeability.
7. Mumford, L. (1937). What is a city. *Architectural Record*.
8. Patel, N. (2007). *Understanding & analysing the dynamics of an urban slum & complexity of its rehabitation: A study of Gulbai Tekra & Odhav, Ahmedabad*. Ahmedabad.
9. Shrivastava, M. (2019, July 28). *Idolising Nomadic Life*. Retrieved from www.deccanherald.com: https://www.deccanherald.com/sunday-herald/sunday-herald-melange/idolising-nomadic-life-749740.html.

Evaluation of the Performance of Permeable and Porous Pavements in the Urban Landscape

Nipun Dinuka Dias Wahalathanthrige and Nandika Miguntanna

Abstract Excess stormwater runoff and the pollution of natural water recourses from the stormwater runoff have become severe global problems nowadays. With the high urbanization, the area of impervious surfaces significantly increases. Consequently, the quantity of stormwater runoff is increasing and therefore the higher loads of pollutants carried by the stormwater runoff to receiving water resources leading to severe degradation of their water quality. The use of pervious pavements in car parks and lightweight driveways in place of traditional impervious surfaces have become one of the innovative, environmental friendly and widely used structural measures globally for past decade. It has a potential to significantly reduce the runoff volume and improve the water quality by trapping the sediments in the infiltrated water. The paper has focused on to investigate the performance characteristics of pervious surfaces and compares their performance against a conventional impervious surface. Three pavement lab models of permeable interlocking concrete pavement (I shaped block), porous interlocking concrete pavement (Grass paving) and impervious interlocking concrete pavement were designed and used for data collection. Data collection was done in two phases: Using stormwater runoff collected from a street and rainwater directly from the sky. Quality analysis were done for both rainwater and stormwater runoff for parameters such as pH, Nitrates, Nitrites, Total Phosphorus (TP), Total suspended solids (TSS) and Electrical conductivity (EC). Quantity analysis was only done for rainwater directly from the sky. Considerable reduction runoff volumes, pollutant concentrations and loads were obtained from the relevant experiments. The findings of this research may contribute to identify the performance of each pervious interlocking concrete pavement and also increase the use of pervious interlocking concrete pavements minimizing the ordinary impermeable pavement surfaces as a stormwater management practice.

Keywords Pervious pavements · Stormwater runoff · Stormwater management

---

N. D. D. Wahalathanthrige (✉) · N. Miguntanna
School of Civil Engineering, Sri Lanka Technological Campus, Padukka, Sri Lanka
e-mail: nipundd1995@gmail.com

A. Jana and P. Banerji (eds.), *Urban Science and Engineering*, Lecture Notes in Civil Engineering 121, https://doi.org/10.1007/978-981-33-4114-2_9

1 Introduction

Many scientists and researchers have found that a rapidly increasing urbanization is the main reason for the excess stormwater runoff. According to the United Nations reports [1] half of the global population already lives in cities, and by 2050 two-thirds of the world's people are expected to live in urban areas. When the amount of the impervious surfaces is increased, it directly affects the quality and quantity of surface runoff. The traditional approach to stormwater management is based on the development of urban drainage networks to convey stormwater away from developed areas quickly. With the high urbanization, the impermeable surfaces are increasing and the problems mentioned above like floods and pollution of water recourses can be occurred due to overloading the existing drainage infrastructure.

Pervious interlocking concrete pavements can be used as an alternative option rather than go for an expensive buildup of a new drainage system. Pervious pavements have been shown to significantly reduce stormwater runoff volumes and increase evaporation compared to conventionally constructed pavements. Moreover, higher rate of filtration through the pervious pavement structure supports to improve the water quality.

Furthermore, researchers [2–10] have noted a considerable reduction of runoff volume and enhancements to water quality parameters in pervious interlocking concrete pavements when compared with impermeable interlocking concrete pavements. According to [7], there are two types of pervious pavements; either can be porous or permeable. Even though they are having similar benefits, they differ significantly in their appearance and in the way they operate. In the South Asian region, there are not many research findings of analyzing the performance of pervious pavements especially experiments' done using local use pervious concrete pavement blocks. The purpose of this research is to study and compare the performance of commercially available permeable and porous interlocking concrete pavements with impermeable interlocking concrete pavement using physical lab models.

2 Literature Review

Stormwater runoff is rainfall that flows over the ground surface. It is created when rain falls on the different paved surfaces and the water does not infiltrate into the ground because of the impermeability. According to [11], stormwater runoff is the primary nonpoint source that pollutes both groundwater and surface water resources. Because the excess surface runoff carries loads of different types of pollutants and adds them to the water bodies. Urbanization occurs due to increase in the population in the cities and towns. Due to the rapid urbanization, the underlying surface condition has changed massively [12]. In addition to that, increasing impervious surfaces are highly affected by the surface runoff and the risk of urban floods adversely.

Due to the infiltration process of pervious surfaces through the soil, it filters out some amount of pollutants and recharges the groundwater table. Impervious surfaces act as solid barriers that do not allow water to penetrate through the surface, so all the stormwater flows over the surface as surface runoff. Pervious surfaces reduce the pressure of the urban drainage system effectively by reducing a part of surface runoff [13]. There are three types of common pervious pavements, namely porous concrete (PC), pervious interlocking concrete pavers (PICP), and porous asphalt (PA) [14].

Pervious interlocking concrete pavements consist of a pervious wearing course and an aggregate subbase course typically built on an open-graded, crushed stone base. Filtering materials such as geotextile can be used in pavement structure depending on the paving type and hydrological conditions [13]. As infiltration trenches, pervious interlocking concrete pavements support pedestrians and vehicles by paving over them [15]. When considering the benefits of pervious interlocking concrete pavers, the main advantage is the ability to reduce the water flow and to improve the water quality properties of natural surfaces and vegetation. Smith [16] shows that the reduction of runoff by as much as 100% from frequent when the runoff consists of low-intensity and short-duration storms. According to [10], heavy metals like Pb, Cu, Cd and Zn are captured by both drainage cells of the layers of pervious pavement structure and geotextile layer below the base material. Pervious pavements are widely used in developed countries but in the South Asian region it is not very common. The main gap has identified as the lack of information and findings of the performance of pervious pavements and their adaptations in local conditions. From these research findings, it is expecting to fill those gaps through a performance analysis of pervious pavements and comparing the results with the impermeable pavement.

3 Methodology

This study evaluated the efficiency of permeable and porous interlocking concrete pavements with their capacity to filter rainfall and stormwater runoff and to compare the efficiency with impermeable interlocking concrete pavement. Three experimental setups were produced, and two pavement models with pervious layers (Porous and permeable) and one with the impervious layer (Impermeable) were tested. Mainly, the data collection was performed in two phases. First, directly using actual rainfall, and then using stormwater runoff collected from a street. Following flowchart (Fig. 1) is a summary of the procedure, which was followed.

The pavement types were selected using a questionnaire survey based on few interlocking manufactures. Preliminary, it was conducted to identify the available types of pavement blocks in Sri Lanka (Pervious and impervious), widely using pavement types locally and to identify problems that occurred when using pervious pavements. I shaped block (Permeable category-Fig. 3), grass paving (Porous category-Fig. 4) and cobble type smooth finish (Impermeable category) were selected. Three glass boxes were used to produce the experimental setups from different interlocking block

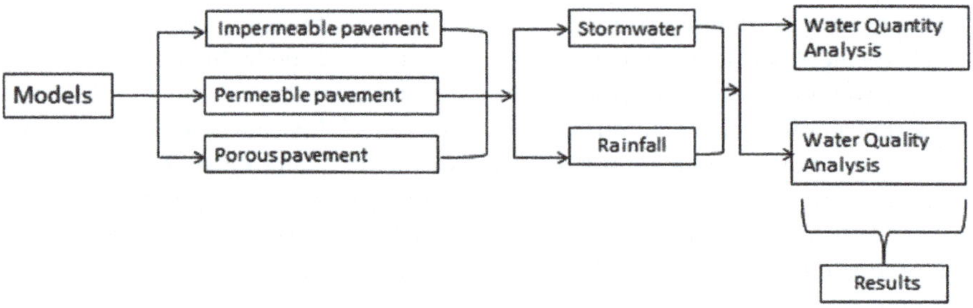

Fig. 1 Summary of the procedure used in this study

types mentioned in above and one empty glass box was used as the control box. Each experimental setup consisted with three aggregate layers and one interlocking concrete block layer at the top as in following Fig. 2 (Following Figs. 2, 3 and 4 are captured during the experiment),

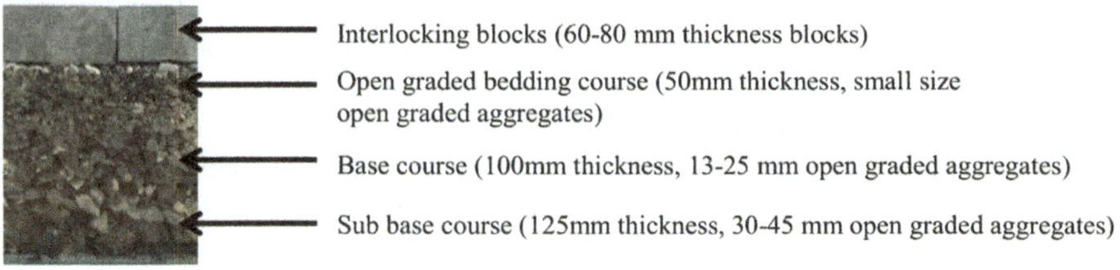

Fig. 2 Structure of the pervious pavement experimental setup

Fig. 3 Experimental setup of I-shaped interlocking concrete block pavement (Permeable pavement)

Fig. 4 Experimental setup of grass paving interlocking concrete block pavement (Porous pavement)

To test with direct rainfall from the sky (Phase 1), the models were kept and exposed to rainfall at the outdoors. For the quantity analysis, for a certain period of time (10–15 min), the amount of rainfall volume, runoff volume and the infiltrated water volume were measured. To measure the rainfall volume at each event, an empty box was used as control. The experiment was repeated for 15 different rainfall events throughout the month of November to get a much accurate and precise value of percentages for each pavement surface.

After the quantity analysis was performed for the direct rainfall, the collected water (Rainwater, Infiltrated rainwater through cobble paving, I-shape and grass paving) was filled into label bottles to transport them to the testing laboratory. The bottles were kept under 4 °C in a thermal box and all the bottles were sealed until they were carried to the testing laboratory. Five infiltrated water samples were collected over three different rainfall events (totally 15 water samples) in each pavement surface type to get an average value for the increase of accuracy.

In phase 2, the Sheet flow method was used to collect stormwater runoff from the streets as it is one of the simplest and standard methods to collect stormwater samples manually. Water sample collection was performed according to the standard guidelines and all the samples were collected within the first 5–10 min (first flush) of the rainfall event. Then the collected water samples were spread over the models using a showerhead, simulating the rainfall falling over the pavement surface. The same procedure was repeated for three different rainfall events to get five infiltrated water samples to analyze (altogether 15 water samples).

4 Sample Analysis

4.1 Quantity Analysis

To evaluate the rainwater infiltration capacity of each pavement model, rainwater infiltrated through each pavement surface was compared with the amount of rainwater collected in the control (Empty) box after each rainfall event. As mention in Fig. 1, quantity analysis is only done for phase 1. The amount of water infiltrated was obtained by using this equation, $E = (V1 - V2)/V1$; here E is the percentage amount of rainwater infiltrate through the pavement models; V1 is the rainwater volume stored in the control box; V2 is the runoff water volume from each surface.

4.2 Quality Analysis

Quality analysis was done for both phase 1 and phase 2. With the limiting resources available, water samples were analyzed for general parameters including pH, Electrical conductivity, and pollutants TSS (Total Suspended Solids), TP (Total Phosphorus), Nitrates and Nitrites. All parameters were measured using the standard analytical methods according to the standard guidelines.

5 Results and Discussion

5.1 Quantity Analysis

It was possible to measure rainwater volumes from 15 rainfall events during the experimental period between the 5 and 30 of November 2019 throughout the month. Because October–November is the seasonal rainy period (second inter monsoon season) of the country, enough rainfall events were occurred during the experimental period to get much accurate and precise data for quantity analysis. The storm that occurred 7 October did not produce any runoff from both grass and I-shaped pavement surfaces, furthermore it did not produce any runoff from the grass paving surface on the 5 October. E values (%) of each surface according to the rainfall event are presented in Table 1.

From Table 1, it is clear that the E values of permeable and porous pavement models are much higher compared with the impermeable pavement model. The higher amounts of voids and infiltration ability of the surfaces are the main reasons for the higher E values of permeable and porous pavement models. If compares the I-shaped (Permeable) and grass paving models, E values are higher in grass paving models. The absorption capability of grass and the higher surface void area

Table 1 E values (%) of each surface relevant to different rainfall events; N/A—not average

Date	Cobble type (Impermeable)	I-shaped (Permeable)	Grass paving (Porous)
05/11/2019	4.9	54.15	N/A
06/11/2019	5.83	62.08	67.083
07/11/2019	4.35	N/A	N/A
09/11/2019	4.27	64.77	72.95
12/11/2019	6.04	55.77	70.60
15/11/2019	N/A	N/A	N/A
16/11/2019	4.71	57.85	71.98
18/11/2019	5.29	59.29	N/A
19/11/2019	N/A	N/A	N/A
20/11/2019	5.15	57.22	N/A
22/11/2019	5.32	63.12	68.44
23/11/2019	4.68	57.31	71.64
25/11/2019	N/A	N/A	N/A
28/11/2019	5.29	62.5	N/A
30/11/2019	5.44	58.17	70.2

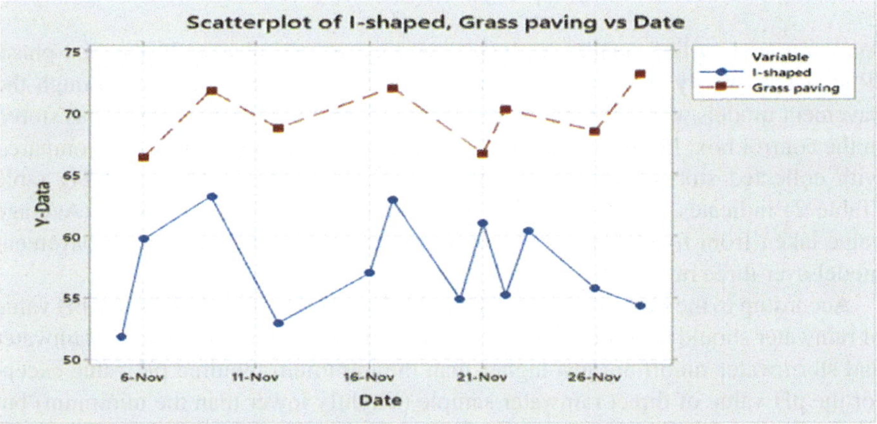

Fig. 5 Percentage of reduction in runoff volumes of pervious pavements compare to impermeable pavement model

of the grass paving model compared with the I-shaped pavement are the possible reasons to higher E values of grass paving model than the I-shaped pavement model.

According to results gained in Table 1, the minimum rainfall required producing runoff from Cobble, I-shaped and grass paving surfaces are 4 mm, 10 mm and 12 mm, respectively. Following chart gives the % reduction in runoff from the I-shaped surface model and the grass paving surface model compared with runoff from the impermeable surface model,

Here, X axis—Date of the rainfall event and Y axis—Infiltration Percentages (%).

Figure 5 indicates that the average percentages of reduction of runoff volume compared with impermeable cobble type pavement, which varied between 51% and 63% for the I-shaped surface model and 65–75% for the grass paving model. From the values presented in Table 1 and Fig. 5, it is verified that pervious pavements are able to manage stormwater flow effectively by infiltrating the surface runoff. Many researchers have found previously [17] that the percentage reductions in the runoff volume through pervious pavements are between 34 and 47%. Therefore, it is confirmed that from this quantity analysis, the results obtained from the I-shaped pavement model and grass paving model within the range or give much better results than the values obtained by previous researchers (even by using the local pervious pavement blocks).

5.2 *Quality Analysis*

As mentioned earlier, quality analysis was done in two phases. In the first phase (Rainwater directly from the sky), the quality of infiltrated rainwater through the pavement models was compared with the quality of the rainwater, which has stored in the control box. In the second phase, infiltrated stormwater runoff was compared with collected stormwater runoff samples from the streets. The following table (Table 2) indicates the average concentration values of each parameter (Average value taken from five collected and infiltrated water samples from each pavement model over three rainfall events for both phases).

According to the rainwater quality guidelines and recommendations, the pH value of rainwater should be within 6.5–8.5. The average pH values obtained for rainwater and stormwater runoff are just higher than the minimum required pH value except for the pH value of direct rainwater sample (Slightly lower than the minimum) but when rainwater and stormwater were infiltrated through the pervious surfaces, the pH values are increased within the allowable limits. On average 8–12% from the I-shaped pavement surface and 12–19% from the grass paving surface are increased. As a reason, it could be assumed that when the water infiltrates through each pavement model aggregate layers added some impurities to the infiltrated water. Therefore, the pH values of infiltrated water are comparatively higher.

Electrical conductivity is a measure of water to conduct electrical current. It directly depends on the total dissolved solids in water. Total dissolved solids can be positive or negative charged ions. In this experiment, when the water is infiltrated through each layer the number of dissolved solids getting reduced. Therefore, the electrical conductivity values are decreased because the ion concentrations

Table 2 Average concentrations of all parameters; R/W—rainwater, S/W—stormwater runoff

Parameter	Collected samples		Infiltrated rainwater samples			Infiltrated stormwater samples		
	R/W	S/W	Cobble	I-shaped	Grass paving	Cobble	I-shaped	Grass paving
pH	6.3	7.2	6.7	7.1	7.7	7.3	7.8	8.1
EC (μS/cm)	36	135.6	34.2	17.5	11.3	126.4	12.4	11.8
TP (mg/L)	0.2	0.92	0.17	0.065	0.075	0.85	0.323	0.374
Nitrite (mg/L)	0.01	0.36	0.009	0.002	0.036	0.29	0.084	0.11
Nitrate (mg/L)	3.94	9.2	3.12	0.886	1.225	8.64	2.488	3.421
TSS (mg/L)	13.1	151.3	12.7	1.143	1.023	149.6	12.452	11.131

decreased, and the electrical conductivities of collected water samples are lesser than the maximum allowable value recommended by rainwater quality guidelines which is 2250 μS/cm. When rainwater and stormwater are infiltrated through the pavement models, both I-shaped and grass paving surfaces decreased the electrical conductivities of the infiltrated water samples.

On average TP, nitrites, nitrates and TSS concentrations reduced between 60%–70%, 65%–75%, 70%–75%, and 90%–95%, respectively, from I-shaped pavement surface model and 52–60%, 60–65%, 60–65%, and 92–95% from grass paving. These removal efficiencies of each water quality parameter were calculated comparing the concentrations of water quality parameters from I-shaped and grass paving pavement surfaces with the cobble (Impermeable) pavement surface. The voids of grass paving model consist with soil and coir dust mixer but in I-shaped the voids are filled with chip aggregates, so it can be assumed that when the water is infiltrated through the grass paving model, it added some extra pollutants or contaminants than in I-shaped model, which might reason to reduce removal efficiency values of grass paving model, and it was identified that the samples taken at the beginning the nitrate and nitrite concentrations are higher. The reason was identified that the leaching of nitrate and nitrite from the subbase of the pavements.

6 Conclusion

This study was carried out to investigate the performance characteristics particularly in terms of reduction of surface runoff volume and pollutant concentrations of commercially available pervious interlocking concrete pavements with impermeable interlocking concrete pavement using physical lab models. Percentage reduction in runoff volume from pervious pavements varied between 50 and 75% when compared with conventional impermeable pavement model. Grass paving pavement model was infiltrated more volume of water than the I-shaped pavement model because it consists of more voids and less impervious surfaces compared with the I-shaped model. Referring to the above results gained, it is verified that both pervious pavements can reduce the stresses on drainage systems by effectively managing the stormwater runoff. The water retained within the pavement structure will evaporate back to the atmosphere.

The results obtained from the experiments verified that the filtration ability of the pervious pavements compared with impermeable pavement. From this pervious pavement experiments, it was observed that the water quality enhancement of the collected infiltration water samples, with reductions in TSS, TP, nitrites and nitrates of around 90%–95%, 50%–70%, 60%–75% and 60%–75%, respectively, confirming findings by previous researchers. The average pH value for the rainwater sample slightly less than the minimum but it increased after the infiltration process and pH values for other samples are also increased after the filtration by the models. Electrical conductivity values are within the allowable limits and it further decreased after infiltrated by the models. If compares to the filtration capacity of the I-shaped pavement model with the grass paving pavement model, the results indicated that the I-shaped surface has more filtration ability to reduce the pollutant concentrations.

From this study, it is also expecting to create awareness among urban designers and planners about the importance of using pervious pavements in place of conventional pavements. The final results of this study provide a great privilege to select a suitable pavement type according to anyone's need (Porous or permeable). As an extension of this research in the field of urban planning, it could be suggested that the filtration water of pervious pavements can be used for rainwater harvesting through a special drainage system or any other useful work. According to the literature review, few researchers have shown that there is a possibility in pervious pavements to filter out heavy metals like Pb, Cu, Cd and Zn. Due to limited resources, the present study was focused only on few water quality parameters (pH, electrical conductivity, TP, nitrate, nitrite, TSS) but it should be recommended to carry out analysis for heavy metals as well in future studies, and also problems that have identified from the literature review like clogging of pervious pavements, maintenance required to facilitate the performance continuously, use the infiltrate water for different useful purposes like etc. are the areas that require further study.

References

1. Zhang, X. Q. (2016). The trends, promises and challenges of urbanisation in the world. *Habitat International, 54,* 241–252.
2. Balades, J. D., Legret, M., & Madiec, H. (1995). Permeable pavements: Pollution management tools. *Water Science and Technology, 32*(1), 49–56.
3. Beecham, S. C., Pezzaniti, D., Myers, B., Shackel, B., & Pearson, A. (2009). Experience in the application of permeable interlocking concrete paving in Australia (Doctoral dissertation, Argentinean Concrete Block Association).

4. Burak, R. J. (2004). Permeable interlocking concrete pavements-selection, design, construction and maintenance. In *2004 Annual Conference and Exhibition of the Transportation Association of Canada-Transportation Innovation-Accelerating the Pace.*
5. Council, R. B. (1998). *Porous asphalt. Highway improvements report.* File ref: hig9750.
6. James, W., & Von Langsdorff, H. (2003, October). The use of permeable concrete block pavement in controlling environmental stressors in urban areas. In *International Conference on Concrete Block Paving (PAVE)* (7th, 12th–17th October, 2003, Sun City, Sur de Africa). Document Transformation Technologies, Sun City, Sur de Africa, Conference Planners.
7. Jayasuriya, N., & Kadurupokune, N. (2013). Comparative performance of permeable and porous pavements.
8. Khaniya, B., Wanniarachchi, S., & Rathnayake, U. (2017). Importance of hydrologic simulation for LIDs and BMPs design using HEC-HMS: A case demonstration. *International Journal of Hydrology, 1*(5), 138–146.
9. Korkealaakso, J., Kuosa, H., Niemeläinen, E., & Tikanmäki, M. (2014). *Review of pervious pavement dimensioning, hydrological models and their parameter needs. State-of-the-art.* Research Report.
10. Legret, M., & Colandini, V. (1999). Effects of a porous pavement with reservoir structure on runoff water: Water quality and fate of heavy metals. *Water Science and Technology, 39*(2), 111–117.
11. Priyadarshana, T., Jayathunga, T., & Dissanayake, R. (2013). Pervious concrete—A sustainable choice in civil engineering and construction.
12. Nguyen, T. T., Ngo, H. H., Guo, W., Wang, X. C., Ren, N., Li, G., & Liang, H. (2019). Implementation of a specific urban water management-sponge city. *Science of the Total Environment, 652,* 147–162.
13. Zhu, H., Yu, M., Zhu, J., Lu, H., & Cao, R. (2018). Simulation study on effect of permeable pavement on reducing flood risk of urban runoff. *International Journal of Transportation Science and Technology.*
14. Razzaghmanesh, M., & Beecham, S. (2018). A review of permeable pavement clogging investigations and recommended maintenance regimes. *Water, 10*(3), 337.
15. Nichols, P. W. B., Lucke, T., & Dierkes, C. (2014). Comparing two methods of determining infiltration rates of permeable interlocking concrete pavers. *Water, 6*(8), 2353–2366.
16. Smith, D. R. (2006). *Permeable interlocking concrete pavements. Selection, design, construction, and maintenance.* Herndon, VA: Interlocking Concrete Pavement Institute.
17. Pratt, C. J., Mantle, J. D. G., & Schofield, P. A. (1989). Urban stormwater reduction and quality improvement through the use of permeable pavements. *Urban discharges and receiving water quality impacts* (pp. 123–132). Pergamon.

Strategic Planning for Greenfield Cities: A Holistic Evaluation of Efficiency and Smartness

Surbhi Bhavsar and Saikat Paul

Abstract This paper aims to develop a framework to evaluate the planning and performance efficiency of greenfield cities as an indicator of their smartness. The contextual prioritization of identified smartness and efficiency indicators and a comparative analysis of proposed landuse alternatives through modeling of city efficiency and smartness have been carried out in this study. This framework could potentially act as a standardized system to gauge the viability of a certain proposal. The preparation of framework has been backed by studies pertaining to the various attributes that are essential indicators of landuse efficiency and smartness in addition to a study of landuse models, trends, modeling tools and techniques, case studies, etc. Quantifiable parameters for both these evaluations have been identified which, if required, could then be used for benchmarking and verification. This method was applied to compare and rank the landuse efficiency and smartness of three landuse plans proposed for the development of Pushpak Nagar, Navi Mumbai as a smart city. This approach recommends physical planning and infrastructure development strategies for a smart and efficient city. The resultant toolkit is useful in determining the effectiveness of development interventions and their implications on policy formulation. It will assist in the identification and addressing inadequacies, prioritization of provisions, and viable allocation of landuses and amenities.

Keywords Smart city evaluation · Landuse efficiency · Transportation · Infrastructure · Environment

1 Introduction

1.1 Overview

Measuring the comparative efficiency levels of cities and urban development projects is an essential prerequisite for creating sustainable urban futures. Comprehensive

S. Bhavsar (✉) · S. Paul
Indian Institute of Technology Kharagpur, Kharagpur 721302, West Bengal, India
e-mail: surbhibhavsar@gmail.com

A. Jana and P. Banerji (eds.), *Urban Science and Engineering*, Lecture Notes in Civil Engineering 121, https://doi.org/10.1007/978-981-33-4114-2_10

development of physical, institutional, social and economic infrastructure is required to cater to the growing urban population, drive economic growth and improve people's quality of life by developing replicable planning strategies. This has engendered the need for employing assessment frameworks [1] as key mechanisms for measuring the impacts of development [2] and as a key policy instrument for supporting the transition to a smart and efficient urban development. Assessment using indicators and indexing methods has gained recognition mainly because of the visualization of a phenomenon and the reliability of quantifiable variables.

1.2 Need for Study

Previously vacant land over 250 acres developed around cities to absorb their growing population, and address their needs constitute greenfield developments. These areas need not be within the limits of an Urban Local Body or an Urban Development Authority but require innovative plan financing, implementation and planning mechanisms [26]. The limited availability of land and resources warrants planning that ensures their optimum utilization. Thus, there is a need for a successful replicable prototype for efficient greenfield smart city development [3, 4], the formulation of which defines the novelty of this paper. With the advent of new technologies, there is a need to analyze and inculcate the implications of the Internet Of Things (IOT) and Information & Communication Technologies (ICT) on the Quality of Life, an integral aspect of the development of Smart Cities. Despite the ongoing discussion of the recent years, there is no agreed definition about smart cities, whereas strategic planning in this field is still largely unexplored [25]. As proposed by several authors, the smartness of a city is not only a function of the technological interventions but also the efficiency of its planning and functioning [5, 6]. For this purpose, especially in developing countries, it is often advisable to rely on spatial reorganization aided by information technology [5, 7, 8], instead of complete dependence on expensive technological tools. This study tries to evaluate the planning and performance efficiency of greenfield cities as an indicator of its smartness.

2 Background and Study Area

Maharashtra is India's third-largest state by area and also the world's second most populated subentity. It is not only the country's trade capital but also the wealthiest and most developed states in India. Navi Mumbai, a planned multinucleated township for about 11 lakh people was developed by City and Industrial Development Corporation Ltd (CIDCO), as a holistic, affordable sister city for Greater Mumbai, to act as a counter magnet for Greater Mumbai to successfully reduce its housing, employment and infrastructural burden, and is now vying to be one of the smart cities. CIDCO also proposed its smart city Pushpak Nagar to be developed as a

Greenfield site. The site with an area of about 230 Ha is supposed to cater to a population of 240,000 and generate about 8.7 lakh high-end jobs [24]. Located toward the southern end of the Navi Mumbai boundary, the area delineated for Pushpak Nagar has the NH4B passing through it (Fig. 1), and is in close proximity to the proposed Navi Mumbai International Airport. The area (Fig. 2) is currently under potentially monocropped uncultivated farmlands acquired under CIDCO's 22.5% and 12.5% schemes. A stream of the Panvel creek delineates its western flank. The primary principle of development is proposed to be Integrated Development.

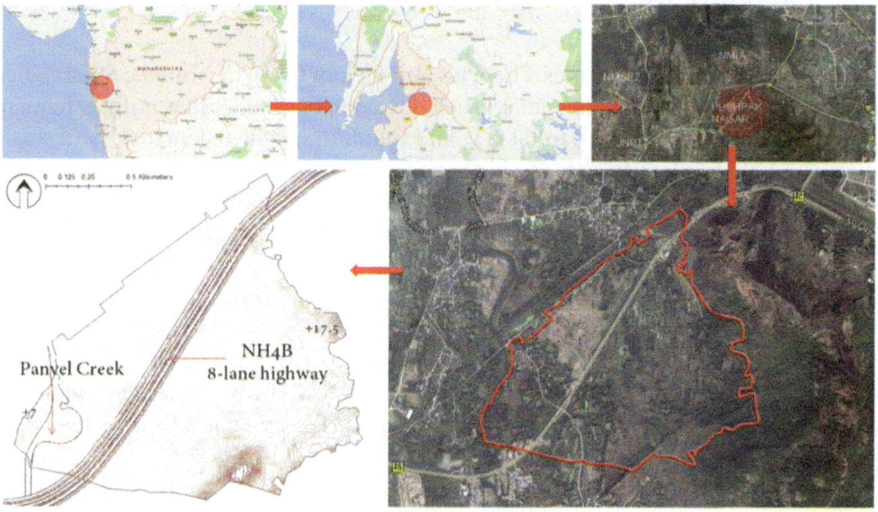

Fig. 1 Pushpak Nagar, site location

Fig. 2 Visual survey of site

3 Methodology

The overall methodology primarily comprised of literature review, data identification, collection and analysis followed by modeling and comparison of three proposed alternatives, culminating with the identification of the most efficient and smart land use alternative.

3.1 Literature Review

Smart Cities, Land Economics Theories and Urban Growth Models: Smart cities are a representation of a conceptual urban development model that utilizes human, collective and technological capital [1]. An exploration of the recent history of smart cities can help to identify certain shaping forces that lead to the conceptualization of "smart city" [23]. For a better understanding of the implications of a landuse proposal, land economics theories [9] and the evolution of urban growth models were studied.

Landuse Efficiency: The planning of cities needs cognizance of several factors to arrive at optimized ideal solutions for a given context. This has prompted considerable research on the various dimensions of landuse efficiency that can be fairly generalized and applied to most planning interventions [9]. Largely, the priorities have included spatial structures [10, 11], energy [12] and transportation efficiency [8], city networks [13], density, size and form as well as policy integration [4].

Modeling Landuse: Logical reasoning may not always suffice to ensure that the landuse plan proposed is the most efficient, ideal solution for a certain context. Thus, quantitative modeling considers land market and ownership, environment, demography, transportation, landuse and urban form, social and institutional amenities, infrastructure and development investment, socioeconomic and biophysical factors, livability, human and technological capital as well as the economic and political environment as some parameters [13, 14].

Modeling Smart Cities: The spurt in smart city initiatives around the world, necessitated the need to set benchmarks and quantify the smartness of a city, based on the extent and efficiency of the initiatives proposed. The parameters considered in research so far have predominantly included environment, governance, civil society, energy, transportation, built form, livability and economy [7].

Assessment Models: Hard and soft models, the revised triple helix model, fuzzy logic, coverage index, Fuzzy Delphi Model, the Analytic Hierarchy Process, the Analytic Network Process, Analysis of Variance, Life Cycle Impact Assessment, Input–Output Analysis, Cost–Benefit Analysis, Entropy Method, PROMETHEE, Planning Support System, Score Transformations, expert opinions analysis and data collection are some of the most commonly used assessment tools for landuse efficiency and smartness [1].

Landuse Efficiency	Smartness
Transit adjacency to residents, services	Punctuality of public transportation & usage, convenience of commuting
Parking supply in employment centers	Comfort level of temperature, water quality
Wastewater, solid waste generation	Travel time between inner-city areas and international airports
Energy use	Level of green coverage
Distance & number of (non) home-based vehicle trips	Public lighting
% Households with flush, net, phone, waste removal access	Transportation fatalities per population
Wastewater treatment state, groundwater availability	Taxi fare
Residential water consumption	Green/renewable energies
Greenhouse emissions generated	Smart grids
Stormwater runoff pollution generated	Waste & water management
Noise pollution generated	No of companies with iso 14001 certification
Distance to nearest transit modes & junctions, nmt coverage	Percentage of renewable energy used, paper recycled

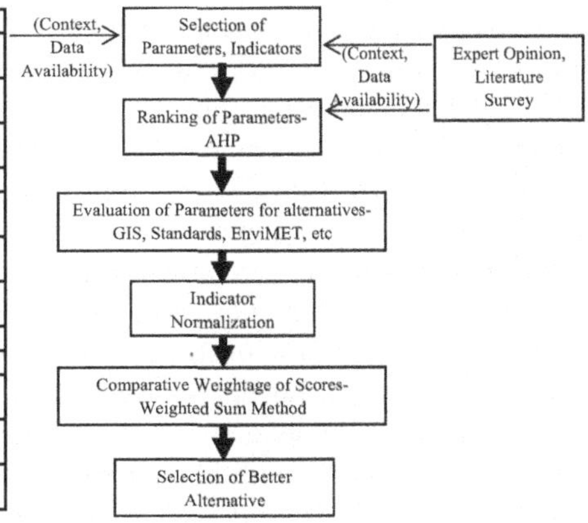

Fig. 3 Landuse efficiency and smartness modeling methodology

The eight smart city projects were studied including the greenfield sites at Naya Raipur (Chhattisgarh, India), Amaravathi (Andhra Pradesh, India), GIFT City (Gujarat, India), Dholera SIR (Gujarat, India), PlanIT (Portugal), Songdo IBD (South Korea), as well smart city interventions in Barcelona (Spain) and Amsterdam (Netherlands). The key findings included the importance of provisions of future growth, justification of need and suitability, significance of technological innovation, the importance of phasing and scaling of interventions, disaster preparedness, public participation, and prioritization of transportation, infrastructure and sustainability as well as efficiency optimization.

3.2 *Modeling Methodology*

The indicators for Evaluation of Landuse Efficiency as well as for Smartness Evaluation were selected based on contextual viability as well as data availability, assisted by primary, secondary and visual surveys carried out as a part of data collection. Figure 3 gives a detailed process for both these evaluations.

3.3 *Landuse Alternatives*

The site is currently completely vacant with no settlements or agricultural land, yet its location being close to the dynamic city of Mumbai, demands a holistic development in sync with the existing urban development, as a proliferating urban continuum. It is a necessity that the development does not occur in isolation, to avoid any future instances of incompatible landuse conflicts.

Alternative 1: The first alternative has been proposed by CIDCO itself. Like most other nodes, the proposal is based on a gridiron pattern. About 48% of the land has been demarcated for residential use, with 5% mixed-use, i.e., residential + commercial. The site slope has been utilized, such that higher ground gets utilized for services such as storage reservoirs, whereas the lower ground close to the Panvel creek can be used for water treatment plants, etc., while the creek and its surrounding area form the primary recreational space.

Alternative 2: This alternative proposes a combination of radial and grid-iron layout for its transit and infrastructure convenience and efficiency in terms of noise mitigation. The core concept is based on compact decentralization, proposed through sectoral divisions, beneficial for administrative purposes with hierarchical distribution of all amenities in addition to parking lots and detention basins. Green corridors connect all the sectors and a multimodal transit hub. A scenic peripheral highway helps to capitalize on resultant higher land values. An industrial hub has been located to the North West, such that pollution is diverted away from the site, while also acting as an economic. Much like the first alternative, the primary recreational area is located close to the Panvel Creek with Treatment Plants following the site slope. There is a seamless integration of all economic groups with suitable FSI provisions to ensure aesthetic integrity in terms of building profiles as well as economic feasibility.

Alternative 3: The third alternative inspired by Indian planning layouts incorporates a Central Business District constituting a stadium, a water body, and primary commercial and institutional spaces close to the Transit Hub. This alternative is based on the concept of sectors and decentralization of commercial and institutional services, such that every sector has optimal facilities to cater to its population. Green corridors connecting the sectors as well as a green buffer between the industrial and residential areas are important features of the alternative. Table 1 gives a summary of the three landuse alternatives.

4 Results and Discussion

The alternatives proposed were analyzed individually for the preidentified parameters. The parameters or indicators identified for analyzing Landuse Efficiency and Smartness as listed in Tables 2 and 3, were evaluated individually under Transportation, Physical Infrastructure, and Environment, based on suitable criteria. Although smartness and efficiency are interrelated, the paper evaluates their indicators separately. This enables independent applicability of the process, as deemed necessary. AHP or Analytical Hierarchical Process [15, 16] employed here for evaluation, is essentially a theory of measurement involving pair-wise comparisons dependent on preferences of one parameter over another. It is a useful multicriteria decision-making tool dependent on consistency, measurement and relationships within and between sets of indicators. The process entails the selection of preferred parameter and assignment of comparative scores by experts, arrival at weightages, evaluation of

Table 1 Landuse alternatives-comparison

<table>
<tr><th></th><th>Landuse plan</th><th>Landuse distribution</th></tr>
<tr><td>Alternative 1</td><td>
</td><td>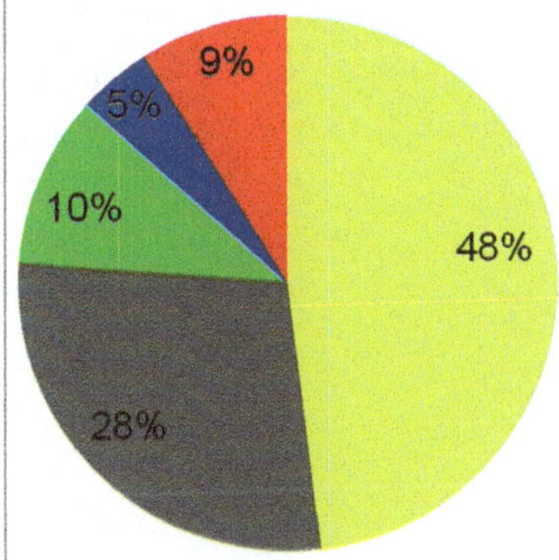
</td></tr>
<tr><td>Alternative 2</td><td>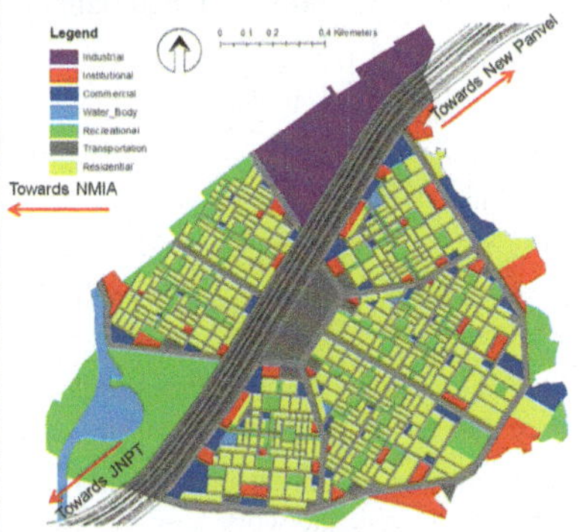
</td><td>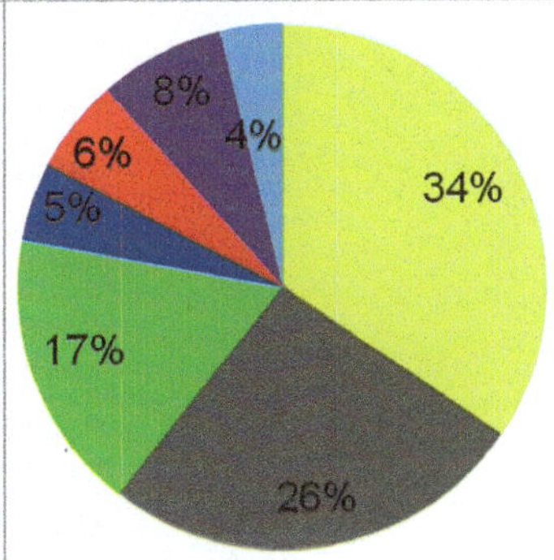
</td></tr>
<tr><td>Alternative 3</td><td>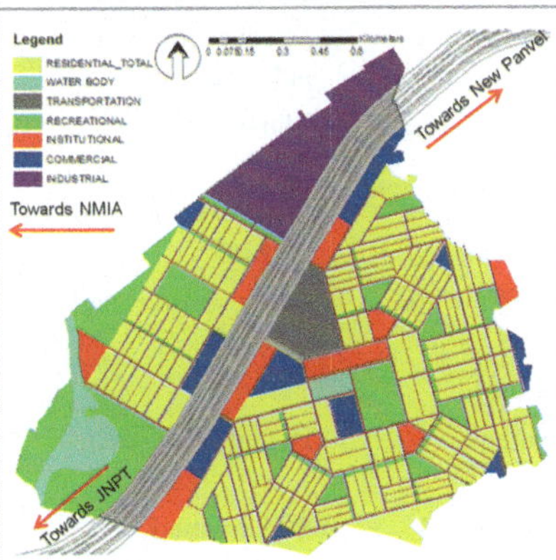
</td><td>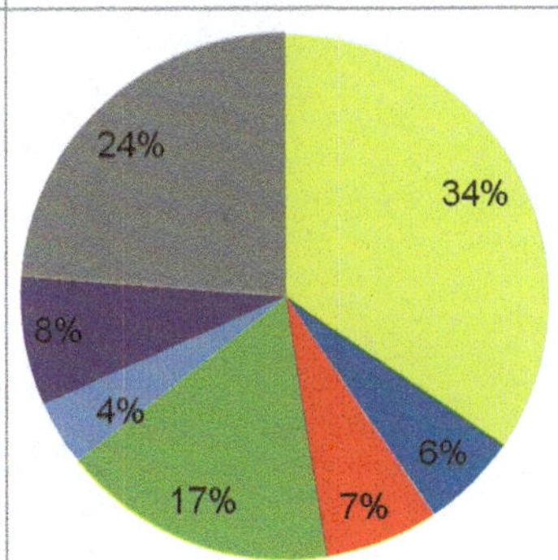
</td></tr>
</table>

Table 2 Landuse efficiency evaluation criteria and weightings

Landuse efficiency	Criteria	Weightage
Transportation		**20.3**
Distance to nearest transit modes	Proximity to 11 m roads	18.7
Transit adjacency to residents	Proximity to 11 m roads	30.4
Transit adjacency to services	Proximity to commercial areas	13.6
Distance and number of vehicle trips	Walkability for work-home-others	7.9
Bicycle Coverage	Streets with bicycle tracks	5.4
Pedestrian coverage	Streets with pedestrian tracks	7.5
Parking supply	Provision of parking lots	3.7
Regional connectivity	Layout	12.8
Infrastructure		**53.4**
Residential water consumption	Management	33
Wastewater generation and management	Street sections and network	23
Solid Waste generation and management	Street sections and network	16
% households with phone access	Street sections and network	11
% households with waste removal access	Street sections and network	4
Wastewater treatment state	Management	3
% households with internet access	Street sections and network	3
% households with toilet access	Street sections and network	7
Environment		**26.3**
Groundwater availability	Detention basin	52
Stormwater runoff pollution	SCS method	13
Energy use	Light and thermal comfort	25
Noise pollution generated	L_{eq} calculations	6
Greenhouse emissions	Air quality index	3

Table 3 Smartness evaluation criteria and weightages

Smartness	Criteria	Weightage
Transportation		**23.5**
Min transport of people and goods	Network	34
Optimum network	Distances	24
Transportation fatalities per population	Major intersections	16
Travel time between inner-city areas	Distance to/from CBD/recreational/institutional	11
Usage of public transportation	Transit node distance (11 m roads)	7
Convenience of commuting	Transit node distance(11 m roads)	5
Taxi fare	Public transportation service	2
Infrastructure		**18.5**
Centralized services efficiency	Management and street network	3
Smart grids	Management	44
Water quality	Management	30
Waste management	Management	11
Public lighting	Management (road length)	8
Cyclic processes	Management	5
Environment		**58**
Level of green coverage	Recreational area	7
Comfort level of temperature	Thermal comfort	32
SPM, SO_2, NO_2, CO_2 emissions	Air quality index	3
Percentage of renewable energy used	Management(solar energy)	24
Green/renewable energies	Management(50% roof area)	17
Percentage of paper recycled	Management	10
ISO 14,001 certification	Management	2
Contextual feasibility	Regional, geographic	5

consistency ratios and generation of an Eigen vector that allows ranking of identified parameters.

A square matrix of size $n \times n$ is constructed for each expert for each category, where, $n =$ number of indicators.

Comparative scores are assigned to each cell (higher values if preferred more, reciprocal values if less)

$$\boldsymbol{P_a = k_a / \Sigma k_a} \tag{1}$$

where, $k_a =$ the nth root of the product of the values in the row for indicator a, $P_a =$ priority vector for each indicator a, $m =$ number of categories

Table 4 Landuse efficiency and smartness category ranking

Rank	Landuse efficiency	Smartness
1	Infrastructure	Environment
2	Environment	Transportation
3	Transportation	Infrastructure

$$\boldsymbol{CI} = \lambda_{\max} - \boldsymbol{n}/(\boldsymbol{n} - 1) \tag{2}$$

where, CI = Consistency Index, $\lambda_{\max} = \Sigma R$, $R = P_a \times S_a$, S_a = sum of the values in the column for indicator a

$$\boldsymbol{CR} = \boldsymbol{CI}/\boldsymbol{RI} \tag{3}$$

where, RI = Random Index (derived from a RI table), CR = Consistency Ratio (Ideally <= 0.01).

A similar procedure is followed for evaluating CR for all categories.

$$\boldsymbol{C_a} = \Sigma(\boldsymbol{P_{ab}} * \boldsymbol{P_b}) \tag{4}$$

where, P_b = priority vector for each category b, P_{ab} = priority vector for indicator a under category b, C_a = Final score for each indicator (forms basis for ranking).

A total of seven experts were referred to for ranking the indicators and assigning weights. The consistency ratio for the scores assigned for Smartness and Efficiency as well as the categories under each of them were under 0.01 and thus were valid, for all seven experts. Table 4 gives the rankings for all three categories of Smartness and Landuse Efficiency, while Tables 5 and 6 give the final rankings for all the indicators under Transportation, Infrastructure, and Environment for Smartness and Landuse Efficiency.

The spatial evaluation was carried out using EnviMET [17] simulations (Fig. 4) and GIS-based analysis (Fig. 5) on a case-specific basis for each indicator. Based on the EnviMET simulation, it can be observed that the third alternative has slightly more cooler areas, whereas the second alternative allows slightly higher wind speeds. From the GIS simulations, it can be observed that the second alternative allows better accessibility to services and streets with public transport accessibility. Contextual feasibility in terms of environment, economy, transportation, livability, and governance was determined based on expert opinion. While some parameters required evaluation based on benchmarks or standard calculations, the following metrics have been used in this study:

Thermal Comfort and Energy Use

ISO 7730 proposes the use of the Predicted Mean Vote (PMV) model [18], which predicts the mean value of thermal assessment of a large group of people exposed

Table 5 Landuse efficiency indicator ranking

Rank	Transportation	Infrastructure	Environment
1	Transit adjacency to residents	Residential water consumption	Energy use
2	Distance to nearest transit modes and junctions	Wastewater generation	Groundwater availability
3	Transit adjacency to services	Wastewater treatment state	Runoff and pollution
4	Regional connectivity	Waste generation	Noise pollution
5	Distance and number of vehicle trips	% households with waste removal	Emissions
6	Pedestrian coverage	% households with toilet access	
7	Bicycle coverage	% households with phone access	
8	Parking supply in employment centers	% households with internet access	

Table 6 Smartness indicator ranking

Rank	Transportation	Infrastructure	Environment
1	Convenience of commuting	Smart grids	Level of green coverage
2	Minimum transportation of people and goods	Water quality	% of Renewable energy used
3	Travel time between inner city areas	Waste management	Green/renewable energies
4	Usage and punctuality of public transportation	Cyclic processes	Comfort level of temperature
5	Optimum network	Centralized service efficiency	% of paper recycled
6	Transportation fatalities per unit population	Public lighting	Emissions
7	Taxi fare		Contextual feasibility
8			ISO 14,001 certification

to the same environment in a 7-point thermal sensation index, rated from −3 to +3 given by the following equation:

$$PMV = f\left(t_a, t_{mrt}, v, p_a, M, I_{cl}\right) \quad (5)$$

where, t_a = air temp, t_{mrt} = Mean Radiant Temp, v = relative air velocity, p_a = Humidity, M = Metabolic Rate, I_{cl} = Clothing Insulation.

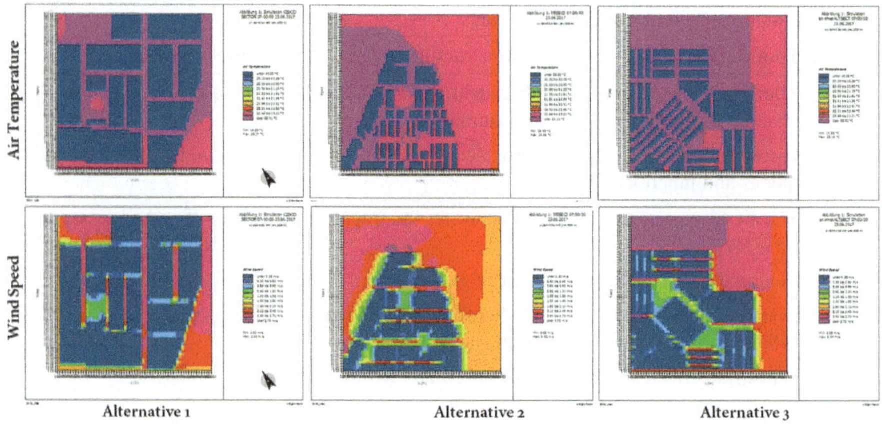

Fig. 4 EnviMET simulations

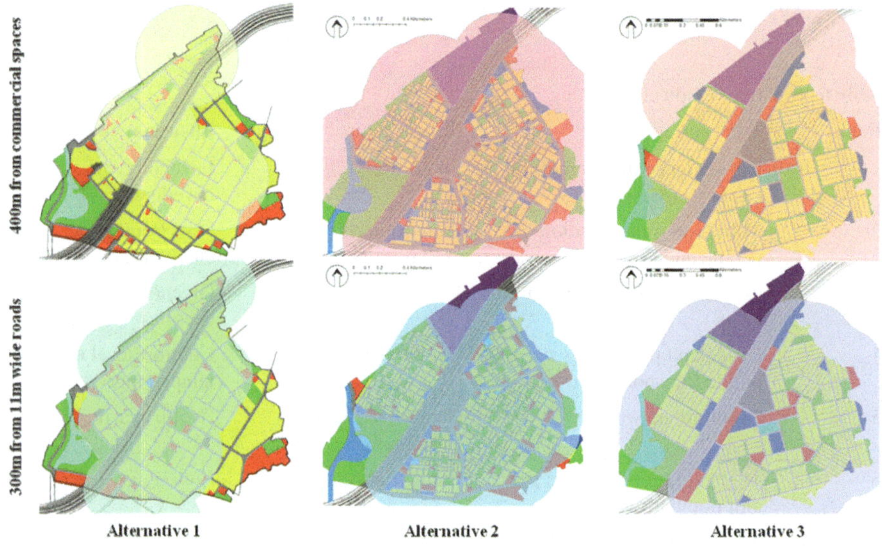

Fig. 5 GIS simulations

The Predicted Percentage of Dissatisfied (PPD) people predict the percentage of thermally dissatisfied people in a given thermal environment. PPD is a function of PMV and is calculated using the following equation:

$$\boldsymbol{PPD} = 100 - 95 * \exp\left(-0.003353 * \boldsymbol{PMV}^4 - 0.2179 * \boldsymbol{PMV}^2\right) \quad (6)$$

Noise Level

Noise levels have been modeled by various researchers [19–22] to assess noise pollution in urban areas due to road traffic and industrial activities. The following equation has been used to quantify the noise levels:

$$L_{eq}Pa(D) = 10\log\left(10L_{eq}^{I(D)/10} + 10L_{eq}^{C(D)/10} + 10L_{eq}^{R(D)/10} + 10L_{eq}^{S(D)/10}\right. \quad (7)$$

where

$L_{eq}Pa(D)$ is total equivalent continuous sound pressure level in dB, $L_{eq}^{I(D)/10}$ is L_{eq} for industrial areas, $L_{eq}^{C\frac{D}{10}}$ is L_{eq} for commercial areas, $L_{eq}^{R(D)/10}$ is L_{eq} for residential areas, $L_{eq}^{S(D)/10}$ is L_{eq} sensitive areas.

Runoff Generated (SCS Method)lx

Runoff curve number (CN) is the main factor in determining runoff in the SCS (Soil Conservation Service) based hydrologic modeling method. SCS-CN method is based on the following relationship between rainfall, P (mm), and runoff, Q (mm):

$$Q = (P - 0.2S)^2/(P + 0.8S) \quad (8)$$

where, variable S (mm) is potential maximum retention after runoff begins, which varies with antecedent soil moisture and other variables, which can be estimated using the following equation:

$$S = (1000/CN) - 10 \quad (9)$$

and

$$CN = CN_p * (1 - F_n) + (CN_n n * F_n) \quad (10)$$

where

F_n = nonpermeable area %, CN_p = CN for permeable surface, CN_n = CN for nonpermeable surface.

CN is a dimensionless catchment parameter having values ranging from 0 to 100. A CN of 100 represents a perfectly impermeable catchment with zero retention and 100% runoff, whereas, a CN value of 0 represents absolutely no runoff irrespective of the rainfall.

The scores arrived at for every indicator had varying units based on the purpose and evaluation criteria. Thus the scores were then normalized and final uniform scores were arrived at through the Weighted Sum Method.

Based on the above results (Table 7), it can be inferred that the second alternative proposed fares better than the other two in terms of landuse efficiency as

Table 7 Final scores

	Alternative 1	Alternative 2	Alternative 3
Efficiency			
Transportation	1039.45	1257.41	1193.44
Infrastructure	2761.68	3457.42	2964.26
Environment	706.77	918.26	865.38
Total efficiency	**4507.90**	**5633.08**	**5023.07**
Smartness			
Transportation	694.52	1027.89	951.72
Infrastructure	751.38	1258.46	1124.34
Environment	1377.97	546.64	655.02
Total smartness	**2823.87**	**2832.99**	**2731.08**
Total	***7331.77***	***8466.07***	***7754.16***

well as smartness. Some of the favorable factors included an integrated grid-iron decentralized design among other strategies.

5 Conclusion

The primary purpose of this study was to put forth a process to evaluate the efficiency of a proposal and to gauge its smartness. It could also be considered as a cumulative summary of relevant existing research. It covers a broad spectrum of indicators that are essential in determining the effect of proposed interventions, such as changes in landuse or FSI, urban design or street configurations or policies for utility provisions. This methodology could prove to be an indispensable tool for planners, not only for physical planning but also as an aid for policy formulation. It may also assist in prioritization of provisions, on a case-by-case basis, as well as the allocation of landuses and amenities, not just for smart cities but also, to ensure efficiency and viability. Further research could entail the implementation strategies of smart city initiatives. Evaluation of Smartness and Landuse Efficiency enables identification and addressing the inadequacies in not just greenfield sites but also existing cities thus facilitating better integration and contiguity.

References

1. Sharifi, A. (2020). A typology of smart city assessment tools and indicator sets. *Sustainable Cities and Society, 53,* 101936.
2. Xu, Z., Zhang, J., Zhang, Z., Li, C., & Wang, K. (2020). How to perceive the impacts of land supply on urban management efficiency: Evidence from China's 315 cities. *Habitat International, 98,* 102145.
3. Breathnach, P. (2009). Greenfield development. In: R. Kitchin & N. Thrift (Eds.), *International encyclopedia of human geography* (pp. 639–643). Elsevier. https://doi.org/10.1016/B978-008044910-4.00841-5.
4. Kane, K., & York, A. M. (2017). Prices, policies, and place: What drives greenfield development? *Land Use Policy, 68,* 415–428.
5. Ninčević Pašalić, I., Ćukušić, M., & Jadrić, M. (2020). Smart city research advances in Southeast Europe. *International Journal of Information Management,* 102127. https://doi.org/10.1016/j.ijinfomgt.2020.102127.
6. Yao, T., Huang, Z., & Zhao, W. (2020). Are smart cities more ecologically efficient? Evidence from China. *Sustainable Cities and Society,* 102008. https://doi.org/10.1016/j.scs.2019.102008.
7. Tang, Z., Jayakar, K., Feng, X., Zhang, H., & Peng, R. X. (2019). Identifying smart city archetypes from the bottom up: A content analysis of municipal plans. *Telecommunications Policy, 43,* 101834.
8. Šurdonja, S., Giuffrè, T., & Deluka-Tibljaš, A. (2020). Smart mobility solutions—Necessary precondition for a well-functioning smart city. *Transportation Research Procedia, 45,* 604–611.
9. Jiao, L., et al. (2020). Assessment of urban land use efficiency in China: A perspective of scaling law. *Habitat International, 99,* 102172.
10. Yang, J., Zeng, C., & Cheng, Y. (2020). Spatial influence of ecological networks on land use intensity. *Science of the Total Environment, 717,* 137151.
11. Chen, H.-C., Han, Q., & De Vries, B. (2020). Modeling the spatial relation between urban morphology, land surface temperature and urban energy demand. *Sustainable Cities and Society* 102246. https://doi.org/10.1016/j.scs.2020.102246.
12. Yu, Y., & Zhang, N. (2019). Does smart city policy improve energy efficiency? Evidence from a quasi-natural experiment in China. *Journal of Cleaner Production, 229,* 501–512.
13. Yu, J., Zhou, K., & Yang, S. (2019). Land use efficiency and influencing factors of urban agglomerations in China. *Land Use Policy, 88,* 104143.
14. Zhu, X., Zhang, P., Wei, Y., Li, Y., & Zhao, H. (2019). Measuring the efficiency and driving factors of urban land use based on the DEA method and the PLS-SEM model—A case study of 35 large and medium-sized cities in China. *Sustainable Cities and Society, 50,* 101646.
15. Leal, J. E. (2020). AHP-express: A simplified version of the analytical hierarchy process method. *MethodsX, 7,* 100748.
16. Wu, Z., Jin, B., Fujita, H., & Xu, J. (2020). Consensus analysis for AHP multiplicative preference relations based on consistency control: A heuristic approach. *Knowledge-Based Systems, 191,* 105317.
17. Tsitoura, M., Michailidou, M., & Tsoutsos, T. (2016). Achieving sustainability through the management of microclimate parameters in Mediterranean urban environments during summer. *Sustainable Cities and Society, 26,* 48–64.
18. Pourshaghaghy, A., & Omidvari, M. (2012). Examination of thermal comfort in a hospital using PMV–PPD model. *Applied Ergonomics, 43,* 1089–1095.
19. Lu, X., et al. (2019). Influence of urban road characteristics on traffic noise. *Transportation Research Part D: Transport and Environment, 75,* 136–155.
20. Vladimir, M., & Madalina, C. (2019). Optimizing urban landscapes in regard to noise pollution. *Procedia Manufacturing, 32,* 161–166.
21. Wang, H., Cai, M., & Luo, W. (2017). Areawide dynamic traffic noise simulation in urban built-up area using beam tracing approach. *Sustainable Cities and Society, 30,* 205–216.

22. Yuan, M., Yin, C., Sun, Y., & Chen, W. (2019). Examining the associations between urban built environment and noise pollution in high-density high-rise urban areas: A case study in Wuhan, China. *Sustainable Cities and Society, 50,* 101678.
23. Angelidou, M. (2015). *Smart cities: A conjecture of four forces.*
24. CIDCO. (2015). *CIDCO Navi Mumbai(South) Smart City Vision and Action Plan.* Navi Mumbai: CIDCO.
25. Hollands, R. G. (2008). *Will the real smart city please stand up?*
26. MoUD. (2015). *URDPFI Guidelines* (Vol. I).

Real-Time Flood Mapping with Temporal SAR Images Using ESA CloudToolbox Service

V. S. K. Vanama and Y. S. Rao

Abstract Flood events are increasing at a drastic rate throughout the globe and became a major threat of increasing concern. The Synthetic Aperture Radar (SAR) system is capable of acquiring images during heavy rains and cloud cover. Many studies have proven the potential of SAR images in disaster management studies. In this study, a recent flood event in Kerala state, India, which occurred during Aug 2018 is analyzed using the European Space Agency (ESA) CloudToolbox service. To map the flood area, the C-band Sentinel-1 temporal SAR images and Sentinel-2 temporal optical images that are available within the duration of flood event are used. To process this massive volume of data, the Sentinel Application Platform (SNAP) with Graph Processing Framework (GPF) is used. The flood mapping approach adopted for each image is automated through XML code which is implemented in GPF. Kittler and Illingworth threshold algorithm is used to obtain the optimum threshold for flood identification. The results are validated against the ground truth data. The results indicate that ESA's CloudToolbox service is quite efficient in handling the big volume of data and real-time information extraction.

Keywords Kerala flood · Kittler · Sentinel · Dual-pol

V. S. K. Vanama (✉)
Centre for Urban Science and Engineering, Indian Institute of Technology Bombay, Mumbai 400076, India
e-mail: vsaikrishna1990@gmail.com
URL: https://sites.google.com/view/venkata-sai-krishna

Y. S. Rao
Centre of Studies in Resources Engineering, Indian Institute of Technology Bombay, Mumbai 400076, India

A. Jana and P. Banerji (eds.), *Urban Science and Engineering*, Lecture Notes in Civil Engineering 121, https://doi.org/10.1007/978-981-33-4114-2_11

1 Introduction

Flood events are increasing at a drastic rate throughout the globe and become a significant threat of increasing concern. Many Indian cities are vulnerable to at least one out six types of natural disasters as per millennium ecosystem assessment report. During the flood disasters in India, organizations like Indian Space Research Organisation (ISRO), India Meteorological Department (IMD) provide the near-real-time information for the end-users. Remote Sensing (RS) technology plays a major role in flood disaster monitoring and management. Optical RS images are often limited by cloud cover, whereas Synthetic Aperture Radar (SAR) signals can penetrate through the clouds and operate day and night particularly the C-band and L-band SAR missions. Thus, a SAR system is capable of acquiring images during heavy rains and cloud cover. Many studies [1–3, 8] have proven the potential of SAR images in disaster management studies, especially in the domain of flood monitoring and management [7, 27]. Flood inundation modelling combined with thresholding the ENVISAT ASAR images is demonstrated on 2008 flood event of Po river, Cremona, Italy [26]. Thresholding combined with region growing, segmentation and change detection methods were implemented on various flood events [9, 20, 21]. Flood area mapping from high incident angle Radarsat-2 images was tested on Kosi basin flood event [16]. A combined approach of thresholding, region growing and fuzzy logic yielded 96% overall accuracy when tested on multiple flood events [18]. The role of interferrometric coherence in accurate flood mapping was also explored [23, 24]. SAR-based flood damage assessment was found to be more reliable [6]. Various change detection algorithms such as Normalize Change Index (NCI) [22, 29], Ratio Index (RI) [10] and Normalized Difference Flood Index (NDFI) [5] are explored for flood area mapping.

With the advancement of technology and many Earth Observation (EO) satellites in space, it is now possible to minimize the intensity of disaster-induced damages. The numerous satellites present in the space are generating tons of data every day [12, 15, 28] confronting us to many challenges like rapid data processing, information extraction and real-time information dissemination to end-users. For India, ISRO's Bhuvan geoportal provides the near-real-time information using state of art technologies for the end-users. However, they utilize the classic algorithms to extract the flood information from the satellite images thus consumes long processing time. The Bhuvan portal only disseminates the information but doesn't allow users to customize the flood mapping procedure. Other cloud processing tools such as Water Mask Processor (WaMaPro) [11], Rapid Mapping of Flooding tool (RaMaFlood) [19], TerraSAR-X Flood Service (TFS) are the in-house tools developed by DLR [17]. The European Space Agency (ESA) provides a CloudToolbox service to process and analyze the SAR images. The end-users can download, process, analyze and extract the useful information from SAR images on the cloud and can transfer the results to their local machines. The advantage of CloudToolbox service is that users can customize and develop their own algorithms.

However, few studies have demonstrated the utility of ESA's CloudToolbox for flood mapping applications. Therefore, in this study, a major disastrous flood event in the history of Kerala state, India, occurred during Aug 2018 is taken to assess the applicability of ESA's CloudToolbox service. Access to the CloudToolbox service is obtained through a research proposal approved by ESA. The multiple tasks involved in the flood mapping procedure are completely processed on CloudToolbox. Section 2 describes the study area and datasets used in the research. The research methodology is given in Sect. 3 followed by results and discussions in Sect. 4. The conclusions of the work are given in Sect. 5.

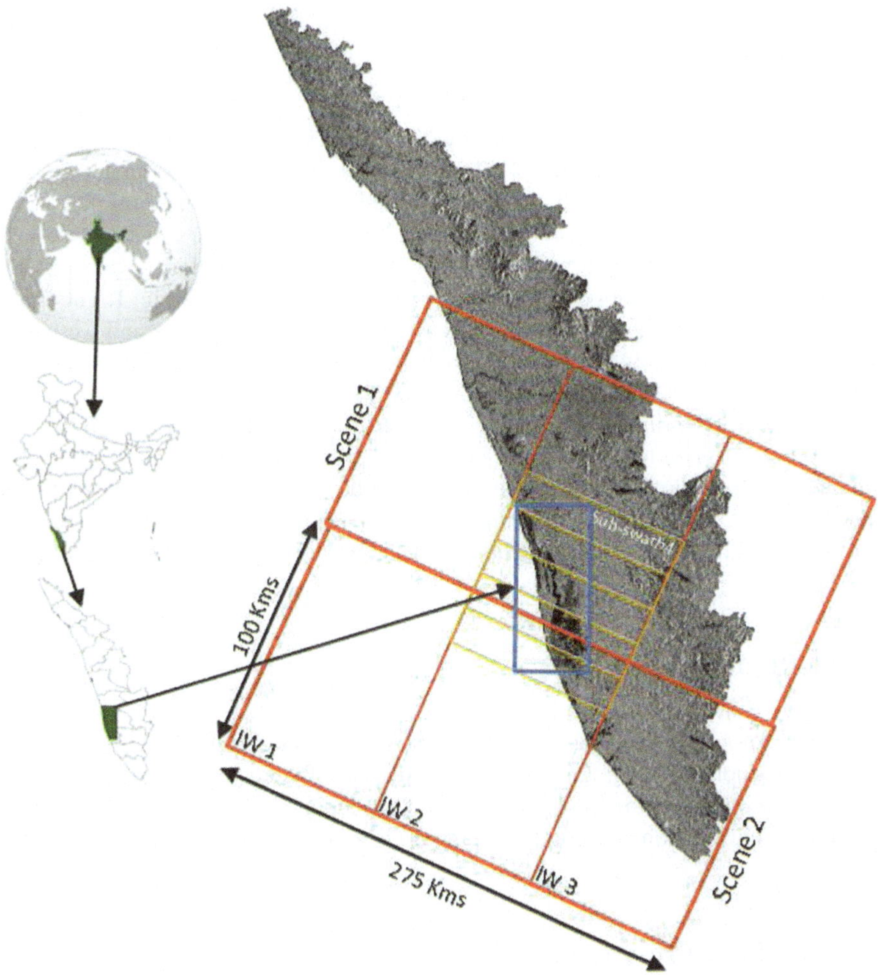

Fig. 1 Study area showing the parts of Alappuzha, Eranakulam and Kottayam districts in Kerala State, India

Table 1 EO datasets used for analyzing the 2018 flood event of Kerala, India

DOA	Satellite	Acq. mode/ Pass	Pol/ Bands	Flood condition
05 Jan 2018	Sentinel-1	IW	VV	Pre-flood
09 Aug 2018	Sentinel-1	IW	VV	During-flood
21 Aug 2018	Sentinel-1	IW	VV	During-flood
22 Aug 2018	Sentinel-2	M1LC	B3/B7/B12	During-flood

DOA Date of Acquisition; *Acq. mode* Acquisition mode; *Pol* Polarization

2 Study Area and Datasets

During Aug 2018, Kerala state of India experienced a disastrous flood resulting in numerous socio-economic losses [4]. Two major rainfall spells occurred during Aug 2018 affected 13 out 14 districts in Kerala, India [13]. Unprecedented rainfall along with landslides and dam water release at many locations made the situation worst. A low-lying region covering parts of Kottayam, Alappuzha, Ernakulam districts which are severely affected by flood is taken as the study area as shown in Fig. 1. The study area is characterized by multiple land uses and is rich in mangroves and wetlands. Also, a major water body Vembanadu lake, a RAMSAR [25] recognized site present in the study area to which many rivers drain into it.

The temporal Sentinel-1 SAR images acquired during the flood are used to analyze the flood event. The temporal SAR images used in study are acquired in level-1 Single Look Complex (SLC) Interferometric Wide (IW) swath mode. For a better understanding of flood event, the Sentinel-2 optical RS images are used for validation purpose. The characteristics of EO datasets used in the study are given in Table 1.

3 Methodology

The methodology of the research is divided into two modules viz., data pre-processing and flood mapping. The first module deals with various steps in data pre-processing and clipping images to the study area extent. The second module deals with optimum threshold identification and flood mapping procedure.

3.1 EO Data Pre-processing

The European Space Agency (ESA) developed Sentinel Application Platform (SNAP) image processing software which is used to process all the temporal Sentinel-1 SAR images. ESA's CloudToolbox service is equipped with SNAP software. The study area (blue rectangle as shown in Fig. 1) is covered in two Sentinel-1 scenes. There-

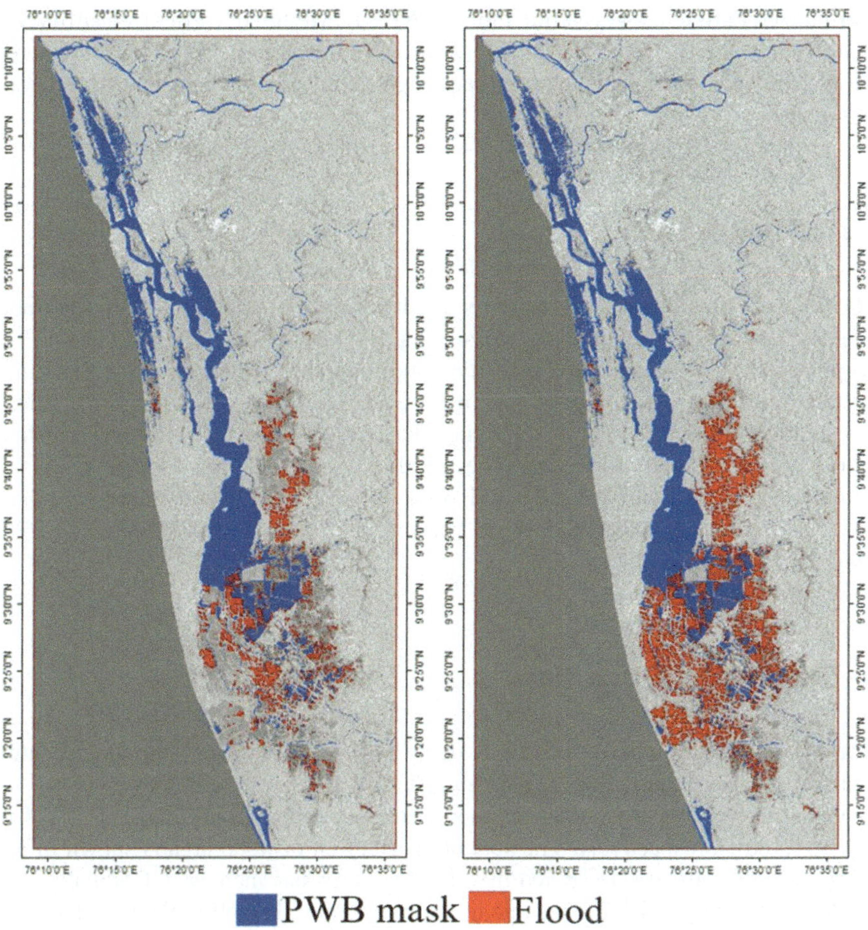

Fig. 2 Spatial distribution of flood area for the Sentinel-1 images acquired on 09 Aug 2018 and 21 Aug 2108

fore, the "Slice Assembly" operator in SNAP is used to assemble (red rectangle as shown in Fig. 1) the two Sentinel scenes. Followed by this "TopSAR Split", "Apply Orbit File", "Calibrate", "TopSAR Deburst" and "TopSAR Merge" SNAP operators are implemented. Rather than processing the entire image, the required sub-swaths (orange rectangle as shown in Fig. 1) are processed to save the computational time and resources. Further, "Subset", "Polarimetric Matrix", "Polarimetric Speckle Filter" and "Terrain Correction" operators (blue rectangle as shown in Fig. 1) are implemented. Finally, the terrain corrected σ^0_{vv} images are produced which will be used for further analysis.

The Sentinel-2 optical images are pre-processed to produce the reflectance values from DN values using Sen2Corr processor operator. The thematic indices like Green

Normalized Difference Vegetation Index (GNDVI), Modified Normalized Difference Water Index (MNDWI) are developed from Sentinel-2 images. A bash file is created in CloudToolbox to automate the pre-processing task for processing the multi-temporal EO images.

3.2 Flood Mapping

The water bodies that behave as the smooth reflecting surface result in low radar backscatter values which helps in classifying flood from other land use classes. The co-polarized (VV) images are used for flood mapping. Kittler's [14] thresholding algorithm is implemented on the terrain corrected σ_{vv}^{0} images to obtain the optimum thresholding algorithm. The pixel values that are less than the optimum threshold values are considered as flooded. The pre-flood Sentinel-1 SAR image is used to generate the Permanent Water Body (PWB). To reduce the over estimation of flood area, the permanent water bodies are masked out from the flood area using PWB mask.

4 Results and Discussions

The optimum backscatter values are −18.7 dB and −21.2 dB for flood identification from the images acquired on 09 Aug and 21 Aug 2018, respectively.

Quantitatively, the flood area obtained from the temporal SAR images is 84.1 km^2 and 151.4 km^2, respectively. The spatial distribution of flood area is shown in Fig. 2. From the results, it is observed that the flood area has increased drastically from 09 Aug to 21 Aug 2018. This is due to the major rainfall spell in the month of Aug 2018. The same can be verified from the optical Sentinel-2 image derived thematic indices like GNDVI, MNDWI as shown in Fig. 3. Although there is a significant time (1 day) gap between Sentinel-1 and Sentinel-2 image acquisitions, the flooded area surrounding the Vembanadu lake is significantly detected.

5 Conclusions

The 2018 flood event of Kerala state of India is analyzed using multi-temporal EO images in ESA's CloudToolbox service. With the launch of future SAR mission, the CloudToolbox service is quite useful for rapid information dissemination during the critical time of disasters. The temporal dynamics of the flood are captured for Kerala flood event. This study can be further extended by identifying the role of interferometirc coherence in flood mapping.

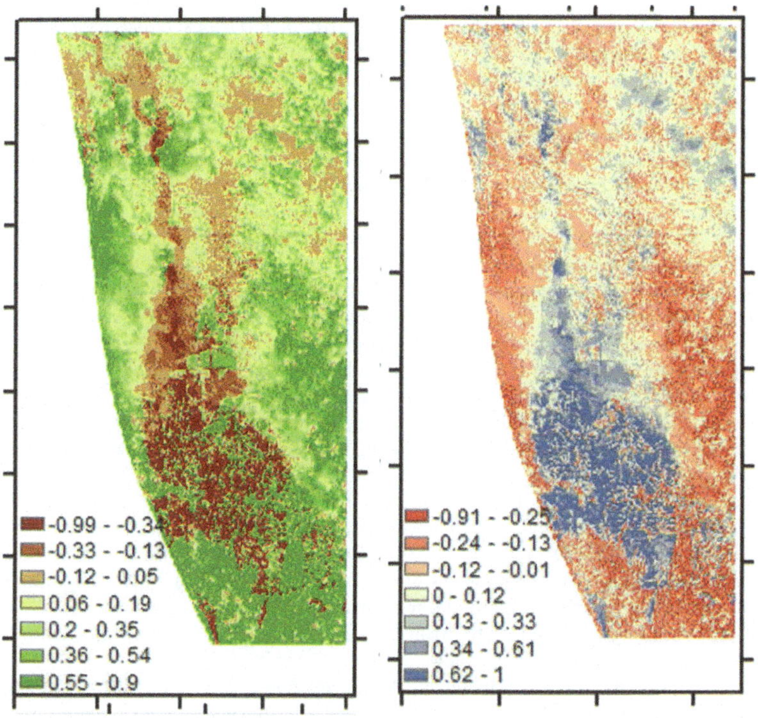

Fig. 3 GNDVI and MNDWI indices developed from Sentinel-2 optical image acquired on 22 Aug 2108

Acknowledgements The first author is thankful to the Ministry of Human Resource Development, GoI for providing necessary computational facilities through the project (MHRD-0A0025) funding. The authors would like to thank the ESA for providing the access to online RSS CloudToolbox.

References

1. Alahacoon, N., Matheswaran, K., Pani, P., & Amarnath, G. (2018). A decadal historical satellite data and rainfall trend analysis (2001–2016) for flood hazard Mapping in Sri Lanka. *Remote Sensing*, *10*(3), 448.
2. Amarnath, G. (2014). An algorithm for rapid flood inundation mapping from optical data using a reflectance differencing technique. *Journal of Flood Risk Management*, *7*(3), 239–250.
3. Amitrano, D., Di Martino, G., Iodice, A., Riccio, D., & Ruello, G. (2018). Unsupervised rapid flood mapping using sentinel-1 GRD SAR images. *IEEE Transactions on Geoscience and Remote Sensing*, *56*(6), 3290–3299.
4. Central Water Commission. (2018). *Study report: Kerala flood of August 2018* (Tech. Rep.). Government of India, Central Water Commission, Hydrological Studies Organisation.
5. Cian, F., Marconcini, M., & Ceccato, P. (2018). Normalized difference flood index for rapid flood mapping: Taking advantage of EO big data. *Remote Sensing of Environment*, *209*, 712–730.

6. Dadhich, G., Miyazaki, H., & Babel, M. (2019). *Applications of sentinel-1 synthetic aperture radar imagery for floods damage assessment: A case study of Nakhon SI Thammarat, Thailand.* International Archives of the Photogrammetry: Remote Sensing & Spatial Information Sciences.
7. Dasgupta, A., Grimaldi, S., Ramsankaran, R. A. A. J., Valentijn, R. N. P., & Jeffrey, P. W. (2018). Towards operational SAR-based flood mapping using neuro-fuzzy texture-based approaches. *Remote Sensing of Environment, 215*, 313–329.
8. Giustarini, L., Hostache, R., Matgen, P., Schumann, G. J.-P., Bates, P. D., & Mason, D. C. (2012). A change detection approach to flood mapping in urban areas using TerraSAR-X. *IEEE Transactions on Geoscience and Remote Sensing, 51*(4), 2417–2430.
9. Giustarini, L., Hostache, R., Matgen, P., Schumann, G. J.-P ., Bates, P. D., & Mason, D. C. (2013). A change detection approach to flood mapping in urban areas using TerraSAR-X. *IEEE Transactions on Geoscience and Remote Sensing, 51*(4), 2417–2430.
10. Gong, M., Cao, Y., & Wu, Q. (2011). A neighborhood-based ratio approach for change detection in SAR images. *IEEE Geoscience and Remote Sensing Letters, 9*(2), 307–311.
11. Gstaiger, V., Huth, J., Gebhardt, S., Wehrmann, T., & Kuenzer, C. (2012). Multi-sensoral and automated derivation of inundated areas using terrasar-x and envisat asar data. *International Journal of Remote Sensing, 33*(22), 7291–7304.
12. Guo, H.-D., Zhang, L., & Zhu, L.-W. (2015). Earth observation big data for climate change research. *Advances in Climate Change Research, 6*(2), 108–117.
13. Indian Meteorological Department. (2018). *Performance of South West Monsoon 2018 Over Kerala.* (Tech. Rep.). Meteorological Centre, Thiruvananthapuram. https://www.imdtvm.gov.in/images/cumulativerainfallforkerala-swmonsoon2018.pdf.
14. Kittler, J., & Illingworth, J. (1986). Minimum error thresholding. *Pattern Recognition, 19*(1), 41–47.
15. Ma, Y., Haiping, W., Wang, L., Huang, B., Ranjan, R., Zomaya, A., et al. (2015). Remote sensing big data computing: Challenges and opportunities. *Future Generation Computer Systems, 51*, 47–60.
16. Manjusree, P., Prasanna Kumar, L., Mohan Bhatt, C., Srinivasa Rao, G., & Bhanumurthy, V. (2012). Optimization of threshold ranges for rapid flood inundation mapping by evaluating backscatter profiles of high incidence angle SAR images. *International Journal of Disaster Risk Science, 3*(2), 113–122.
17. Martinis, S., Kuenzer, C., Wendleder, A., Huth, J., Twele, A., Roth, A., et al. (2015). Comparing four operational SAR-based water and flood detection approaches. *International Journal of Remote Sensing, 36*(13), 3519–3543.
18. Martinis, S., & Rieke, C. (2015). Backscatter analysis using multi-temporal and multi-frequency SAR data in the context of flood mapping at River Saale. *Remote Sensing, 7*(6), 7732–7752.
19. Martinis, S., Wendleder, A, Künzer, C., Huth, J., Twele, A., Roth, A., & Dech, S. (2016). Four operational SAR-based water and flood detection approaches: A comparison. In *Proceedings of ESA Living Planet Symposium 2016 (SP-740)*. Spacebooks Online.
20. Mason, D. C., Schumann, G. J.-P., Neal, J. C., Garcia-Pintado, J., & Bates, P. D. (2012). Automatic near real-time selection of flood water levels from high resolution Synthetic Aperture Radar images for assimilation into hydraulic models: a case study. *Remote Sensing of Environment, 124*, 705–716.
21. Matgen, P., Hostache, R., Schumann, G., Pfister, L., Hoffmann, L., & Savenije, H. H. G. (2011). Towards an automated SAR-based flood monitoring system: Lessons learned from two case studies. *Physics and Chemistry of the Earth, Parts A/B/C, 36*(7–8), 241–252.
22. Nico, G., Pappalepore, M., Pasquariello, G., Refice, A., & Samarelli, S. (2000). Comparison of SAR amplitude vs. coherence flood detection methods-a GIS application. *International Journal of Remote Sensing, 21*(8):1619–1631.
23. Pulvirenti, L., Pierdicca, N., Chini, M., & Guerriero, L. (2011). An algorithm for operational flood mapping from Synthetic Aperture Radar (SAR) data based on the fuzzy logic. *Natural Hazard and Earth System Sciences.*

24. Pulvirenti, L., Chini, M., Pierdicca, N., & Boni, G. (2016). Use of SAR data for detecting floodwater in urban and agricultural areas: The role of the interferometric coherence. *IEEE Transactions Geoscience and Remote Sensing*, *54*(3), 1532–1544.
25. Ramsar Sites Information Services. (2018). *Annotated list of wetlands of international importance*. https://rsis.ramsar.org/sites/default/files/rsiswp_search/exports/Ramsar-Sites-annotated-summary-India.pdf?1538638066.
26. Schumann, G., Di Baldassarre, G., Alsdorf, D., & Bates, P. D. (2010). Near real-time flood wave approximation on large rivers from space: Application to the river PO, Italy. *Water Resources Research*, *46*(5)
27. Schumann, G. J. P., Bates, P. D., Apel, H., & Aronica, G. T. (2018). *Global flood hazard: Applications in modeling, mapping, and forecasting* (Vol. 223). Wiley & Sons.
28. Yang, C., Huang, Q., Li, Z., Liu, K., & Fei, H. (2017). Big data and cloud computing: Innovation opportunities and challenges. *International Journal of Digital Earth*, *10*(1), 13–53.
29. Yulianto, F., Sofan, P., Zubaidah, A., Sukowati, K. A. D., Pasaribu, J. M., & Khomarudin, M. R. (2015). Detecting areas affected by flood using multi-temporal ALOS PALSAR remotely sensed data in Karawang, West Java, Indonesia. *Natural Hazards*, *77*(2), 959–985.

Evaluation of Hedonic Price Models that Explain Transit Induced Impact on Housing Prices

Karan Barpete and Arnab Jana

Abstract This review article evaluates housing price models and their configurations across 21 housing projects in different parts of the world that have a transit project being in close vicinity. These projects try to explain the impact of transit projects on the housing prices. The housing projects and their respective hedonic price models are studied to compile all the dependent and explanatory variables used in them. The explanatory variables are then classified into six categories, namely (i) Proximity variables, (ii) Proximity premium, (iii) Land/Structural variables, (iv) Neighborhood variables, (v) Accessibility variables, and (vi) Temporal variables. In most of these studies, the dependent variable is a variation of price (rent price, sale price, land price, etc.). The differences in the dependent variables are noted and highlighted in the evaluation. The comparison of different functional form of HPM is done to establish the difference in results and the applicability of them in appropriate models.

Keywords HPM(Hedonic price Model) · Transit · Housing

1 Introduction

1.1 Effect of Metro Rail on Real Estate in Its Vicinity

Although results are somewhat heterogeneous across study areas, many researchers have identified a positive effect of metro or transit station proximity on housing price [11, 13]. While some results have shown nonsignificant results, no studies were found showing a negative impact of metro rail proximity to housing prices. In research assessing the effect of metro lines on housing prices, a wide range of proximity premiums were found: from as low as 0.3% in Seoul [4] to as high as 75% in London (Banister 2007).

K. Barpete (✉) · A. Jana
Centre for Urban Science and Engineering, IIT Bombay, 400076 Mumbai, India
e-mail: b.karan@iitb.ac.in

A. Jana
e-mail: arnab.jana@iitb.ac.in

A. Jana and P. Banerji (eds.), *Urban Science and Engineering*, Lecture Notes in Civil Engineering 121, https://doi.org/10.1007/978-981-33-4114-2_12

In Bangalore, the Metro system was found to uplift property values by 10.7% within a 500 m catchment area [38]. Gadziski and Radzimski [15] examined the effect of a new transit line on three variables: travel behavior, housing choices, and property prices. They found a significant influence of the transit line on at least one of the three studied components in every examined region. Metro rail provides quicker and easier access to jobs and can increase housing demand, and thus prices, in a given region. In a Bangkok study, researchers found a significant impact of proximity to mass transit stations, along with proximity to other transportation infrastructure such as arterial roads, on housing prices [2]. Interestingly, it seems that even the announcement of a planned transit line can have an impact on rental prices [17].

Many articles call attention to the positive impacts of land value increase due to metro rail—e.g., opportunities for city governments to earn revenue through land value capture [38]. However, some authors have highlighted the possible negative effects of proximity to transit stations. For example, in the USA, new transit stations (particularly "walk and ride" transit stations) can lead to gentrification or displacement of residents by increasing rent in the catchment area of the station [26]. Additionally, proximity to transit can be a disamenity due to additional noise or crime risk [37]. However, generally speaking, results have indicated positive relationships between access to metro stations and land value.

While many articles have been published quantifying the effect of metro rail on housing prices in urban areas, the present research has much to add to the literature. First, as Mumbai currently has one operational metro line and multiple planned lines, the present study can test the effect of both operational and nonoperational metro lines on housing prices. Second, a majority of published articles assess metro systems in the global north. This article is one of few addressing a metro system in a rapidly developing nation, and one of even fewer addressing an Indian city. The nature of Mumbai, particularly its dynamic and rapidly expanding real estate market, makes this study extremely relevant. Third, this paper features several different data sources and two methodologies. This multiple-methods approach allows for internal validation of results.

Some of the studies [5] that focus on the large-scale operational and established urban rail/metro investments study the impact on property values. However small-scale mode like LRT/trams are not studied as much. There lies a research gap for these modes. Multiple studies [13, 14, 20] have confirmed similar positive effect of having urban rail transit system closer on the property prices. Although there are other studies that undermine the effect and call it marginal or only important when the supporting conditions are favorable. Out of 24 North American cities studied in 19 different studies, 13 tested positive correlation in property prices, while 4 were found to have no effect of LRT infrastructure. However, 7 cities showed signs of decline in property prices. The results are not ambiguous because the negative correlations have an explanation. The cities where increase in socioeconomic inequality was triggered by the LRT infrastructure, the benefits of accessibility and proximity to LRT are outweighed by negative externalities like rise in crime, noise levels.

The conclusions of these studies suggest that the significance of the positive effect of LRT on real estate can be guided by appropriate and supportive policies. This will require context-specific empirical studies because the generalized solution from the case studies of the Anglo-Saxon countries and Western Europe will not be suitable in India. The conclusion of these reviews hangs in the form of three primary questions:

1. How does the proximity to the new transportation infrastructure affect travel behavior in the residents of locality?
2. How does the transportation infrastructure affect the quality of life environment of the locality neighborhoods? How satisfied are the residents with their housing choices?
3. How does the proximity to public transportation network influence the property prices in the study area?

2 Hedonic Price Modeling

"Hedonics" comes from a Greek word hedonikos, which means pleasure, and in economics it refers to the satisfaction one gets from consumption of goods and services, i.e. utility [10]. HPM is extensively used in housing value and real estate research and even though the accuracy of the results may sometime be off the mark, but it continues to be valid for empirical research in the real estate market [10]. HPM analysis does economic analysis with following five assumptions:

- Homogenous land/housing market.
- Perfect competition in the market.
- Consumers and suppliers are free to enter and exit the market.
- Consumers and suppliers are perfectly informed about the products and prices.
- Market is at equilibrium and prices and attributes have no inter-relationship.

HPM applies least squares regression analysis, and there is a linear relationship between the dependent variable and explanatory variables. In case of housing/land market, the observed price (P) is explained using the following parametric land price equation.

$$P_i = f\left(X_j; \beta_j\right) + \varepsilon_i \tag{1}$$

- P_i is the assessed land/residential property price of the ith observation,
- X_j is a vector of quantitative and qualitative attributes of land/residential property,
- β_j is the unknown hedonic hidden price, of the land/residential property for attribute *j*, and
- ε_i is the stochastic error term.

2.1 Functional Form of Hedonic Price Models

As the association between the explanatory and dependent variables in HPM of Housing markets is mostly nonlinear, there are different configuration of HPM functional form for Housing, land and real-estate models to bypass this lack of linearity, which assume that explanatory variables are continuous, not binary in nature. Hannonen [19] proposes that the methodology choice of the correct functional form for the HPM decides the accuracy of estimation process, and an inappropriate choice can make the subsequent analysis invalid. In parametric research, it is essential to work on a variety of alternative model configuration to decide, which suits the land, housing market or its submarket being analyzed. An incorrect selection of functional form can result in unreliable estimates [6, 10]. And even though, it has been practiced for a long time, the theory lacks guideline on the decision of choosing correct functional form for varying application. Among the variety of hedonic price models, the mostly used ones are (i) Linear HPM, (ii) Log-linear HPM, (iii) Linear-Log HPM, and (iv) Double-Log HPM.

The following tables (Tables 1 and 2) summarize the different forms of hedonic price models used in the 21 projects around the world. They also tell us the degree of success these models had in describing the pricing of these housing projects. Tables 3, 4, 5 and 6 list down the respective independent variables as used in these models. The ID column in the latter tables relates with the Tables 1 and 2.

3 Conclusion and Identified Research Gaps for Further Study

Some of the studies [5] that focus on the large-scale operational and established urban rail/metro investments study the impact on property values. However small-scale modes like LRT/trams are not studied as much. There lies a research gap for these modes. Multiple studies [13, 14, 20) have confirmed similar positive effect of having urban rail transit system closer on the property prices. Although there are other studies that undermine the effect and call it marginal or only important when the supporting conditions are favorable. Out of 24 North American cities studied in 19 different studies, 13 tested positive correlation in property prices, while 4 were found to have no effect of LRT infrastructure. However, 7 cities showed signs of decline in property prices. The results are not ambiguous because the negative correlations have an explanation. The cities where increase in socio-economic inequality was triggered by the LRT infrastructure, the benefits of accessibility and proximity to LRT are outweighed by negative externalities like rise in crime, noise levels.

The conclusions of these studies suggest that the significance of the positive effect of LRT on real estate can be guided by appropriate and supportive policies. This will require context-specific empirical studies because the generalized solution from the

Table 1 Hedonic price models studying impact of BRT and LRT projects

Id.	Author	Location and transit system	HPM form	HPM # Obs. (model R^2)	Dependent variable
BRT1	Rodriguez and Targa [35]	Bogota, Colombia TransMilenio BRT	OLS—linear, log/linear, log/log	494 (0.71)	Linear, log (rental cost)
BRT2	Rodriguez and Mojica [34]	Bogota, Colombia Trans Milenio BRT	OLS WLS—log/linear	3976 (0.694)	Ln (Advert. Sale Price)
BRT3	Perk and Catala (2009)	Pittsburgh, USA, MLK, Jr East Busway	Robust LS—linear	128,717 (0.8)	Praised value (fair market value)
BRT4	Cervero and Kang [7]	Seoul, South Korea, Seoul BRT	Multi-level logit	25,410 (0.992)	Land value
BRT5	Mulley and Tsai [32]	Sydney Australia Liverpool-Parramatta BRT	ANOVA & OLS	1167 (0.67)	Ln (sale price)
LRT1	Golub et al. [17]	Phoenix, USA, Phoenix LRT	OLS—log/log	88,308 (0.533)	Ln (adjusted sale price)
LRT2	Atkinson-Palombo [3]	Phoenix, USA, rezoning around the phoenix LRT	GLS log/linear	9177 (0.76)	Ln (sales price)
LRT3	Du and Mulley [12]	England, UK, tyne & wear light rail	OLS & GWR log/linear	1700 (0.38)	Ln (house price)
LRT4	Cervero and Duncan [8]	San Diego, USA LRT	OLS—linear	14,756 (0.605)	Sale price
LRT5	Garrett (2004)	Missouri, USA St. Louis Metrolink LRT	OLS log/linear	1516 (−)	House price

case studies of the Anglo-Saxon countries and Western Europe will not be suitable in India. The conclusion of these reviews hangs in the form of three primary questions:

(1) How does the proximity to the new transportation infrastructure affect travel behavior in the residents of locality?
(2) How does the transportation infrastructure affect the quality of life environment of the locality neighborhoods? How satisfied are the residents with their housing choices?
(3) How does the proximity to public transportation network influence the property prices in the study area?

Table 2 Hedonic price models studying impact of metro & commuter rail projects

ID.	Author	Location & transit system	HPM form	HPM # Obs. (model R^2)	Dependent variable
Metro 1	Banister (2007)	London, UK, London metro Jubilee line	GWR	–	Land and property valuations
Metro 2	Gatzlaff and Smith [16]	Miami, USA heavy rail/metro	OLS linear log/linear exp. log/log	912 (0.72–0.84)	Sale price
Metro 3	Laakso [28]	Helsinki, Finland Helsinki Metro	OLS log/linear	6732 (0.940)	Ln (sale price)
Metro 4	Bae et al. [4]	Seoul, South Korea Heavy Rail	GLS Log/Linear	956 (0.9542)	Ln (sales price)
Metro 6	Celik and Yankaya [33]	Izmir, Turkey Izmir Metro	OLS Linear Log/linear Log/Log	360 (0.83)	Sale price
Metro 6	Modelewska and Medda [31]	Warsaw, Poland Warsaw Metro	OLS Log/Linear	1130 (0.696)	Sale price
CR1	Cervero and Duncan [8]	San Diego, USA Commuter Rail	OLS	25,923 (0.7)	Sales price
CR2	Sedway Group [36], Mathur and Ferrell [30]	San Francisco USA Bay Area Rapid Transit (BART)	OLS Log/Log	2133 (0.74)	Ln (Sales Price)
CR3	Gruen [18],Chaney [9]	Chicago, USA METRA, Commuter Rail	OLS Log/linear	796	Property value
CR4	Voith [39]	Pennsylvania & New Jersey, USA Commuter Rail	OLS	571 (0.711)	Property value
CR5	Lochl and Axhausen [29]	Zurich, Switzerland, commuter rail	OLS, spatial autoregressive model, GWR, Log/Log	8592 (0.85)	Ln (rent)

Table 3 Explanatory variables in HPM models with BRT

ID	Proximity variable	Proximity premium	Land/structural variables	Neighborhood variables	Accessibility variables	Time-based variables
BRT1	5 min walk	6.8% to 9.3%	Property Area Beds Baths Living Room Age	Socioeconomic Conditions. Population Density Employ. Density % Diff. Land Uses Crime Poverty 400 m Busway	Distance to BRT Ped. time to BRT BRT travel time Distance to CBD	
BRT2	150 m	13% to 15%	House/AptAge Bedroom Bath Garage Area	Socio Economic Conditions. Population Density Employ. Density % Diff. Land Uses Crime	Prox. 150 m BRT 500 m BRT	Year Dummies Interaction terms
BRT3	Distance to BRT	Significant and +ve	Lot Area Living area size Beds Bath 1/2 Bed Age	Income Socio Economic Conditions Population Density	Distance to BRT	
BRT4	90 m to 300 m of BRT stop	5% to 10%	Land use Building coverage ratio floor area ratio % Age Demo. % College degree	Population Density Employment Density Distance to River % Park % Land Developed Road area ratio % Res & Comm. Develop. capacity	Distance to Freeway ramp Distance to BRT Distance to CBD Distance to Subway Distance to Major Rd. Distance to Bus Job Accessibility by Car	

(continued)

Table 3 (continued)

ID	Proximity variable	Proximity premium	Land/structural variables	Neighborhood variables	Accessibility variables	Time-based variables
BRT5	400 m	Up to 3.3%	Bed. Bath Parking House/Apt	% Eng. Language Unemployed Income	Within 50 m of BRT stop	Time dummies & interaction terms;preconstruction during const. & operations

Table 4 Explanatory variables in HPM models with LRT

ID	Proximity variable	Proximity premium	Land/structural variables	Neighborhood variables	Accessibility variables	Time-based variables
LRT1	200ft	25%	Living size, Lot size, Age, #Patios, #Bath, #Floors, Pool, TOD Zoning		Dist. to LRT Stn., Dist. to LRT Alignment, Dist. to CBD, Dist. to Airport	Time dummies Prior NEPA, During NEPA Review, Planning & Design, Construction, Operations
LRT2	1/2 mile	17% Transit 34% Transit + TOD Overlay	Lot Size House size Pool Age	Socio Economic Data TOD Overlay Zoning	LRT Ped Catchment Dist. to Fwy Dist to CBD	Pre and Post dates from the introduction of the TOD overlay
LRT3	200 m	17.1%	House Type, #Bedroom,	Local School Indicator, % unemployed, %Higher Profession Occupation	PT Access (School, College…), Car Access (School College…), Dist. to LRT	
LRT4	400 m	3.8% to 17.3%	Size, #Units #Bath, Bed, Age	Housing Density Income Race Profile %Senior %Vacant Land	½ Mile LRT Dist. to Hwy/Fwy Dist. to Fwy Ramp	Time Dummies Monthly to reflect different sale times
LRT5	700 m	32%	#Bed #Bath #Stories Garage Pool Age Lot Size House size	Dist. to Hwy interchange %Res. With College Education Income Property Tax rate School District Test Scores Does nearest LRT have P&R?	Dist. to nearest LRT Stn Noise impact from LRT by Dist. to LRT	–

Table 5 Explanatory variables in HPM models with metro rail

ID	Proximity variable	Proximity premium	Land/structural variables	Neighborhood variables	Accessibility variables	Time-based variables
Metro 1	2000 m Access to metro	75%		Comm. & environ. amenity Car ownership Socio. economic	Access to shops Dist. to School Access to metro	–
Metro 2	Dist. to metro	Mixed between stations	House area Lot size Age	Est. house price index	Dist. to metro	Construction announcement dummy
Metro 3	250 m	3.5% to 6%	Ln (age) Ln (area) Terrace house Pool Indoor sports Health Stn Library Day care	Ln (%Park) Ln (income quartile) Dist. to Coast Ln (Dist. to CBD)	Metro station dummies Feeder bus dummies Commuter rail dummy Shopping center dummy	Transaction time dummies
Metro 4	400 m	0.3% to 2.6%	Apart. size, Age, #Houses block #Parking Heating Type Dist. to Park	Dist. from Han River School District Pop. Density Job Density	Dist. to Subway Dist. to CBD Dist. to Sub center	Time dummies Sales in 1995 Sales in 1997 Sales in 2000
Metro 5	500 m	0.7% to 13.7%	House size #Apt in Bldg. #Apts. in Floor Age #Bed #Storey of Bldg. Corner location Parking Heating	Location Type of ground	Dist to Subway Dist. to Bus Dist to Shop	
Metro 6	1000 m	6.7%-7.13%	Area #Rooms #Floors in Bldg. Age Parking	School District	Dist to Hospital Dist. to Green Area Metro Catchment dummy	Time Dummy for year of sale

Table 6 Explanatory variables in HPM models with commuter rail projects

ID	Proximity variable	Proximity premium	Land/structural variables	Neighborhood variables	Accessibility variables	Time-based variables
CR1	½ mile	−7.1 to 46.1%	House size Lot size #Bath #Bed Age	Housing Density Income %White Neighbourhood	½ Mile Commuter Rail Dist. to Hwy Ramp Job Access Hwy Job Access Transit	Time Dummies Monthly to reflect different sale times
CR2	½ mile	20% 1.5%	House size Lot size #Bath #Bed Age	Income %Hispanic Neighbourhood	Dist. to BART Dist. to Bus Dist. to Hwy/Fwy	Time Dummies for years 1995–2002
CR3	400 m	14.5 to 20%	House Size Lot Size #Bath Age Furnished Garage Fireplace House Type		Dist. from Station Dist. from Hwy Squared Dist. from Station Squared Dist. from Hwy	
CR4	½ mile	6 to 10%	Size Detached Age #Rooms	%Black Neighbourhood	Auto Commute Station Rail Commute	–
CR5	500 m	4 to 8%	House Size Lift Balcony #Bath Age Furnished Garage Fireplace Single house View	Within 100 m Autobahn Air Noise Job within 1 km Pop. Density per hectare %Foreigners per Hectare Local Tax level Slope	Dist. to CBD Car Access time to employment PT Access to employment Rail stn. catchment	Transaction time dummies

References

1. Alleviation, M. O. (2015). *National urban rental housing policy*. Government of India.
2. Anantsuksomsri, S., & Tontisirin, N. (2015). The impacts of mass transit improvements on residential land development values: Evidence from the Bangkok metropolitan region. *Urban Policy and Research, 33*(2), 195–216. https://doi.org/10.1080/08111146.2014.982791.
3. Atkinson-Palombo, C. (2010). Comparing the capitalisation benefits of light-rail transit and overlay zoning for single-family houses and condos by neighbourhood type in metropolitan

phoenix Arizona. *Urban Studies, 47*(11), 2409–2426. https://doi.org/10.1177/0042098009357963.
4. Bae, C.-H. C., Jun, M.-J., & Park, H. (2003). The impact of Seoul's subway Line 5 on residential property values. *Transport Policy, 10*(2), 85–94. https://doi.org/10.1016/s0967-070x(02)00048-3.
5. Banister, D., & Thurstain-Goodwin, M. (2011). Quantification of the non-transport benefits resulting from rail investment. *Journal of Transport Geography, 19*(2), 212–223. https://doi.org/10.1016/j.jtrangeo.2010.05.001.
6. Blomquist, G., & Worley, L. (1981). Hedonic prices, demands for urban housing amenities, and benefit estimates. *Journal of Urban Economics, 9*(2), 212–221. https://doi.org/10.1016/0094-1190(81)90041-3.
7. Cervero, R., & Kang, C. D. (2011). Bus rapid transit impacts on land uses and land values in Seoul, Korea. *Transport Policy, 18*(1), 102–116. https://doi.org/10.1016/j.tranpol.2010.06.005.
8. Cervero, R., & Duncan, M. (2002). Residential self selection and rail commuting: a nested logit analysis.
9. Chaney, H. (2005). *Evaluating the capitalization effects of metra commuter rail transit upon land values in the suburban Chicago municipality of Arlington heights: A tale of two stations.* The University of North Carolina at Chapel Hill University Libraries. https://doi.org/10.17615/FFTG-PQ41.
10. Chau, K. W., & Chin, T. L. (2003). A critical review of literature on the hedonic price model (June 12, 2002). *International Journal for Housing Science and Its Applications, 27*(2), 145–165. Available at SSRN: https://ssrn.com/abstract=2073594.
11. Debrezion, G., Pels, E., & Rietveld, P. (2007). The impact of railway stations on residential and commercial property value: A meta-analysis. *Journal of Real Estate Finance and Economics, 35,* 161–180. https://doi.org/10.1007/s11146-007-9032-z.
12. Du, H., & Mulley, C. (2007). Transport accessibility and land value: A case study of tyne and wear. *RICS Research Paper Series, 7*(3), 52
13. Dubé, J., Thériault, M., & Des Rosiers, F. (2013). Commuter rail accessibility and house values: The case of the Montreal South Shore, Canada, 1992–2009. *Transportation Research Part A: Policy and Practice, 54*, 49–66. https://doi.org/10.1016/j.tra.2013.07.015.
14. Efthymiou, D., & Antoniou, C. (2013). How do transport infrastructure and policies affect house prices and rents? Evidence from Athens, Greece. *Transportation Research Part A: Policy and Practice, 52*, 1–22. https://doi.org/10.1016/j.tra.2013.04.002.
15. Gadziski, J., & Radzimski, A. (2015). The first rapid tram line in Poland: How has it affected travel behaviours, housing choices and satisfaction, and apartment prices? *Journal of Transport Geography*. https://doi.org/10.1016/j.jtrangeo.2015.11.001.
16. Gatzlaff, D. H., & Smith, M. T. (1993). The Impact of the Miami metro rail on the value of residences near station locations. *Land Economics, 69*(1), 54. https://doi.org/10.2307/3146278.
17. Golub, A., Guhathakurta, S., & Sollapuram, B. (2012). Spatial and temporal capitalization effects of light rail in phoenix. *Journal of Planning Education and Research, 32*(4), 415–429. https://doi.org/10.1177/0739456x12455523.
18. Gruen, A. (1997). *The effect of CTA and metra stations on residential property values.* Report to the regional transportation authority, Chicago, Ill.
19. Hannonen, M. (2014). Urban housing policy considerations: perspectives from the finish housing market. *Journal of Heterodox Economics, 1*(2), 114–130. https://doi.org/10.1515/jheec-2015-0007.
20. Hess, D. B., & Almeida, T. M. (2007). Impact of proximity to light rail rapid transit on station-area property values in Buffalo, New York. *Urban Studies, 44*(5–6), 1041–1068. https://doi.org/10.1080/00420980701256005.
21. Hills, J. (2007). *Ends and means: The future roles of social housing in England.* ESRC Research Centre for Analysis of Social Exclusion.
22. Hui, E. C., & Yue, S. (2006). Housing price bubbles in Hong Kong, Beijing and Shanghai: A comparative study. *Journal of Real Estate Finance and Economics, 33,* 299–327. https://doi.org/10.1007/s11146-006-0335-2.

23. India, R. B. (2018). Affordable housing in India. *RBI Bulletin January 2018*, 13.
24. Jones, C., & Macdonald, C. (2004). Sustainable urban form and real estate markets. *European Real Estate Conference, Milan*, 2–5.
25. Journal, A. T. (2001). Impossibility of competitive equilibrium in the real estate brokerage industry.
26. Kahn, M. E. (2007). Gentrification trends in new transit-oriented communities: Evidence from 14 cities that expanded and built rail transit systems. *Real Estate Economics*, *35*(2), 155–182. https://doi.org/10.1111/j.1540-6229.2007.00186.x.
27. Kaplanski, G., & Levy, H. (2012). Real estate prices: An international study of seasonality's sentiment effect. *Journal of Empirical Finance, 19,* 123–146. https://doi.org/10.1016/j.jempfin.2011.11.004.
28. Laakso, S. (1992). Public transport investment and residential property values in helsinki1. *Scandinavian Housing and Planning Research*, *9*(4), 217–229. https://doi.org/10.1080/02815739208730308.
29. Löchl, M., & Axhausen, K. W. (2010). Modelling hedonic residential rents for land use and transport simulation while considering spatial effects. *Journal of Transport and Land Use*, *3*(2), https://doi.org/10.5198/jtlu.v3i2.117.
30. Mathur, S., & Ferrell, C. (2013). Measuring the impact of sub-urban transit-oriented developments on single-family home values. *Transportation Research Part A: Policy and Practice*, *47*, 42–55. https://doi.org/10.1016/j.tra.2012.10.014.
31. Modelewska, M., & Medda, F. (2011). *Land value capture as a funding source for urban investment: The Warsaw metro system.*
32. Mulley, C., & Tsai, C. (2013). The impact of liverpool-parramatta transitway on housing price: A repeat sales approach. In: *Australasian Transport Research Forum 2013.* [online] Available at: [Accessed 1 January 2020].
33. Murat Celik, H., & Yankaya, U. (2006). The impact of rail transit investment on the residential property values in developing countries. *Property Management*, *24*(4), 369–382. https://doi.org/10.1108/02637470610671604.
34. Rodríguez, D., & Mojica, C. (2008). *Land value impacts of bus rapid transit: the case of Bogota's Transmilenio.* [online] Lincolninst.edu. Available at: https://www.lincolninst.edu/sites/default/files/pubfiles/1359_680_Bus%20Bogota.pdf. [Accessed 1 January 2020].
35. Rodríguez, D. A., & Targa, F. (2004). Value of accessibility to Bogotá's bus rapid transit system. *Transport Reviews*, *24*(5), 587–610. https://doi.org/10.1080/0144164042000195081.
36. Sedway Group. (1999). *Regional impact study.* A report commissioned by the bay area rapid transit system.
37. Seo, K., Golub, A., & Kuby, M. (2014). Combined impacts of highways and light rail transit on residential property values: a spatial hedonic price model for Phoenix, Arizona. *Journal of Transport Geography*, *41*, 53–62. https://doi.org/10.1016/j.jtrangeo.2014.08.003.
38. Sharma, R., & Newman, P. (2018). Can land value capture make PPP's competitive in fares? A Mumbai case study. *Transport Policy, 64,* 123–131. https://doi.org/10.1016/j.tranpol.2018.02.002.
39. Voith, R. (1991). Transportation, sorting and house values. *Real Estate Economics*, *19*(2), 117–137. https://doi.org/10.1111/1540-6229.00545.

Strategies for Sustainable, Efficient, and Economic Integration of Public Transportation Systems

Challa Prathyusha, Sandeep Singh, and P. Shivananda

Abstract In India, the city and intercity State Road Transport Undertakings (SRTUs) cater 6.8 million passengers/day over a length of 1.48 billion passenger-km/day during 2015–2016. The average occupancy ratio of 47 SRTUs has decreased by 1.56%, from 70.74 in 2014–2015 to 69.65 in 2015–2016. The traditionally followed operational plan for the public transportation system in urban areas results in a financial loss of 7% in SRTUs; from Rs. 10,587.98 crore in 2014–2015 to Rs. 11,349.78 crore in 2015–2016. In order to achieve the equalization of public service, there is a necessity for operational integration of the public transportation system to reduce the financial losses to public buses. The public transportation system productivity and efficiency could be maximized through the components of the Intelligent Transportation System (ITS) like Electronic Ticketing Machines (ETMs) and Global Positioning System (GPS) equipped with Vehicle Tracking Unit (VTU), which helps in the collection of data for a network. In this paper, the bus transit frequency is evaluated for one route with even headway by point check method. The bus service frequency software (Visual Basic-VB6) is developed to estimate user and operator costs for economic analysis. The results show that the waiting time cost of the passenger and the fleet size is minimized using deficit function as the solution for timetable synchronization and vehicle scheduling.

Keywords Operational integration · Frequency · Intelligent transportation system (ITS) · Electronic ticketing machines (ETMs) · And vehicle tracking unit (VTU)

C. Prathyusha (✉) · P. Shivananda
REVA University, Yelahanka, Bangalore, India
e-mail: prathyushareddy1990@gmail.com

S. Singh
National Institute of Technology Tiruchirappalli, Tiruchirappalli, India
e-mail: sandeepsingh.nitt@gmail.com

A. Jana and P. Banerji (eds.), *Urban Science and Engineering*, Lecture Notes in Civil Engineering 121, https://doi.org/10.1007/978-981-33-4114-2_13

1 Introduction

In the context of Indian cities, the dependence of the urban trips on public transport is based on numerous parameters like the city size, geographical considerations, land use, and functional segregation of activities over the city. The Multiple Criteria Decision-Making (MCDM) method enhances the traveling behavior, and the provision of better transit service is challenging in the cities [1]. In India, the city and intercity State Road Transport Undertakings (SRTUs) cater 6.8 million passengers/day over a length of 1.48 billion passenger-km/day during 2015–2016. The average occupancy ratio of 47 SRTUs has decreased by 1.56%, from 70.74 in 2014–2015 to 69.65 in 2015–2016. The traditionally followed operational plan for the public transportation system in urban areas results in the financial loss of 7% in SRTUs, from Rs. 10,587.98 crore in 2014–2015 to Rs. 11,349.78 crore in 2015–2016 [2]. In urban areas, public transit remains to be a challenge due to an increase and variation in travel demand [3]. There is a necessity for public transport transit integration: organizational, physical, and operational, which can be achieved through push and pull measures [4]. If the demand for public transportation is less, then flexible service is desirable and conventional service method during high demand. The integration of these two methods develops a variable-type service [5]. In order to achieve the equalization of public service, there is a necessity for operational integration of the public transportation system to reduce the financial losses to public buses.

2 Literature Review

The public transportation system productivity and efficiency could be maximized through the components of the Intelligent Transportation System (ITS) like Electronic Ticketing Machines (ETMs) and Global Positioning System (GPS) equipped with Vehicle Tracking Unit (VTU) which helps in the collection of data for a network [6]. Public transportation performance is checked through various service level benchmarks: headway, service coverage area, transit-accessible area, average waiting time, high-frequency accessible area, boarding, % transit ridership, organized Public Transportation (PT), peak hour, and availability of PT. The modified indicator was operating ratio, which is defined as the ratio between cost/km and earning/km [7]. According to [8], the SRTU performance indicators in the Indian context were (1) Capacity-fleet size and vehicle seat capacity; (2) Serviceability-revenue per km, passengers per km, load factor, passengers carried, and number of routes; (3) Safety-number of accidents, fatalities, and injuries; (4) Productivity-operating cost, cost per km, maintenance cost, traffic revenue, non-traffic revenue, operating ratio, and cost recovery; (5) Reliability-trip scheduled, regularity, actual trips operated, departure and arrival timings, number of breakdowns, and rate of breakdown; and (6) Comfort-average age fleet were considered in the frequency and time table scheduling of the bus. The various factors influencing the efficiency of a bus operating system are reliability of service, frequency, route coverage, excessive transfers, fare and low profit [9]. Reliability of travel time is one of the concerned aspects for both user and operator and the bus arrival time will be influenced by various factors such as boarding and alighting of passengers, weather, and congestion [10]. However, to predict the travel time, weekly, daily, and hourly patterns need to be analyzed. The different arrival time prediction models are time series, regression, Kalman filtering, artificial neural networks, and super vector machines [11]. The time series model is used for the travel time of the bus at the stops. The bus service reliability is evaluated to improve the operational efficiency in two stages: strategic and tactical planning [12]. The 12 cities have been surveyed to know the satisfaction and dissatisfaction levels of the user from the bus service (primary mode-53.6%), and overcrowding is the most critical aspect of dissatisfaction; this can be improved through higher frequency [13]. The strategic-network design and the tactical-frequency setting and timetabling are given more importance in public transportation planning. The objective function, constraints, and decision variables of user and operator are used to optimize public transportation [14].

3 Aims and Objectives

In Indian context, the public transport system productivity and efficiency could be maximized through the basic operational planning process. In this paper, the bus transit frequency is evaluated for one route with even headway by point check method. The bus service frequency software (Visual Basic-VB6) is used to estimate user and operator costs for economic analysis. The main objectives are as follows:

- To develop a model for frequency setting for a route.
- To develop a model for scheduling of bus for a particular route.

4 Model Formulation

The four basic operational planning processes include network route design, setting timetables, scheduling vehicles to trip, and assignment of drivers [15]. A cost-effective frequency setting is one of the most complicated strategic planning for public transportation due to variations in travel patterns along the routes. The major service attributes route, operation, vehicle, user-level, and cost are considered with transit operators. The optimal frequency for different time periods of a single route for a single time period is determined considering user and operator attributes and parameters in "Bus Service Frequency Setting Software," which is mathematical programming in Visual Basic (VB6). The service efficiency constraints in the transit assignment such as cost per passenger, passenger per vehicle-kilometer, cost recovery ratio, number of trips per vehicle, number of passengers per trip per vehicle, operation cost per trip per passenger per vehicle with the objective to minimize the fleet size, and maximize the patronage. The regional bus scheduling problem [RBSP] is a network problem [NP], which is solved in this paper using a bi-level programming model [16] and which helps in integrating the frequency and timetable of a single route [14].

4.1 Assumptions

1. In frequency setting for a bus, any size of the time period for a single day (service period ≤24 h), stop to stop operation cost, for each stop minimum dwell time, for every passenger lost-penalty cost, passengers in-vehicle travel time and waiting time, seat availability, and standing cost are considered.
2. Different bus sizes and capacity, access, and egress time are not considered.
3. In the sequence of bus departures, bus bunching considered were short turning, overtaking, and stop skipping not allowed.

The simulation executed under the following conditions:

1. Buses dispatched for a given headway.
2. The constraints of the user and operator are followed while assigning the passengers for each bus.
3. Historical boarding and alighting of passengers are applied for each bus.

Let on route k, $i = 1, 2, 3 \ldots$ V denotes a set of vehicles dispatched from the stops $j = 1, 2, 3 \ldots$ S, vehicle type Z with a seating capacity Q_i^Z. The arrival, departure, and dwell time are denoted as $t_{ij}^{arr}, t_{ij}^{dep} and\ t_{ij}^{dwell}$, respectively, for a given period $h = 1, 2, 3 \ldots$ H. To set vehicle frequency, boarding and alighting of passenger count, passenger arrival rate, and stop to stop travel time are necessary from the operator's point of view [17].

4.2 Cost Minimization

The user and operator cost attributes influence the bus operating system. The overall cost is minimized considering two objective functions, minimum user and operator cost for a given route. The cost function is estimated using the handbook on feasible service delivery ranges for bus transit in the Indian Context part 2-operator perspective, chapter-4 for a given service period [17] as follows:

$$\text{Overall Cost } (C_S) = \text{ User Cost}(C_U) + \text{ Operator Cost } (C_O) \tag{1}$$

$$\begin{aligned}\text{User Cost } (C_{\mathrm{U}}) = {} & \text{Waiting Time Cost } (C_{\mathrm{W}}) + \text{In - vehicle Seating Time Cost } (C_{\mathrm{Tseat}}) \\ & + \text{In - vehicle Standing Time Cost } (C_{\mathrm{Tstand}}) \\ & + \text{Passenger Inconvenience Cost } (C_{\mathrm{P}}) \end{aligned} \quad (2)$$

$$\begin{aligned}\text{Operator Cost } (C_{\mathrm{O}}) = {} & \text{Fuel cost } (C_{\mathrm{F}}) + \text{Depreciation Cost } (C_{\mathrm{D}}) \\ & + \text{Crew Cost } (C_{\mathrm{crew}}) + \text{Maintenance Cost } (C_{\mathrm{M}}) \\ & + \text{Operator Penalty Cost } (C_{\mathrm{P'}}) \end{aligned} \quad (3)$$

4.3 Frequency

Frequency is defined as the minimum number of vehicles requires for a given period [18]. In this work, the frequency is estimated for the baseline scenario using an hourly load point.

$$\mathrm{F}_{hk} = \max\left[\mathrm{P}_{mh}/\gamma_h \mathrm{Q}, \mathrm{F}_{mhk}\right] \quad (4)$$

F_{hk} Frequency (Number of vehicles for a given period on route k).
P_{mh} Total number of passengers at maximum load point.
γ_h Load Factor.
Q Capacity of bus.
F_{mhk} Minimum required frequency.

The constraint is formulated on route k subject to the minimum objective function $\mathrm{C_S}$ for departures:

$$1.\, \alpha_k^{max} < \alpha_k^{route} < \alpha_k^{min} \quad (5)$$

$$2.\, hw_k^{max} < hw_k^{route} < hw_k^{min} \quad (6)$$

where α = Crowding level, h = Headway.

5 Solution Method

5.1 Theorem 1

The set of all trips arrives and departs within a scheduling horizon and does not consider Dead Heading (DH); the minimum number of vehicles required to serve all the trips is equal to the sum of all deficits. This is known as the fleet size theorem.

$$\text{MinimumV} = \sum_{j=1}^{S} \text{D(S)} = \sum_{j=1}^{S} \text{Maximum}(j, h)|h = 1, 2, 3, \ldots H| \tag{7}$$

5.2 *Deficit Function*

Deficit function (DF) is defined as the necessity of a minimum number of vehicles required to complete N trips in the scheduling. In the scheduling, the change of the departure timings reduces the fleet size and waiting cost of the passenger. The failure of this method results in unbalanced loads, overcrowding, empty, and bus bunching problems, so DF is studied carefully while assigning vehicles to the trips.

If the precedence relation R is satisfied, then vehicle is serviced sequentially as follows:

$$R{:}\, i \;<\; (i-1) \rightarrow t_{ij}^{arr} \le t_{(i-j)j}^{dep} \text{ and } j_{arr}^{i} = j_{dep}^{i-1} \tag{8}$$

A deficit function is also known as step function changes by −1 and + 1 for every arrival and departure times of each trip during the service period.

$$\text{Minimum V} = \sum_{j=1}^{S} \text{Maximum}\{\text{DF(j, h)}|\text{h} = 1, 2, 3 \ldots.\text{H}| \tag{9}$$

5.3 *Maximum Flow Fleet Theorem*

$$\min N = V - \max C_{\text{O}} \tag{10}$$

N Chain of the trip.
V_z Number of vehicles of size Z.

Proof Given $i = 1, 2, 3, 4 \ldots$ V, each trip is assigned to i vehicles for Up and Down by the vehicle i, the required fleet size can be reduced where the flow C_{O} can be saved by linking the trips together. Therefore, the minimum fleet size is required to perform all the trips.

5.4 Passenger Boarding and Alighting Prediction Algorithm

The historic data of passengers and capacity of the bus is considered to predict the passengers boarding and alighting.

The predicted number of boarding passengers:

$$Pb_{j+1}^{i+1} = a_{j+1}(PT_{j+1,A}^{i+1} - t_{j+1,A}^{i}) + R_{j+1}^{i} \tag{11}$$

Pb_{j+1}^{i+1} Number of boarding passengers for bus $(i + 1)$ at stop $(j + 1)$.
λ_{j+1} Number of passenger arrival rate at stop $(j + 1)$.
$PT_{j+1,A}^{i+1}$ Arrival time of bus $(i + 1)$ is predicted at stop $(j + 1)$.
$t_{j+1,A}^{i}$ Arrival time of bus (i) at stop $(j + 1)$.
$PT_{j+1,A}^{i+1} - t_{j+1,A}^{i}$ Headway for the bus (i) is predicted at the stop $(j + 1)$.
R_{j+1}^{i} After the departure of the bus (i), remain passengers from stop $(j + 1)$.

The number of passenger's in-vehicle:

$$N_{j+1}^{i} = N_{j}^{i} + b_{j+1}^{i} - a_{j+1}^{i} \tag{12}$$

N_{j+1}^{i} In-vehicle passengers for bus (i) after departure from stop $(j + 1)$.
N_{j}^{i} In-vehicle passengers for the bus (i) after departure from the stop (j).
b_{j+1}^{i} Boarding passengers for bus (i) at stop $(j + 1)$.
a_{j+1}^{i} Alighting passengers for bus (i) at stop $(j + 1)$.

Check for bus capacity:

$$C_{j+1}^{i} = P_i - N_{j+1}^{i} \tag{13}$$

C_{j+1}^{i} saturation of capacity for bus (i) after departure from stop $(j + 1)$.

If $C_{j+1}^{i} \leq 0$ it presented that the number of passengers in the bus i reached the capacity.

C_{j+1}^{i} After departure from stop $(j + 1)$, in-vehicle passengers for bus (i).
P_i Maximum number of passengers can be in the bus i.

If the vehicle capacity of the bus is exceeded, then the remaining passengers are supposed to wait for the next bus. The remaining number of passengers can be calculated as

$$R_{(j+1)}^{i} = \begin{array}{ll} Pb_{(j+1)}^{i} - b_{(j+1)}^{i}, & C_{(j+1)}^{i} \leq 0 \\ 0, & C_{j+1}^{i} > 0 \end{array} \tag{14}$$

where

R^{i}_{j+1} After the departure of the bus (i) from the stop ($j + 1$) remaining passengers.
Pb^{i}_{j+1} Passengers boarding for bus ($i + 1$) at stop ($j + 1$).
b^{i}_{j+1} Passengers boarding for bus (i) at stop ($j + 1$).
C^{i}_{j+1} After departure from the stop ($j + 1$), a saturation of capacity for the bus (i).

The number of alighting passengers can be estimated by passenger's in-vehicle:

$$pa^{i+1}_{j+1} = \sigma_{j+1} C^{i+1}_{j+1} \tag{15}$$

where

pa^{i+1}_{j+1} predicting alighting passengers for bus ($i + 1$) at stop ($j + 1$).
C^{i+1}_{j+1} In-vehicle passengers for bus (i) after departure from stop ($j + 1$).
σ_{j+1} % passengers alights at stop ($j + 1$).

6 Case Study

Bengaluru Metropolitan Transportation Corporation (BMTC) is one of the most dominant modes with a fleet size of about 6165 buses, including feeder services to metro covering 53,984 km catering 45 lakh commuters per day along 2194 routes with 43 depots during 2016–2017 [19]. The fleet utilization of BMTC buses is decreased by 7.1% from 2013–2014 (91.2%) to 2018–2019 (84.1%). The revenue of BMTC for the non-air conditioned bus is Rs. 41.48/km, whereas the operating cost is Rs. 57.88/km [20]. Hence, there is a necessity for operational integration to reduce the loss of BMTC buses.

The selected bus route-290E goes along the route Yelhanka to Shivajinagar, located in the northern part of Bangalore city. It has a route length of around 20.1 km with 13 major bus stops. The input details for morning peak hour 08:00:00 a.m. to 08:59:00 a.m. in the software are given in the tables as follows (Tables 1, 2, 3 and 4):

The model and software developed balances the bus service operation for different service periods considering upper and lower constraints. Figures 1 and 2 illustrates the developed time table for up and down directions, respectively and for a given service duration, the performance details are listed below:

Total Number of buses Yelahanka Old Town to Shivajinagar (UP) = 5.
Total Number of buses Shivajinagar to Yelahanka Old Town (Down) = 5.
Crowd = 1.5, Departures from A −12, Departures from B = 12.
Minimum Fleet Size = 12, Maximum Fleet size = 15.
Vehicle–Kilometer = 482.4, Fuel Consumption = 96.95 L.
Fuel Cost = Rs. 5817.18, Vehicle Depreciation Cost = Rs. 31963.47.
Crew Cost = Rs. 1,44,000.00, Maintenance Cost = Rs. 2412.00.
User Cost = Rs. 192,313, Operator Cost = Rs. 29660.08.

Table 1 Distance and travel time between stops

Origin	Destination	Distance (kms)	Travel time (min) (Up)	Travel time (min) (Down)
Yelahanka	Maruthinagar	2.2	4.9	4.5
Maruthinagar	Kogilu	1.5	4.5	4.1
Kogilu	Ittige Factory	1.3	6.1	5.7
Ittige Factory	R.K. Hedge Nagar	3.7	17.55	16.2
R.K. Hedge Nagar	Thanisandra	1.8	5.7	5.3
Thanisandra	Nagavara	2.3	10.6	9.8
Nagavara	Arabic College	1.0	6.9	6.4
Arabic College	Kodugondanahalli	1.3	12.2	11.31
Kodugondanahalli	Periyar Nagar	1.3	2.0	1.9
Periyar Nagar Circle	Coles Park	1.5	4.1	3.8
Coles Park	CSI Park	1.5	10.6	9.8
CSI Park	Shivajinagar	0.7	5.3	4.9

Table 2 Boarding and alighting of passengers up and down

Place	Passenger arrival rate (Pass/min) (Up)	Passenger arrival rate (Pass/min) (Down)	Boarding		Alighting	
			Up	Down	Up	Down
Yelahanka	6.5	0.0	44	0.0	0.0	23
Maruthinagar	4.5	1.9	2	5	12	11
Kogilu	3.9	4.5	7	4	13	16
Ittige Factory	11.0	4.1	16	5	24	18
R.K Hedge Nagar	5.4	3.5	9	18	3	14
Thanisandra	5.7	4.9	24	5	18	17
Nagavara	4.2	3.8	4	4	19	5
Arabic College	3.9	5.1	9	18	7	11
Kodugondanahalli	3.1	4.9	30	10	15	2
Periyar Nagar Circle	4.5	10.0	14	6	8	6
Coles Park	3.9	3.5	3	5	8	4
CSI Park	2.1	4.1	8	8	13	3
Shivajinagar	0.0	5.9	0	42	30	0

Table 3 Service level constraints

Service level constraints	**Vehicle specifications**
1. Headway range: minimum—2 and maximum —5 2. Crowding level range: minimum—1.2 and maximum—1.5 3. Increment factor: 1 min 4. Maximum waiting time: 25 min 5. Maximum number of unserved buses: 4	1. Bus seat capacity (C): 65 2. Vehicle life span: 15 years 3. Vehicle life span: 800,000 vehicle-km 4. Vehicle fuel consumption: running-5 kmpl, engine idle-2 L/h 5. Number of doors: 2
Operation specifications	**Operation cost specifications**
1. Minimum layover time: 0 min 2. Average boarding time: 5 s 3. Average alighting time: 3 s 4. Minimum dwell time: 0 s 5. Number of crew on board: 2 6. Bus bunching threshold: 1 min 7. Minimum bus holding time: 1 min	1. Cost of vehicle: 7,000,000 INR 2. Fuel cost: 60 INR/Lit 3. Vehicle maintenance cost: 5INR/km 4. Crew cost: 300,000 INR/Month 5. Operator penalty: 10 INR/Lost-Passenger **User cost specifications** 1. Waiting time cost: 0.8 INR/Min 2. Travel time cost: seating—0 INR/min and standing—0.4 INR/min 3. User inconvenience cost: 0 INR/Pass

Table 4 Route fare chart

Stops	1	2	3	4	5	6	7	8	9	10	11	12	13
1	0	5	10	15	19	20	20	22	22	22	23	23	24
2	5	0	5	10	15	19	20	20	22	22	22	23	23
3	10	5	0	5	10	15	19	20	20	22	22	22	23
4	15	10	5	0	5	10	15	19	20	20	22	22	22
5	19	15	10	5	0	5	10	15	19	20	20	22	22
6	20	19	15	10	5	0	5	10	15	19	20	20	22
7	20	20	19	15	10	5	0	5	10	15	19	20	20
8	22	20	20	19	15	10	5	0	5	10	15	19	20
9	22	22	20	20	19	15	10	5	0	5	10	15	19
10	22	22	20	20	20	19	15	10	5	0	5	10	15
11	23	23	22	22	22	20	20	19	10	5	0	5	10
12	23	23	22	22	22	20	20	19	15	10	5	0	5
13	24	23	23	22	22	20	20	20	20	15	10	5	0

7 Conclusion

The study results point out the frequency of a single route for a single time period considering user and operator costs. The simulation-based frequency through mathematical programming considering operator and user service constraints caters to the sustainable, efficient, and economical public transportation system, which saves resources and provides an environment-friendly transportation system in urban areas. The waiting time cost of the passenger and the fleet size is minimized using the Deficit function as the solution for timetable synchronization and vehicle scheduling. The objective function is the same for pre- and post-COVID-19. However, the desired occupancy constraint values vary under social distancing, frequency, and timetable synchronization of a particular route that can be evaluated using this model based on demand variation. Thus, the routes are rationalized based on the variation in boarding and alighting of the passengers between the origin and destination.

Appendix

Terminal A
===========

Dep.	Pool	Dep_bus	Dep. time	Arr_bus	Arr_time
1	0	A 1	08:00:00	---	--------
2	0	A 2	08:05:00	---	--------
3	0	A 3	08:10:00	---	--------
4	0	A 4	08:15:00	---	--------
5	0	A 5	08:20:00	---	--------
6	0	A 6	08:25:00	---	--------
7	0	A 7	08:30:00	---	--------
8	0	A 8	08:35:00	---	--------
9	0	A 9	08:40:00	---	--------
10	0	A 10	08:45:00	---	--------
11	0	A 11	08:50:00	---	--------
--	1	---	--------	B 1	09:38:09
--	2	---	--------	B 2	09:43:06
--	3	---	--------	B 3	09:48:06
--	4	---	--------	B 4	10:04:06
--	5	---	--------	B 5	10:02:09
--	6	---	--------	B 6	10:04:58
--	7	---	--------	B 7	10:08:02
--	8	---	--------	B 8	10:09:38
--	9	---	--------	B 9	10:11:38
--	10	---	--------	B 10	10:13:38
--	11	---	--------	B 11	10:15:38

Fig. 1 Time table for up direction

Terminal B
===========

Dep.	Pool	Dep_bus	Dep. time	Arr_bus	Arr_time
1	0	B 1	08:00:00	---	--------
2	0	B 2	08:05:00	---	--------
3	0	B 3	08:10:00	---	--------
4	0	B 4	08:15:00	---	--------
5	0	B 5	08:20:00	---	--------
6	0	B 6	08:25:00	---	--------
7	0	B 7	08:30:00	---	--------
8	0	B 8	08:35:00	---	--------
9	0	B 9	08:40:00	---	--------
10	0	B 10	08:45:00	---	--------
11	0	B 11	08:50:00	---	--------
12	0	B 12	08:55:00	---	--------
--	1	---	--------	A 1	09:58:16
--	2	---	--------	A 2	10:03:11
--	3	---	--------	A 3	10:08:11
--	4	---	--------	A 4	10:13:11
--	5	---	--------	A 5	10:35:37
--	6	---	--------	A 6	10:34:05
--	7	---	--------	A 7	10:36:05
--	8	---	--------	A 8	10:38:05
--	9	---	--------	A 9	10:40:05
--	10	---	--------	A 10	10:42:05
--	11	---	--------	A 11	10:44:05
--	12	---	--------		10:46:05

Fig. 2 Time table for down direction

References

1. Fierek, S., & Zak, J. (2012). Planning of an integrated urban transportation system based on macro—simulation and MCDM/A methods. *Procedia - Social and Behavioral Sciences, 54*, 567–579.
2. MoRTH. (2017). Review of the performance of state road transport undertakings for Apr 2015–Mar 2016 (p. 65).
3. Rasouli, S., & Timmermans, H. (2012). Uncertainty in travel demand forecasting: Literature review and research agenda. *Transportation Letters, 4*(1), 55–73.
4. Saliara, K. (2014). Public transport integration: The case study of Thessaloniki, Greece. *Transportation Research Procedia, 4*, 535–552.
5. Myungseob Kim, F. A., & Schonfeld, P. (2012). Conventional, flexible, and variable-type bus services. *American Society of Civil Engineers, 138*(March), 263–273.
6. Patnaik, J., Chien, S., & Bladikas, A. (2004). Estimation of bus arrival times using APC data. *Journal of Public Transportation, 7*(1), 1–20.
7. Center of Excellence in Urban Transportation. (2013). Service level benchmark in urban transport for Indian cities volume-I-benchmarking manual (pp. 1–72).

8. Jain, M., Jain, H., Tiwari, G., & Rao, K. R. (2016). Indicators to measure performance efficiency of bus systems final report Grant Agreement Ref: G15 SSEF-178, Sept 2016.
9. Perceived Problems. Retrieved May 18, 2020, from https://ppiaf.org/sites/ppiaf.org/files/documents/toolkits/UrbanBusToolkit/assets/1/1b/1b.html.
10. Ma, Z. L., Ferreira, L., Mesbah, M., & Hojati, A. T. (2015). Modeling bus travel time reliability with supply and demand data from automatic vehicle location and smart card systems. *Transportation Research Record, 2533*, 17–27.
11. Yin, T., Zhong, G., Zhang, J., He, S., & Ran, B. (2017). A prediction model of bus arrival time at stops with multi-routes. *Transportation Research Procedia, 25*, 4623–4636.
12. Mahdavi Moghaddam, S. M. H., Rao, K. R., Tiwari, G., & Biyani, P. (2019). Simultaneous bus transit route network and frequency setting search algorithm. *Journal of Transportation Engineering, Part A: Systems, 145*(4), 1–14.
13. Gupta, N. (2017). User satisfaction with city bus public transport in India.
14. Ceder, A. (2009). Public-transport automated timetables using even headway and even passenger load concepts. In *32nd Australasian Transport Research Forum*, ATRF, 2009, no. October (pp. 1–17).
15. Chen, W., Yang, C., Feng, F., & Chen, Z. (2012). An improved model for headway-based bus service unreliability prevention with vehicle load capacity constraint at bus stops. *Discrete Dynamics in Nature and Society*.
16. Chien, S., Dimitrijevic, B., & Spasovic, L. (2003). Optimization of bus route planning in urban commuter networks. *Journal of Public Transportation, 6*(1), 53–79.
17. Krishanu Santra, D. P. (2019). Handbook on feasible service delivery level ranges for bus transit in Indian context part 2 : Operator perspective.
18. Ceder, A. (1984). Bus frequency determination. *Transportation Research Part A: Policy and Practice, 18*(516), 439–453.
19. Viswanathan, S., Nair, A., Raghupathy, S., & Murali, A. (1997). Financial brief on Bangalore metropolitan.
20. Record loss in FY19 raises doubts about BMTC's future. *The Economic Times*. Retrieved May 13, 2020, from https://economictimes.indiatimes.com/news/politics-and-nation/record-loss-in-fy19-raises-doubts-about-bmtcs-future/articleshow/69572872.cms.

Do Socioeconomic Characteristics Affect Travel Time and Transport Perception? Insights from Mumbai, India

Neenu Thomas, Arnab Jana, and Santanu Bandyopadhyay

Abstract The influence of socioeconomic condition and perception on travel behavior has been a subject of research for the past few decades. However, both travel behavior and transport perception may have an association with the socioeconomic characteristics of an individual, which is rarely explored. In developing countries like India, where the socioeconomic disparity is significant, understanding the relation of socioeconomic characteristics with travel behavior and perception is relevant for transport policies. Hence, the study aims to investigate the association of travel time (in different modes) and perception on sustainable transportation, among the individuals in Greater Mumbai, India, through a Socioeconomic Index representing different socioeconomic characteristics. The study uses the data collected from 722 individuals residing in Greater Mumbai. Initially, the study focuses on the formulation of the Socioeconomic Index using Principal Component Analysis. Consecutively, the variation in mode-based travel time and transport perception among individuals with different index scores are analyzed separately to capture the behavioral differences. Ridit analysis is used to interpret the perception of sustainable transportation. The results reveal that with the increase in the socioeconomic index, overall travel time and time spent on walking and public transport are reducing significantly, while time spent on private and paratransit modes are increasing. Besides, the perception study indicates that the tolerance for sustainable transportation is weaker among high index scored individuals. The study identifies the disparity in perception and behavior, which can benefit policymakers in distinguishing the different ways to provide and encourage sustainable transportation specific to an individual's socioeconomic condition.

Keywords Socioeconomic index · Travel time · Transport perception · Principal component analysis · Ridit analysis

N. Thomas (✉) · A. Jana
Centre for Urban Science and Engineering, IIT Bombay, Mumbai, India
e-mail: thomas_neenu@iitb.ac.in

S. Bandyopadhyay
Department of Energy Science and Engineering, IIT Bombay, Mumbai, India

A. Jana and P. Banerji (eds.), *Urban Science and Engineering*, Lecture Notes in Civil Engineering 121, https://doi.org/10.1007/978-981-33-4114-2_14

1 Introduction

Individuals spend a fraction of the day traveling in different transport modes for completing their daily activities. However, choice of transport mode, time spent on traveling, and perception of different transport settings may vary among different individuals. The socioeconomic condition of an individual affects travel behavior [1], as well as transport perception [2]. Besides, the studies on transport poverty show their association with the individual's socioeconomic status [3, 4]. The lower socioeconomic condition may cause restricted access to different transport options [4], which even becomes a hindrance to several opportunities in life [4]. Nevertheless, the reality of transport poverty and variations in travel behavior among individuals from cities in a developing country, which has a vast disparity in socioeconomic conditions among the population, are explored relatively lesser.

While providing safe, affordable, and accessible transportation to the urban population itself is a strenuous task for most developing countries, enforcing control measures for automobile penetration is also relevant for achieving sustainable transport goals. The count of two-wheeler and four-wheeler vehicles is increasing drastically in the densely populated cities of developing countries, such as India and China [5], affecting transport efficiency and the environment negatively [5]. Hence, an investigation of the travel behavior such as choice of mode and time spent for travel to complete their daily activities, and perception on the use of sustainable transport options, through socioeconomic perspective, can guide planning authorities and policymakers in understanding the transport preference and transport disparity. Thus, the study can help in achieving sustainable transport goals persuasively for a heterogeneous society like India.

The study aims to explore the association of (i) travel time (in different modes) and (ii) perception on sustainable transportation, among the individuals in Greater Mumbai, India, through a single indicator representing different socioeconomic characteristics. The research contains three sections. The initial section of the study focuses on the formulation of the single indicator (named as Socioeconomic Index), representing the socioeconomic characteristics of an individual, using Principal Component Analysis. The second and third sections focus on the association of the socioeconomic index with mode-wise travel time and transport perceptions, respectively.

2 Data Collection and Description

The study uses the data collected through a computer-assisted personal interview survey, which was conducted at the household level in the administrative area of Municipal Corporation of Greater Mumbai (MCGM) between the period of April 2016 and July 2016. The interview was applied to individuals aged more than five years. The survey captured the revealed and stated information related to an individual's travel behavior and transport perception along with the socioeconomic characteristics. The samples for the survey were chosen cautiously to capture the representative information of people from different behavioral segments (formal and informal settlements) residing across Mumbai.

Based on the objectives of the study, necessary data cleaning was done, and a sample of 722 individuals was used. Three sections of the questionnaire were used for the study; (i) socioeconomic conditions, (ii) travel diary of the weekday trips, and (iii) the perception of individuals towards traveling and transport conditions. Household income, residence type, and individual literacy are the variables considered for the representation of the Socioeconomic Index. Individuals' perception of public transport, shared rides, and environment-friendly modes is considered the transport perception study.

The data gathered for the study shows trips made on foot (walk) share maximum travel time among weekday trips, which is about 41.54%. Mumbai's travel pattern shows that public transport plays a vital role in the city's urban transportation. The share of public transport in total travel time is 77.64% among the motorized trips. However, the share of people who use public transport for at least one trip in a weekday is not more than 54.7% of the sample size. Besides, the data indicates that 77% of the public transport users are either residing in informal settlements or Slum Rehabilitation Authority (SRA) housing. Evidently, travel characteristics are dissimilar among individuals from different socioeconomic conditions.

3 Research Methodology

The research focuses on the exploration of the association of travel behavior and transport perception with socioeconomic characteristics. The study comprises of three sections: (i) Developing the socioeconomic index to capture the behavioral differences among individuals, (ii) Analyzing the association of mode-based travel time with socioeconomic index, and (iii) Perception study to identify the perceptional differences between the individuals with varied characteristics using the index score. The research framework is depicted in Fig. 1.

3.1 Socioeconomic Index

The travel behavior studies have found a significant association between transport choices and individual-specific characteristics such as socioeconomic conditions. Generally, such variables become transport dis/advantage factors that are affecting transport choice behavior. In this study, a socioeconomic index (SEI), a single index to represent different socioeconomic characteristics, is developed. The motive of developing a socioeconomic index is that the individuals can be ranked based on the

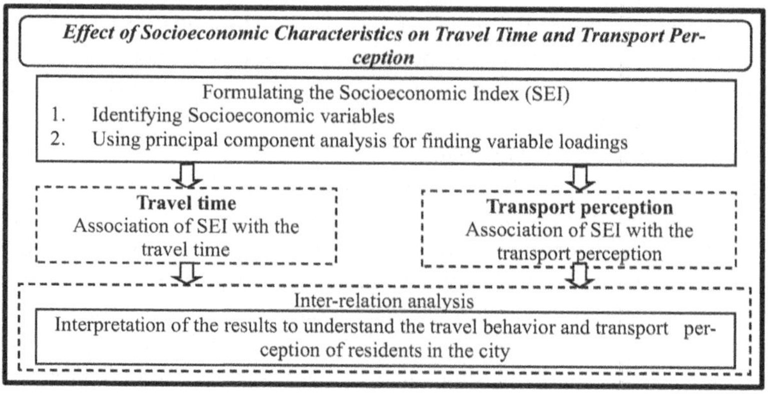

Fig. 1 Research framework

index score obtained by them. Subsequently, the index helps in exploring the change in travel behavior and transport perception with the change in socioeconomic conditions. Hence, SEI is a quantity to indicate the socioeconomic status of an individual, and it is a composite index expressed with the different socioeconomic variables. Ten unidentical variables were selected based on the literature and the correlation test was conducted to identify the variables with medium to high correlation. Income, residence typology, and literacy are the socioeconomic variables selected for the study (see Table 1).

SEI is developed using Principal Component Analysis (PCA), a dimension reduction technique that can be used to formulate the composite index. That is, the index is developed with the weighted aggregation of the observed variables corresponding to an index. Income, residence typology, and literacy are the observed variables in the study. The mandatory steps such as check for suitability of data, check for the percentage of the variance captured in the first loading, and identification of the weight of variables are considered for the development of the index formula [6]. The equation for SEI calculation is expressed as below.

$$SEI = a_1 V_1 + a_2 V_2 + \cdots + a_n V_n \quad (1)$$

Table 1 Description of variable considered for SEI calculation

Index	Individual-specific variables	Variable description
Socio-economic index (SEI)	Income (*I*)	Household income
	Residence typology (*R*)	Number of separate bedrooms in the residence
	Literacy (*L*)	Whether the individual is literate; 1 = if yes; else 0

where a_n is the weight of the variable, and V_n is the normalized value of the nth variables corresponding to an individual. The value of SEI can range from 0 to 1 where 1 implies that the individual is among the topmost category with best socioeconomic characteristics and 0 implies that the individual is among the bottom-most category with rest to socioeconomic characteristics values.

3.2 *Association of Mode-Based Travel Time with the Socioeconomic Index*

Travel choices of individuals are considered to be a rational decision made subjective to their capabilities and constraints. Socioeconomic index (SEI) is capturing the disparity prevailing among the individuals in Mumbai city. The travel time variation of individuals with different index score is analyzed to observe the behavioral changes in using the different transport modes available in the city to complete the daily trips. A significant association between the time spent on different modes and the index score is anticipated.

3.3 *Ridit Analysis for Perception Study*

The study explores the perception of different transport settings using ridit analysis. The data for perception studies was in the format of ordered categorical responses. Ridit analysis is the method developed to analyze such data that can neither be treated as ordinal data nor as nominal data [7, 8]. Ridit analysis helps comparing individuals with different SEI and the analysis is explained in terms of a reference group. The steps involved in the ridit analysis is explained below.

Step 1: Prepare the frequency distribution table: Identify the frequency f_{ij} of each ordered category j (strongly disagree, somewhat disagree, neutral, somewhat agree and strongly agree) corresponding to every obtained value of index score, i. Similarly, identify the frequency f_{sj} of each ordered category j for the total sample, s. The total sample is the reference group in this study.

Step 2: Calculate the proportion P_{ij} and P_{sj} of each ordered category j.

$$\text{Proportion of category } j,\ P_{ij} = \frac{f_{ij}}{\sum_j f_{ij}} \forall i \tag{2}$$

$$\text{Proportion of category } j \text{ in the total sample, } P_{sj} = \frac{f_{sj}}{\sum_j f_{sj}} \tag{3}$$

Step 3: Calculate the ridit value R_{sj} of each ordered category j for the reference group.

Table 2 Statements analyzed for perception study

	Statements analyzed for perception study
a	Feels safe traveling in public transport
b	Tired of the crowd in the sub-urban train
c	Environmental factors play a major role in the mode of travel
d	Ready to share rides

$$\text{Ridit value of category } j \text{ for the reference group, } R_{sj} = \sum_{j=1}^{j} P_{sj} - \tfrac{1}{2} P_{sj} \quad (4)$$

Step 4: Comparison of the index groups with the reference group (total sample): Calculate the mean ridit score S_i corresponding to each index group *i*.

$$\text{Mean ridit score of index group } i,\ S_i = \sum_{j=1}^{n} (P_{ij} \times R_{sj}) \forall i \text{ where } j = 1, 2, \ldots, n \quad (5)$$

The mean ridit score of the reference group must be 0.5. If the mean ridit score of an index group is above 0.5, the result indicates there is a higher probability that a random individual chosen from the index group agrees to the perception statement than a random individual from the reference group. Besides, the study also looks separately into the behavior of individuals who use public transport for their daily trips. The statements analyzed for the perception study are mentioned in Table 2. The selected statements can reveal the attitude of people towards public transport and environment-friendly modes and that can be analyzed along with the time spent on different transport modes.

4 Results

4.1 Formulating the Socioeconomic Index

The socioeconomic index (SEI) is formulated using PCA, and this index is the measuring tool used to categorize individuals based on their socioeconomic characteristics. Data suitability for the PCA method was checked prior to the calculation of variable weight. The correlation of parameters was considered for choosing the variables for the PCA technique. In addition, a sample of 722 is a very good sample size for the application of the PCA technique [9]. Subsequently, PCA was performed for the data set to compute the variable weightage. The KMO test for the measure of sampling adequacy gave the values above 0.5, which is in the mediocre category. However, Bartlett's test of sphericity shows a significant p-value which confirms the

Table 3 Result of principal component analysis

Index		Socioeconomic index (SEI)		
Suitability of data for PCA	KMO test	0.547**		
	Bartlett's test	0*		
	PC1 Variance	49.44%		
Component loading of variables	Factors	Income (I)	Residence typology (R)	Literacy (L)
	Loading	0.796	0.806	0.447

**Adequate, *Significant p-value

acceptability of PCA for the data tested. The variable weights are obtained from the component loading of variables in the first principal component. For both the indices, the total variance of data captured by the first principal component is above 40% (see Table 3).

Based on the obtained component loadings, the socioeconomic index score of an individual can be calculated as expressed below in Eq. 6. The component loading of Residence typology and Income is approximately about 0.8. In contrast, loading of Literacy is lesser, representing approximately 64% of the variance (square of loading value) of the two variables are explained by the index. In comparison, only 20% of the variance of the variable Literacy is being explained by the index developed.

$$SEI = 0.796 \times \mathrm{I} + 0.806 \times R + 0.447 \times L \tag{6}$$

4.2 *The Link Between the Mode-Based Travel Time and the Index*

The association of travel time in different types of mode and index developed for the study is examined graphically by plotting the average travel time with the index scores of the individuals. The graph is plotted for index scores with a significant number of samples (index score representing less than three samples are not included). The trendline of the travel time variation corresponding to different mode types is depicted in the graph (see Fig. 2). Furthermore, the likelihood of the association between travel time and index was also estimated.

The graph shows that the total travel time is gradually decreasing with the increasing socioeconomic index. Similarly, the average travel time in active modes decreases with the increase in the socioeconomic index (SEI). While the average travel time of active mode trips for individuals with the least SEI score is about 55 min, it has decreased to 20 min for high SEI score individuals. Travel time on public transport is also decreasing after the average score of SEI. However, the time

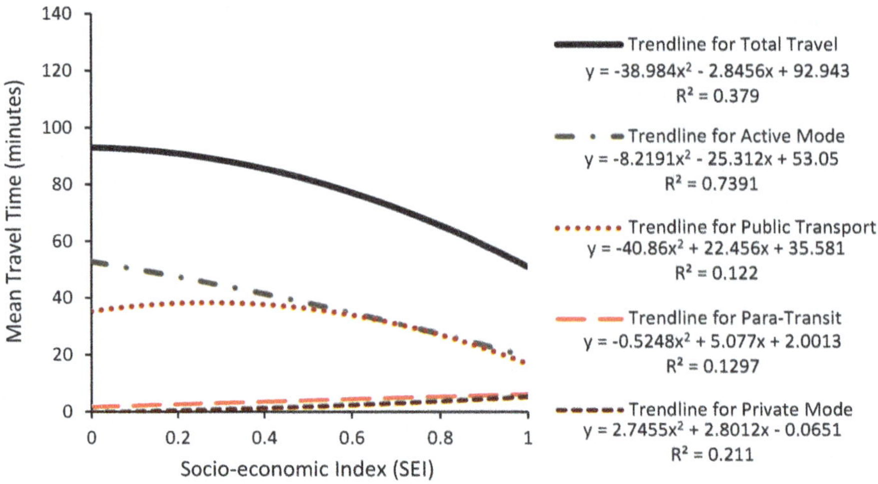

Fig. 2 Association of mode-based travel time with socioeconomic index

spent on paratransit and private modes are comparatively high for individuals with higher SEI scores.

4.3 Ridit Analysis for Perception Study

The study uses ridit analysis to explore the perceptional differences among individuals with different index score. Unlike the typical application of ridit analysis with a limited number of groups (2–5 groups) compared to a reference group, this study has several SEI groups that are compared with a total population. Therefore, we choose to represent the measurement of perceptional differences in the graphical format (see Fig. 3). In addition, the figure also represents perceptional variation between the whole sample and the individuals using public transport.

The graphs indicate that an individual with a low socioeconomic index (SEI) is more open to sustainable transport options than the high SEI score individual even though low SEI groups are tired of traveling on a crowded train. Among the four perception statements tested, ride-sharing is the option with the least variability within the sample, and higher SEI groups least prefer ride-sharing. Interestingly, the low SEI group up to a score of 0.2 does not show variation in their preference for ride-sharing, and slowly the ridit score decreases below the average for high SEI group with a score more than 0.6. Besides, the graph also shows that environmental factors play less role in mode choice for higher SEI groups compared to other populations. Although there is no considerable perceptional difference between the public transport users and non-users from higher SEI groups, among the same lower SEI score individuals, public transport users feel safer and tired in public transport than

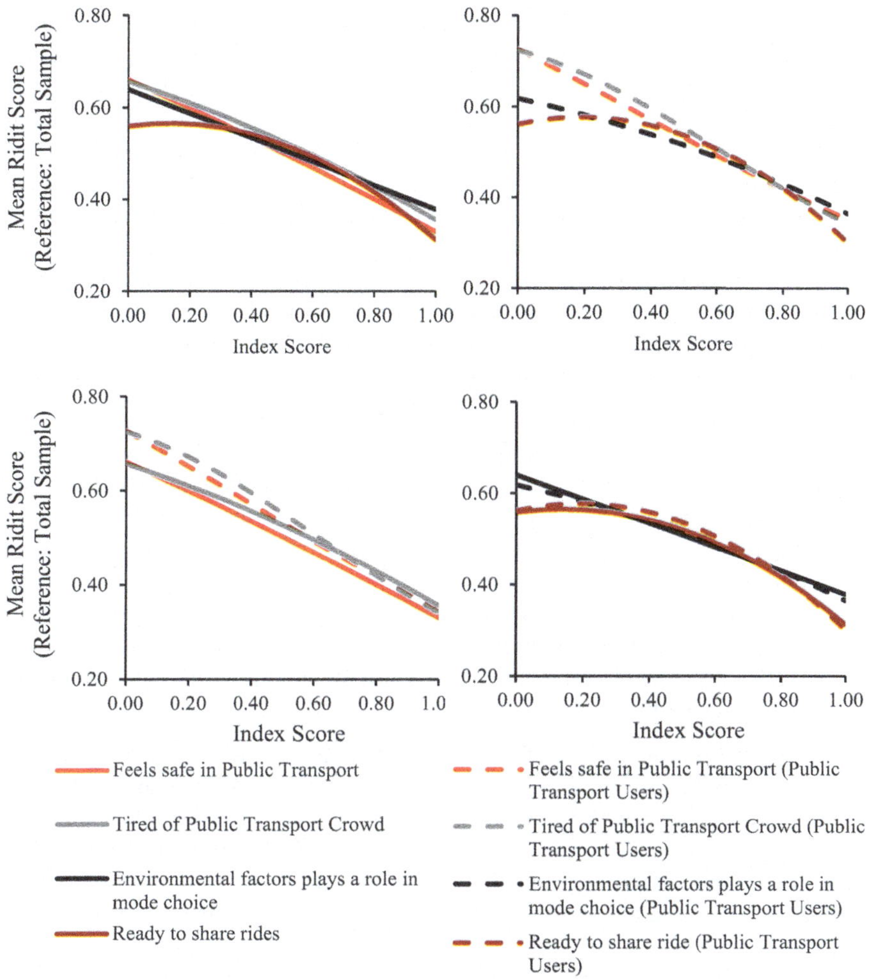

Fig. 3 Perception of transport condition varying with an index score

the non-users. These results show lower socioeconomic groups' inclination to use sustainable transport options like public transport and ride-sharing.

5 Discussion and Conclusions

The socioeconomic condition of an individual does affect their travel behavior and transport perception. The study highlights the time spent traveling in a day, choice of transport modes, and the perception of transport settings of an individual vary significantly among individuals from different socioeconomic conditions. The present study was undertaken in Mumbai, the densest city in India, where a significant proportion of the population stays in informal housing settlements. The study gives limelight to the existing transport disparity in developing nations. On a regular day, an individual with better socioeconomic conditions tends to travel lesser time compared to an individual with a lower socioeconomic condition. The average time spent walking by an individual with the lowest socioeconomic condition is more than twice that of the highest. Besides, the choice of lesser sustainable transport options (such as private transport and paratransit) are majorly made by the individuals with higher socioeconomic conditions, while the lower socioeconomic group prefers traveling more in public transport or walking. The perception study also reveals that the tolerance to public transport and ride-sharing decreases with increasing socioeconomic conditions. The observations discussed indicate;

(i) Higher travel time and more active mode trips among lower socioeconomic groups might be highlighting the disparity in transport accessibility and affordability prevailing in the Mumbai city. This result could be an indication that individuals with a reduction in socioeconomic status tend to use slow modes which are cheap and need to spend more time traveling to achieve the work opportunities.
(ii) People tend to feel public transport is less safe, and the use of private transport and paratransit increases with the increase in socioeconomic conditions. Thus, the efficiency and quality of public infrastructure might be affecting the quality of travel in the city.

The observations are relevant for the planning authority and the policymakers to provide better public transportation, accessibility to the workplace, and containing the use of private modes. A limitation of the study is that weekend trips were not considered for the study. Further research can include this into consideration. The major limitations in the data are that the sample contains a smaller number of high net worth households, and the number of variables for the perception study was limited.

Acknowledgements The authors thank the Ministry of Human Resource Development, India and Industrial Research and Consultancy Centre, IIT Bombay, for providing the funds necessary for conducting this data collection. The title of the grant received is "Frontier Areas of Science and Technology (FAST), Centre of Excellence in Urban Science and Engineering' (grant number 14MHRD005).

References

1. Lu, X., & Pas, E. I. (1999). Socio-demographics, activity participation and travel behavior. *Transportation Research Part A: Policy and Practice, 33A*(1), 1–18.
2. Meng, M., Rau, A., & Mahardhika, H. (2018). Public transport travel time perception: Effects of socioeconomic characteristics, trip characteristics and facility usage. *Transportation Research Part A: Policy and Practice, 114*, 24–37.

3. Lucas, K., & Jones, P. (2012). Social impacts and equity issues in transport: An introduction. *Journal of Transport Geography, 21,* 1–3.
4. Lucas, K., Phillips, I., Mulley, C., & Ma, L. (2018). Is transport poverty socially or environmentally driven? Comparing the travel behaviours of two low-income populations living in central and peripheral locations in the same city. *Transportation Research Part A: Policy and Practice, 116,* 622–634.
5. Thomas, N., Varghese, V., & Jana, A. (2019). Two-wheeler vehicle ownership, transport disadvantage, and mode choice: a case study of different behavioral segments from Mumbai, India. In: 98th Annual Meeting of the Transportation Research Board. Washington, DC.
6. Thomas, N., Jana, A., & Bandyopadhyay, S. (2019). Investigating the association of individual and neighborhood characteristics with travel time: Empirical evidences from Mumbai, India. In: 13th International Conference of Eastern Asia Society for Transportation Studies. Colombo, Sri Lanka.
7. Bross, I. D. J. (1958). How to use ridit analysis. *Biometrics, 14*(1), 18–38.
8. Beder, J. H., & Heim, R. C. (1990). On the use of ridit analysis. *Psychometrika, 55*(4), 603–616.
9. Comrey, A. L., & Lee, H. B. (1992). *A first course in factor analysis.* Hillsdale, NJ: Lawrence Erlbaum Associates.

Understanding the Household Transportation Characteristics in Calicut City Using System Dynamics

P. Deepa and P. Bimal

Abstract System dynamics is an effective tool to model dynamic relationships between various parameters that interact, interrelate, and influence each other. This paper attempts to model transportation characteristics of a household in terms of vehicular ownership and its daily usage, which are influenced by socioeconomic parameters such as household size, household income, household structure, age categories, educational status, and working status which are interrelated with each other. Once a model is generated at the household level, the number of trips generated and the traffic volume generated from the neighborhood can be calculated. A model can be generated using the system dynamics technique. This study conducted at Calicut city has used primary data collected from 103 households. Which are considered as the agent module, which can be replicated to higher scales of an urban system. Data collected has been analyzed using statistical and a causal feedback diagram is generated using these parameters and their inter relationship with one another at a household level has been established. Further, a system dynamics model is built using STELLA. It is found that socioeconomic parameters such as age of the household member and education status highly influence the income of the HH and their readiness to spend for a comfortable and convenient travel mode. Similarly, vehicular ownership was also found to be a function of the number of graduates, postgraduates, retired members, and household income. The integrated model of all these parameters provides household travel characteristics in terms of quantifiable data of vehicular ownership and daily usage of vehicles.

Keyword Transportation modeling · Sustainable planning · System dynamics · STELLA

P. Deepa (✉) · P. Bimal
National Institute of Technology, Calicut, India
e-mail: deepa_p190026ar@nitc.ac.in

P. Bimal
e-mail: bimalp@nitc.ac.in

A. Jana and P. Banerji (eds.), *Urban Science and Engineering*, Lecture Notes in Civil Engineering 121, https://doi.org/10.1007/978-981-33-4114-2_15

1 Introduction

Transportation is one of the vital infrastructures which enable the functioning of the cities efficiently. In the recent past, there is a tremendous increase in the use of personalized motor vehicles for transportation, especially the use of two-wheelers and cars [1]. This has led to various transportation and traffic woes [2]. Planning interventions for improving the transportation scenario in many cities could not find the result, as transportation planning has been carried out in isolation where the planning was done to anticipate the future travel demand. Increasing the transportation infrastructure as the demand grows is not a sustainable solution for transportation problems. A few attempts have been made to integrate transportation planning with land use planning where the attempt was to reduce the need for traveling [3]. Other than land use there are lots of other factors that influence the travel behavior in a city. Household socioeconomic characters are one of these factors particularly income [4]. It is very clear that household vehicular ownership plays role in the personalized vehicle usage on the roads [4]. In this context, this paper tries to answer the question of whether or not increased household ownership of the vehicles has a significant impact on the personalized trip generated in the urban areas. This study presents the development of a system dynamics model for household vehicular ownership using household demographic and socioeconomic characteristics.

System dynamics is a holistic approach [5] to understand the behavior of a complex system over a period of time. It involves modeling, simulation, and validation based on the system concept. A system functions as a whole with the several interactions of the subsystems. All the subsystems are interlinked and interdependent to each other and function as a whole. If one of the subsystems of the system or combination of few subsystems defunct or take a lead role during its function, its effect can be seen in the entire system over a period of time. In system dynamics, the structure and function of the system is explained by causal loop diagrams that explain the interactions and feedback mechanisms within a system. Further, these diagrams are developed as stock–flow diagrams and system dynamics models area built. Thus, built models are simulated, validated, and policy recommendations are made. Thus, this technique enables to analyze dynamics systems of various complexities [6]. In this study, a household will be considered as a system with various parameters interlinked and interrelated to each other which affect the functioning of the household as a whole.

2 Research Background

This study is an enquiry into the conviction that increased personalized vehicular ownership leads to increased personalized trips on the road which probably cannot be accommodated by augmenting the road infrastructure. The root cause of this could be either the inefficiency of the existing public transportation system or the socioeconomic and demographic characteristics of the household [7]. This study attempts to identify these characteristics of the household which leads to the increased personalized vehicle ownership and its usage on daily commutes. This scenario can be best approached through a system concept where all socioeconomic and demographic parameters as the parameters that are interrelated and influence each other and as a whole contribute to the household as a system. So the aim of this study is to develop a system dynamics model for vehicular ownership and usage at the household level using household demographic and socioeconomic characteristics. Challenge lies in identifying the parameters influencing the vehicular ownership at the household level and establishing interactive relationships among these parameters. The model thus developed may be used for simulating future household car and two-wheeler ownership, usage, and the number of daily trips generated.

2.1 Methodology and Tools

A primary data collection is conducted from 103 households in Calicut city, Kerala, to gather socioeconomic and demographic parameters at the household level. These parameters include household size, staying status, age of members, gender, marital status, education status, working status, household income, household expenditure profile, ownership of the car and two-wheeler, usage of the car and two-wheeler, distance to workplace, travel time to workplace, travel distance, trip cost. Further, the relationships between various parameters are identified using statistical methods. Initially, the parameters that are significantly correlated are identified using correlation. Further, a regression analysis is done with the variables of interest as the dependent variables and household characteristics as the independent variables. Multiple linear regression and binary logit regression has been employed for this. Once the relationships have been established between various parameters, a causal feedback diagram has been generated which then further developed into a system dynamics model. SPSS software has been used for statistical analysis, ANYLOGIC for causal feedback diagram and STELLA has been used for system dynamics modeling.

2.2 Data Sources

The study has been conducted in the Calicut Corporation, Kerala. A cross-sectional data has been collected from 103 households. Random sampling method has been used for this. Calicut is the third largest urban agglomeration in Kerala, spreading over an area of 125 km^2 with a total population of 4.3 lakh and density 3464 persons per sq.km. 48% of the total population of the Calicut corporation constitute male and the remaining 52% is female population. Literacy rate of the study area is 97% which higher than the state average of 93.91% and national average of 74.04%. 32% of the population belongs to working-class and is mainly engaged in tertiary sector. 53% of the total male population is working and only 13% of the female population is employed.

2.3 Data Analysis

In this study, parameters and interrelationships has been identified and established through statistical analysis. Correlation analysis has been done to find out the significant parameters that are directly interrelated to and influence the variables of interest viz. monthly income, travel expense, car ownership, car usage, two-wheeler ownership, and two-wheeler usage. The parameters that are found to be significantly correlated are graphically represented in the Fig. 1. The dependent variables are represented in color red in the figure. They are monthly_income (monthly household income), monthly_expense_travel (monthly travel expense of household), TW_usage (daily two-wheeler usage), TW-ownership (two-wheeler ownership), car_ownership (car ownership of the household), car_usage (daily car usage).

These five interesting variables are assumed to be dependent on the socioeconomic characteristics of the household. From the correlation analysis, the probable parameters that influence monthly income, vehicular ownership, and its daily usage has been identified. Further, regression analysis has been done to identify the exact parameters and establish a mathematical relationship among these variables, which can be modeled using system dynamics.

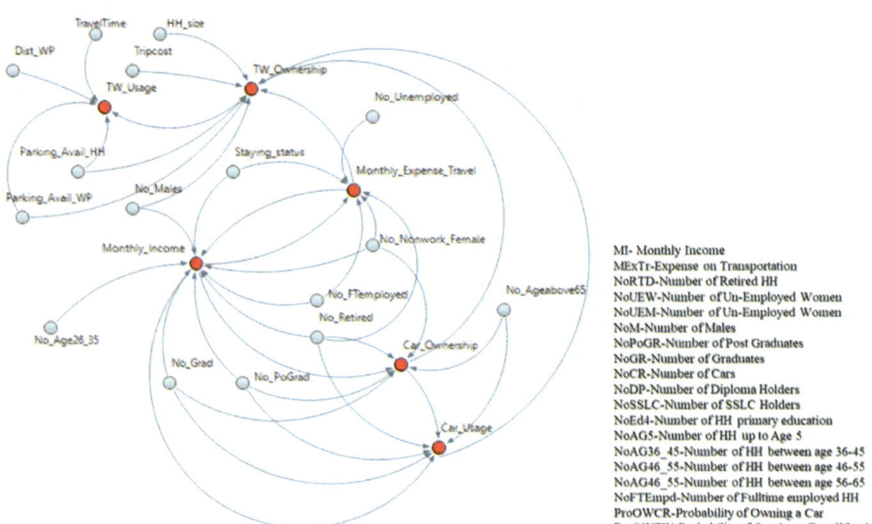

Fig. 1 Interrelationships among various parameters (*Source* Author)

3 Modeling

3.1 Regression Model

Multiple linear regression models are formulated for dependent variables monthly income and travel expense using all the parameters as the independent variables, while a binomial logit regression is used to represent vehicular ownership and its daily usage.

Monthly income: Total monthly income can be expressed in terms of number of postgraduates, number of graduates, number of unemployed women, number of retired persons, and number of males. The average household income is found to be 24886.527 and it increases by Rs. 25,267, 8651, 15,455, and 6260 by the presence of a post-graduate, graduate, retired, and males, respectively. The household income is also found to be less by 19,911 than of an average household if the household has unemployed women. Regression model follows as below.

$$\begin{aligned}\text{Monthly income} = {} & 24886.527 + 25266.935(\text{number of post graduates}) \\ & + 8650.848(\text{number of graduates}) \\ & - 19911.277(\text{number of unemployed women}) \\ & + 15454.791(\text{number of retired}) + 6259.528(\text{number of males})\end{aligned} \tag{1}$$

The regression was carried out by entering all logically related socioeconomic parameters as independent variables and the model has automatically omitted the insignificant variables. It is found that independent variables *number of postgraduates, number of graduates, number of unemployed women, number of retired members, and number of males* are the significant predictors. The R value of 0.533 indicates these independent variables collectively have a good correlation with dependent variable income. R^2 value of 0.509 indicates that 50.9% variation in income is explained by this model. Statistical significance of this model is tested and found that F value is 22.173 with a P value less than 0.005, and thus rejects the null hypothesis. Therefore, the model is statistically significant.

Monthly travel expense: Monthly travel expense is found to be a function of household characteristics such as household monthly income, number of unemployed men, number of HH who belongs to education category up to 4, and the number of HH belonging to age category 36–45 years. Average monthly expense is Rupees 1008, which increases by rupees 0.44, 1018, and 627 as per unit increase in monthly income, number of HH of education up to 4 and age category 36–45, where it is less by rupees 2336 as per unit number of the unemployed men. Regression equation follows as below.

$$\begin{aligned}&\text{Monthly expense on transport}\\&\quad = 1008.486 + 0.044(\text{monthly income}) - 2336.007(\text{number of unemployed men})\\&\qquad + 1017.720\,(\text{number of HH of education upto 4})\\&\qquad + 627.007(\text{number of HH of age } 36-45)\end{aligned} \tag{2}$$

The regression was carried out by entering all logically related socioeconomic parameters as independent variables and the model has automatically omitted the insignificant variables. It is found that independent variables monthly income, number of unemployed men, number of households with education status up to 4, number of households of age category 36–45 are the significant predictors. The R value of 0.430 indicates that these independent variables collectively have a good correlation with the dependent variable, i.e., monthly travel expense. R^2 value of 0.398 indicates that 39.8% variation in monthly travel expense is explained by this model. Statistical significance of this model is tested and found that F value is 16.17 with a P value less than 0.005, and thus rejects the null hypothesis. Therefore, the model is statistically significant.

Probability of owning a car: A binary logistic regression was performed on car ownership as the outcome and educational status, working status, and family structure as the predictors. Insignificant predictors were automatically removed and the number of graduates, postgraduates, and retired members is found to be the strong predictors. A test of the full model with all these 3 significant predictors (Significance less than 0.005) against a constant only model was statistically significant with Hosmer and Lemeshow Chi-square 2.665 and rejects the null hypothesis that adding variables does not significantly increase the models ability to predict the outcome. Table shows the regression coefficients Walds statistics and odds ratio. Among these predictors, the number of postgraduate is found to be a strong predictor with a Walds chi-square of 13.59 and with a p value less than 0.001. It is clear from the table that the presence of postgraduate member and retired member increases the probability to own the car by 5 times, whereas the presence of a graduate member in the family increases this probability only by 2 times compared to that of households without these category members.

$$\begin{aligned}&\text{Probability of owning car}\\&\quad = \frac{e^{(-1.216-1.927*\text{MILs20000}-1.029*\text{MI20000\_40000}+1.182*\text{NoPoGR}+0.597*\text{NoGR}+1.509*\text{NoRTD})}}{1+e^{1.216-1.927*\text{MILs20000}-1.029*\text{MI20000\_40000}+1.182*\text{NoPoGR}+0.597*\text{NoGR}+1.509*\text{NoRTD}}}\end{aligned} \tag{3}$$

Probability of owning a Two-wheeler: Similar analysis to that of car ownership was performed on two-wheeler ownership as the outcome and educational status, working status, and family age structure as the predictors. Insignificant predictors were automatically removed and only the age structure of the family is found to be the strong predictors. They are the number of members belonging to age group up to 5,

Table 1 Identified socioeconomic parameters

Socioeconomic characteristic of households	Parameters
HH members	Number of males
Age structure	Upto 5, 36–45, 46–55, 56–65
Education status	Number of HHs with education up to 4, number of graduates, number of postgraduates
Working status	Retired, number of full time employed, number of unemployed men, number of unemployed women
Economic	Total monthly income, monthly travel expense

between age 36–45, 46–55, and 56–65. A test of the full model with all these 4 significant predictors shows that this model has 80.6% predictability against a constant only model. Model is statistically significant with Hosmer and Lemeshow Chi-square 11.268 and rejects the null hypothesis that adding variables does not significantly increase the models ability to predict the outcome. Table shows the regression coefficients Walds statistics and odds ratio. Among these predictors, number of members between ages 36–45 is found to be a strong predictor with a Walds value of 7.535. It is clear from the table that the presence of family members belonging to age group 36–45 and age up to 5 increases the probability by 3 times, whereas age members in the age category 46–55 and 56–65 increases this chance by 2.5 times compared to that of households without these category members.

$$\text{Probability of owning two} - \text{wheeler} = \frac{e^{(0.867+1.376*\text{NoAG5}+1.242*\text{NoAG36\_45}+1.035*\text{NoAG46\_55}+0.954*\text{NoAG56\_65})}}{1+e^{(-0.867+1.376*\text{NoAG5}+1.242*\text{NoAG36\_45}+1.035*\text{NoAG46\_55}+0.954*\text{NoAG56\_65})}} \quad (4)$$

From regression analysis it is found that the following are the household socioeconomic and demographic parameters that influence a Households vehicular ownership and its daily usage. They can be summarized as Table 1 and the relationships are as shown in Eqs. 1–4.

3.2 System Dynamics Model

A causal feedback diagram has been developed with a polarity which represents whether they are directly or inversely related to each other. This casual feedback diagram has been further developed into a system dynamics model (Fig. 2).

In the system dynamic model, there are stock variables and flow variables. In this model-dependent variables viz. monthly income, monthly travel expense, probability of owning the car and daily use of the car, and probability of owning a two-wheeler and its daily usage has been taken as stock variables. Knowing the demographic characteristics of Calicut city such as total population, average household size; the

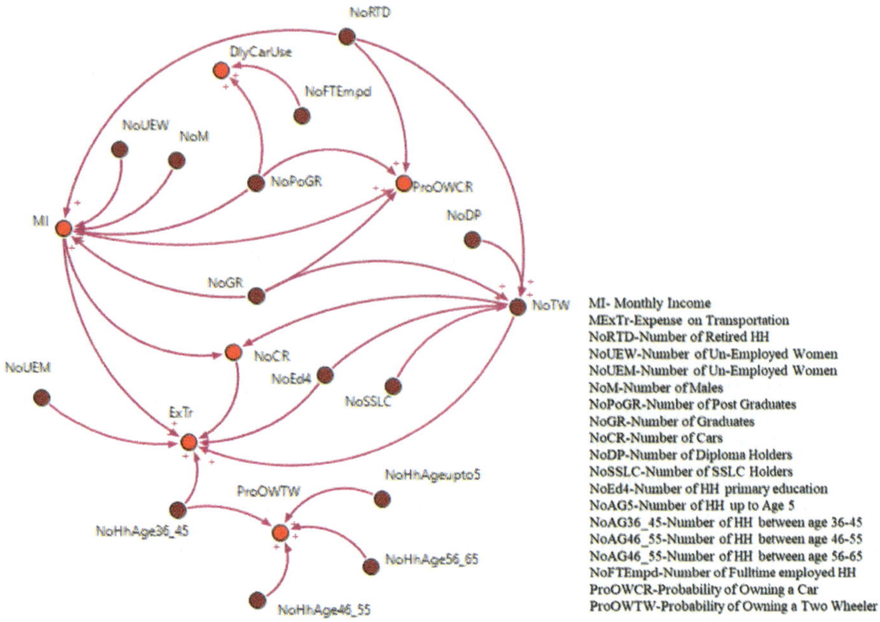

Fig. 2 Causal feedback diagram for household transportation dynamics

total vehicular ownership and its contribution to daily trips can be simulated using this model (Fig. 3).

3.3 Simulation

In order to demonstrate how the developed system dynamics model can be applied to future predictions, simulations are conducted to predict car ownership, two-wheeler ownership, and daily trips generated under different hypothetical scenarios. The simulation was conducted for Calicut city demographic data. In all scenarios, the average household size is taken as 5 and a combination of parents, 2 children, and a grandparent, and the possibilities are explored if the parents belong to various categories of age groups. Model has been simulated for different scenarios as follows.

Scenario 1: Changing the education status of the household members and the rest of the parameters as constant.

Scenario 1a: Parents belong to the age category of 36–45 and grandparent in the age of 56–65 and children are not likely to fall in any of the significant age categories (Table 2).

Scenario 1b: Parents belong to the age category of 46–55 and grandparent and children are not likely to fall in any of the significant age categories (Table 3).

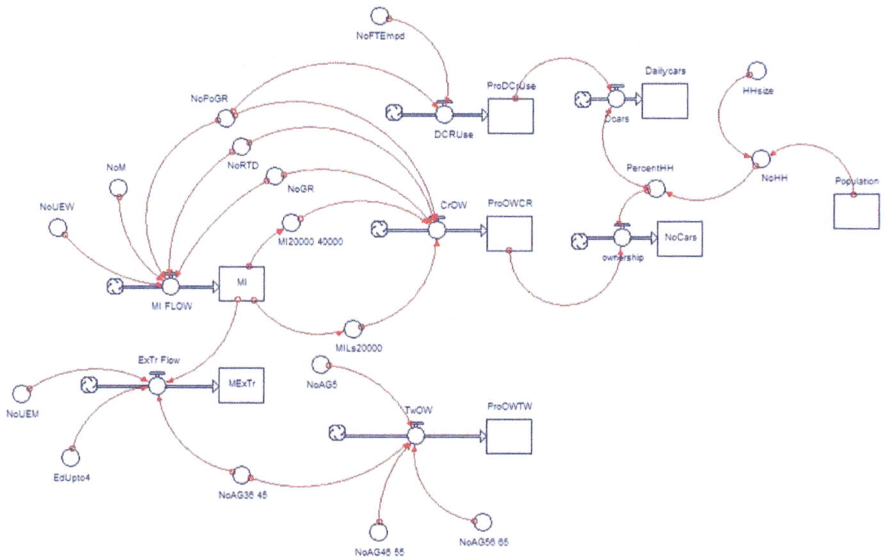

Fig. 3 System dynamics model for household transportation dynamics (*Source* Author)

Scenario 1c: Parents belong to the age category of 56–65 and children at the age of 36–45 (Table 4).

Scenario 2: Changing the working status of the household members and the rest of the parameters as constant.

Scenario 2a: Parents belong to the age category of 36–45 and grandparent in the age of 56–65 and children are not likely to fall in any of the significant age categories (Table 5).
Scenario 2b: Parents belong to the age category of 46–55 and grandparents and children are not likely to fall in any of the significant age categories (Table 6).
Scenario 2c: Parents belong to the age category of 56–65 and children I the age of 36–45 (Table 7).

4 Results and Discussion

4.1 Scenario 1: Changing the Education Status of the Household Members and the Rest of the Parameters as Constant

It is observed from the simulation statistics that as monthly income and car ownership increases as the education status of the household members are high whereas car

Table 2 Representation of scenario 1a

Fulltime employed	Unemployed male	Unemployed female	Retired	Up to 5	**36–45**	46–55	**56–65**	Up to 4	**Graduates**	**Post graduates**
1	0	1	0	0	**2**	0	**1**	0	**0**	**0**
1	0	1	0	0	**2**	0	**1**	0	**1**	**0**
1	0	1	0	0	**2**	0	**1**	0	**2**	**0**
1	0	1	0	0	**2**	0	**1**	0	**0**	**1**
1	0	1	0	0	**2**	0	**1**	0	**1**	**1**
1	0	1	0	0	**2**	0	**1**	0	**0**	**2**

Table 3 Representation of scenario 1b

Fulltime employed	Unemployed male	Unemployed female	Retired	Up to 5	36–45	**46–55**	56–65	Up to 4	**Graduates**	**Postgraduates**
1	0	1	0	0	0	**2**	0	0	**0**	**0**
1	0	1	0	0	0	**2**	0	0	**1**	**0**
1	0	1	0	0	0	**2**	0	0	**2**	**0**
1	0	1	0	0	0	**2**	0	0	**3**	**0**
1	0	1	0	0	0	**2**	0	0	**4**	**0**
1	0	1	0	0	0	**2**	0	0	**1**	**1**
1	0	1	0	0	0	**2**	0	0	**2**	**1**
1	0	1	0	0	0	**2**	0	0	**3**	**1**
1	0	1	0	0	0	**2**	0	0	**1**	**2**
1	0	1	0	0	0	**2**	0	0	**1**	**3**
1	0	1	0	0	0	**2**	0	0	**2**	**2**
1	0	1	0	0	0	**2**	0	0	**0**	**4**

Table 4 Representation of scenario 1c

Fulltime employed	Unemployed male	Unemployed female	Retired	Up to 5	**36–45**	46–55	**56–65**	Up to 4	**Graduates**	**Postgraduates**
3	0	1	0	0	**2**	0	**2**	0	**0**	**0**
3	0	1	0	0	**2**	0	**2**	0	**1**	**0**
3	0	1	0	0	**2**	0	**2**	0	**2**	**0**
3	0	1	0	0	**2**	0	**2**	0	**3**	**0**
3	0	1	0	0	**2**	0	**2**	0	**4**	**0**
3	0	1	0	0	**2**	0	**2**	0	**1**	**1**
3	0	1	0	0	**2**	0	**2**	0	**2**	**1**
3	0	1	0	0	**2**	0	**2**	0	**3**	**1**
3	0	1	0	0	**2**	0	**2**	0	**1**	**2**
3	0	1	0	0	**2**	0	**2**	0	**1**	**3**
3	0	1	0	0	**2**	0	**2**	0	**2**	**2**
3	0	1	0	0	**2**	0	**2**	0	**0**	**4**

Table 5 Representation of scenario 2a

Fulltime employed	Unemployed male	Unemployed female	Retired	Up to 5	36–45	46–55	56–65	Up to 4	Graduates	Postgraduates
0	0	1	0	0	2	0	1	0	1	1
1	0	1	0	0	2	0	1	0	1	1
0	0	0	0	0	2	0	1	0	1	1
1	0	0	0	0	2	0	1	0	1	1
1	0	1	1	0	2	0	1	0	1	1
1	0	0	1	0	2	0	1	0	1	1
0	0	1	1	0	2	0	1	0	1	1
0	0	0	1	0	2	0	1	0	1	1
1	0	0	1	0	2	0	1	0	1	1

Table 6 Representation of scenario 2a

Fulltime employed	**Unemployed male**	**Unemployed female**	Retired	Up to 5	36–45	**46–55**	56–65	Up to 4	Graduates	Postgraduates
0	**0**	**1**	0	0	0	**2**	0	0	2	1
1	**0**	**1**	0	0	0	**2**	0	0	2	1
0	**0**	**0**	0	0	0	**2**	0	0	2	1
1	**0**	**0**	0	0	0	**2**	0	0	2	1
1	**0**	**1**	1	0	0	**2**	0	0	2	1
1	**0**	**0**	1	0	0	**2**	0	0	2	1
0	**0**	**1**	1	0	0	**2**	0	0	2	1
0	**0**	**0**	1	0	0	**2**	0	0	2	1
1	**0**	**0**	1	0	0	**2**	0	0	2	1

Table 7 Representation of scenario 2c

Fulltime employed	**Unemployed male**	**Unemployed female**	Retired	Up to 5	**36–45**	46–55	**56–65**	Up to 4	Graduates	Postgraduates
1	**0**	**1**	0	0	**2**	0	**2**	0	2	1
2	**0**	**1**	0	0	**2**	0	**2**	0	2	1
1	**0**	**0**	0	0	**2**	0	**2**	0	2	1
2	**0**	**0**	0	0	**2**	0	**2**	0	2	1
2	**0**	**1**	0	0	**2**	0	**2**	0	2	1
2	**0**	**0**	0	0	**2**	0	**2**	0	2	1
2	**0**	**1**	1	0	**2**	0	**2**	0	2	1
2	**0**	**0**	1	0	**2**	0	**2**	0	2	1
3	**0**	**1**	1	0	**2**	0	**2**	0	2	1
3	**0**	**0**	1	0	**2**	0	**2**	0	2	1

usage and travel expense does not increase drastically (Figs. 4, 5 and 6). From the model value, it is clear that in scenario 1a, only 15.8% are likely to own the car and only 5.7% of the HHs are likely to use the car on daily basis, but for scenario 1b, 27.7% of the HHs are likely to own the car and 18.5% are likely to use it on a daily basis, whereas in scenario 1c, 27.6% of the HHs are likely to buy a car and 27.5% of the HHs are likely to use it on daily basis. This result implies that as the educational qualification of the HH members increases, they tend to possess a car. Only in HH where the majority of the members are highly qualified are likely to use it on a daily basis. In a city of households with a combination of these in equal proportion, 71.2% of the HHs are likely to own a car and 51.8% are likely to use it on a daily basis.

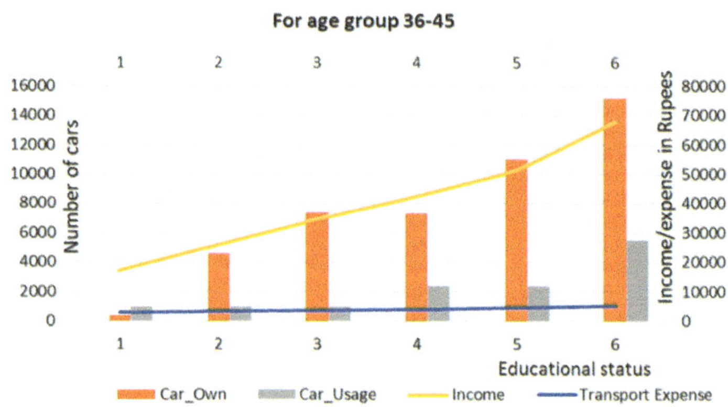

Fig. 4 Simulation result for scenario 1a

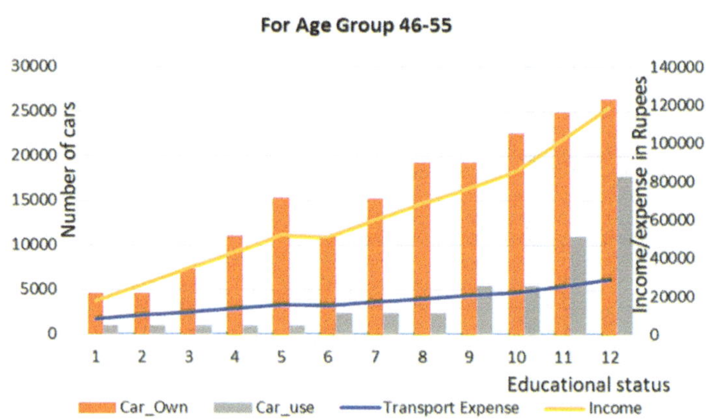

Fig. 5 Simulation result for scenario 1b

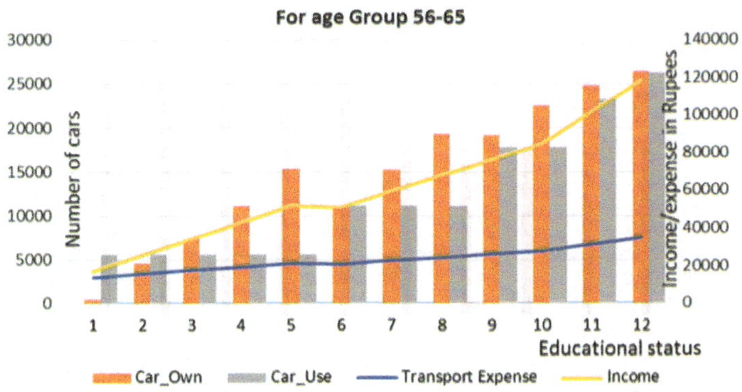

Fig. 6 Simulation result for scenario 1c

4.2 *Scenario 2: Changing the Working Status of the Household Members and the Rest of the Parameters as Constant*

Simulation 2 statistics show that as monthly income and car ownership increases as the working status of the household members are high whereas car usage and travel expense does not increase drastically (Figs. 7, 8 and 9). From the model value, it is clear that in scenario 2a, only 15.9% are likely to own the car and only 5.8% of the HHs are likely to use the car on daily basis. But for scenario 1b, 25.1% of the HHs is likely to own the car and 11.6% are likely to use it on a daily basis, whereas in scenario 1c, 22.8% of the HHs are likely to buy a car and 5.8% of the HHs are likely to use it on daily basis. This result implies that as the working status of the HH member's improves, they tend to poses a car. But their daily usage is not significant compared to that of scenario 1. In a city of households with a combination of these

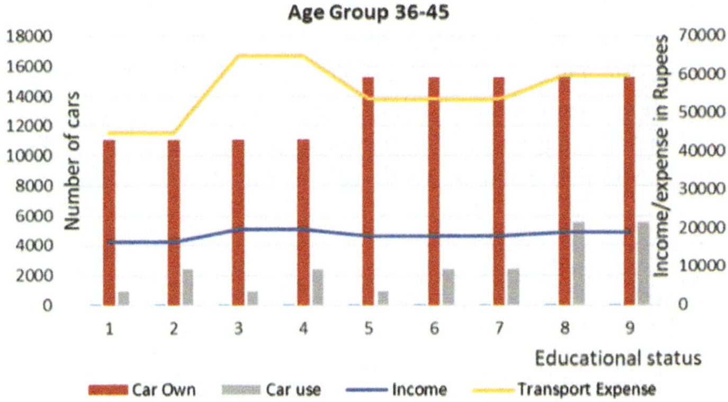

Fig. 7 Simulation result for scenario 2a

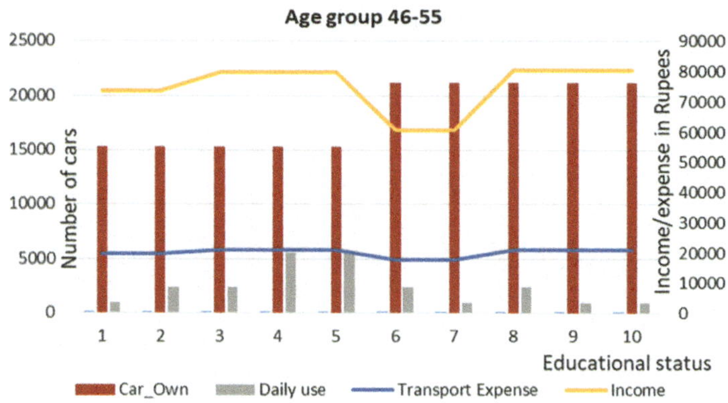

Fig. 8 Simulation result for scenario 2b

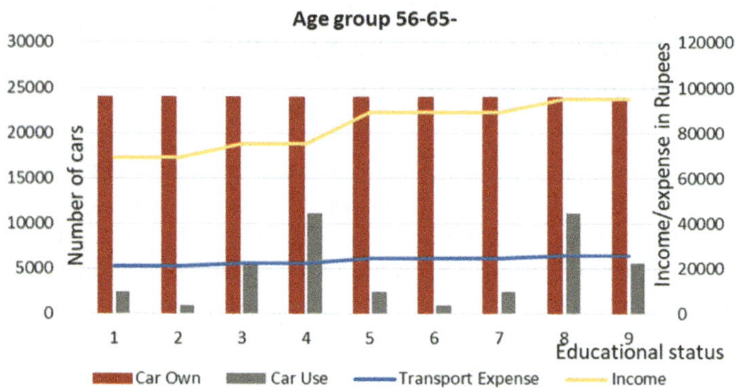

Fig. 9 Simulation result for scenario 2c

in equal proportion, 63.3% of the HHs are likely to own a car and 23.3% are likely to use it on a daily basis.

If we compare both scenarios, the education status of the household plays a major role in car ownership and car usage. Employment status of the household influences car ownership, however, it doesn't lead to daily car usage. Employment status of the household in combination with high educational qualification leads to daily car usage. Similarly, simulation can be done for two-wheelers as well.

5 Conclusion

The simulated model in the context of Calicut city suggests that as the socioeconomic status of the household increases in terms of higher educational qualifications, working status and income, the vehicular ownership, and usage also increases.

In both scenarios, more than 60% of the HHs in Calicut city (71.2% in scenario 1 and 63.3% in scenario 2) own a car at home and its usage is for daily commuting is 51.8% and 23.3% for scenario 1 and 2, respectively. Here it indicates that owning a private vehicle is more than a means for daily commutation, it is an asset. It is a symbol of status and their newly attained affluence. In the case of Calicut city, it is unlikely that these cars significantly contribute to the daily trips generated in the city. Most of the people use either public transportation or two-wheelers for their daily commutation. But there is a high chance of daily usage of cars in a situation where public transportation fails to provide a safe, convenient, and comfortable travel experience to the users. Measures shall be taken to ensure an efficient public transportation system with well networked last mile connectivity.

Household demographic and socioeconomic characteristics are significant parameters that influence car ownership and travel behavior of a household. The interactions among these parameters can be modeled using the system dynamics technique as discussed. Such a model developed at the household level can be used as an agent representing a typical household in a city. This model may be calibrated for different classes of households in the city. Thus, it may be used in conjunction with a suitable agent-based modeling platform to predict the resulting vehicular ownership and usage for a given context. The model would give a specific conclusion about the transport characteristics of the cities based on the socioeconomic parameters.

References

1. Hussein, R. M. R. (2013). *Towards sustainable urban transportation case studies* (Vol. 7, no. 9, pp. 2511–2519).
2. Black, W. R. (1996). Sustainable transportation: A US perspective. *Journal of Transport Geography, 4*(3), 151–159.
3. Dean, E. (2009). *Changing course: A new paradigm for sustainable urban transport* (Vol. 26, no. 2).
4. Borgoni, R., & Prskawetz, A. (2002). *How important are household demographic characteristics to explain private car use patterns ? A multilevel approach to Austrian data* (Vol. 49, p. 26).
5. Chen, M., Ho, T., & Jan, C. (2006). *A system dynamics model of sustainable urban development: Assessing air purification policies at Taipei City* (Vol. 4, no. 1).
6. European Commission. (2001). European transport networks: Results from the transport research programme. *Reproduction.*
7. Kitamura, R. (2010). *Trip generation, and modal split: Model development* (Vol. 14, no. 2009, pp. 711–732).

Informal Settlements and Workplace Spatial Dynamics—A Case Study of Vijayawada

Ekta and Faiz Ahmed Chundeli

Abstract In the recent past, the policy response to informal settlement planning has predominately been to evict and relocate in the form of resettlement, rehabilitation, and redevelopment. Since the inception of the master plan preparation process, planning for informal settlements is still unaddressed or under-addressed. In addition, informal settlements and their workplace spatial dynamic for drawing planning guidelines are often not carried out. This is also evident from the mass exodus of millions of migrant laborers across India, after the Covid19 lockdown. In this research, spatial relation between informal settlement and their workplace dynamics with respect to landuse types has been exercised, with an intent to draw directions for inclusive informal settlement planning. Multi-regression analysis has been carried out between slum location (slum population) and landuse types for the entire Vijayawada Municipal Corporation area. Further, five identified informal settlements in Vijayawada, India, is studied as pilot cases. The study includes mapping of all the five informal settlements and extraction of their spatial correlation with workplaces using GIS. Overlay analysis and origin-destination matrix for pilot cases were performed to understand the spatial dynamics and location preferences of slum dwellers. Besides, random surveys from pilot cases were interviewed to comprehend and ascertain the socioeconomic status and the travel pattern of slum dwellers. The results indicate and reiterate that socioeconomic conditions of informal settlements force people to live as close as possible to their workplace and the need for inclusive policy and urban planning measures at a zonal scale. The study draws the attention of researchers, planners, and policy makers to carry out further research to include informal housing into city development plans.

Keywords Informal settlement · Spatial relation · Land use dynamics, workplace · Geospatial planning

---

Ekta (✉)
Department of Planning, School of Planning and Architecture, Vijayawada, India
e-mail: ekta@spav.ac.in

F. A. Chundeli
Department of Architecture, School of Planning and Architecture, Vijayawada, India

A. Jana and P. Banerji (eds.), *Urban Science and Engineering*, Lecture Notes in Civil Engineering 121, https://doi.org/10.1007/978-981-33-4114-2_16

1 Introduction

1.1 Urbanization and Growth of Informal Settlements

Urbanization is taking place at an unprecedented rate. Thirty percent of the world population was urban in 1950, and within a span of 100 years, it is expected to reach sixty-eight percent by 2050. Approximately half of the world's urban dwellers reside in the small settlement of less than 5,00,000 inhabitants, while around one-eighth live in 33 megacities with more than 10 million inhabitants. According to the report, slum dwellers account for almost one-third of the world's population [15]. Further, the United Nations millennium program recommends the government to strategize the issues of urban poverty, slums, and informal settlements at the national level. The need for inclusion of such national development strategies in India, can be seen in the eleventh five-year plan that highlights the strategy of "inclusive growth" as a key paradigm, where the inclusion of poor and disadvantaged sections in the planning and growth processes is deliberated [9]. As the demand for inclusive planning and development increases, a relook into the missing gaps in our current policy approach demands thorough investigation. The focus of our policy and urban planning measures must shift towards the disadvantaged and lower income people to achieve the goal of inclusive sustainable development.

1.2 Need for a New Policy Approach

The Government of India and the local state government across the country have been exercising various approaches/schemes towards providing houses for the low-income group from industrial housing to low-income affordable housing, from eviction and relocation to resettlement and rehabilitation. Narayan [10] in his study, groups the government policies towards slums and squatters in three categories: First, "Laissez-faire Policies" where existing slums and informal squatter settlements are ignored over the development of other sectors. Second, "Restrictive or Preventive Policies" where the focus of the planning policy has been eliminating or reducing the low-income settlements. In such policy measures, basic urban infrastructure, social, physical, and economic aspects within the slums are neglected [8]. This is another exclusion strategy adopted by the authority for evicting or relocating the residents outside the core city. The third, "End Supportive Policies" where deliberate policy measures are taken to exclude slums and informal squatter settlements from the national policy on the development of urban land and housing policy [10]. However, the housing policies which have been adopted in India, since the first five-year plans (1951–61) have exercised various concepts of a subsidized housing scheme for industrial workers and the economically weaker section, for instance, The Low-Income Housing Scheme of 1954, Slum Clearance and Improvement programme of 1956, etc. These schemes, however, were not as effective as expected because of a conceptual flaw in the policies. The reasons for failure are the unaffordability and poor location choices of relocated slums. As a result, it is inevitable for policy and decision makers to consider and understand the spatial relationship dynamics of job and slum locations [4]. During the 1970–90's, major programs such as the Environmental Improvement of Urban Slums (EIUS) in 1972 and the Urban Basic Services (UBS) Scheme launched in 1986, were initiated focusing on the provision of basic infrastructural facilities for social services. These schemes include the initiatives such as learning opportunities for women, vocational training, pre-school programs for children, and the setting up of community organizations [4]. Further, in the 1990s Urban Basic Services for the Poor (UBSP) and Nehru Rozagar Yojana tried to integrate employment and housing. In the twenty-first Century, within The Jawaharlal Nehru National Urban Renewal Mission (JNNURM), schemes of Basic Services of Urban Poor and the Integrated Housing and Slum Development Programme (IHSDP) were launched to tackle infrastructural and housing challenges. This may be noted that Rajeev Awas Yojana (RAY) aimed for slum free India. In recent years, Pradhan Mantri Awas Yojana (PMAY) has been launched with an intent to provide housing for all [4]. This is clear from the above policy measures and planning initiative that a number of schemes have been experimented so far, including concepts of relocation slums, rehabilitation, and subsidies on loans, etc. Traditionally, slum demolition and resettlement of slum populations were the dominant methods implemented by local and national governments to deal with the "slum problem" [3]; Macharia, 1992). However, the slums continue to exist and they keep growing. To summarize, our approach towards slums and informal housing so far has been treating slums as problems and trying to solve them. However, informal settlements have become an integral part of our urban systems that cannot be negated. The city cannot afford to function without the support services extended by people living in informal settlements. Therefore, in recent initiatives, steps are taken to integrate informal settlements within the urban planning processes. New schemes in the form of 35% reservations for low income and informal housing at zonal level plan is being experimented with. However, such integration poses a number of novel questions that need to be resolved before new policy and urban planning measures are derived. One such question is being investigated in this paper, i.e., the relationship between the informal settlements and workplace dynamics, to develop new policy and planning measures.

1.3 Location Theories and Workplace Relation of Informal Settlements

To explain the slum location and workplace dynamics, the following parameters are used, i.e., private-sector housing location, travel-cost minimization, the travel-cost/housing-cost trade-off, and maximum housing expenditure. A detailed summary of these theories is given by Balchin et al. [2]. Of these theories, one of the major observations drawn is that, given an opportunity, a perfect mobile household would move to a plot where it can satisfy its spatial requirements while paying acceptable transport costs [11]. According to Harvey [5], there are three determinants of residential location, namely accessibility, environmental characteristics, and rent. There is a trade-off of accessibility and environmental characteristics against rent [7]. Urban poor prefer to stay in unplanned settlements, often against the rules and stay close to their employment opportunities to avoid the burden of expenditure on transportation [1]. Further, the finding of Tiwari [14] suggests that the lowest income group relies heavily on walking—indeed, more than half of all of the trips by this group are conducted on foot. Tiwari [14] also points out that trip distances are typically low for low-income groups as the poor tend to live as close as possible to their workplaces, often in sub-standard housing in unplanned settlements, because of long working hours and the need to minimize transport expenditure. Although, broad policy measures on workplace dynamics and slum locations are widely discussed, however, their impact on spatial planning is less explored. In this study, an attempt is made to explore the linkages between the spatial land-use plan and slum locations.

2 Methodology

In this study, a three-layered analytical approach is adopted. First, the existing policies on slums have briefly been studied to understand the current procedures, policy measures, and planning approaches in practice for addressing the issues of informal settlements in cities. Secondly, the residential location theories (in general) have been explored to understand the logic of slum location choices. Further, an attempt has also been made to understand in specific, the criteria used by the slum dwellers of low-income households in Vijayawada, on housing location choices. Thirdly, landuse intensity analysis for commercial, mixed, public, semipublic, and residential is done for all the slums in Vijayawada Municipal Corporation Area (VMCA) to establish a relationship between the land-use and slum locations. Land-use intensity maps for various land-use typologies were derived for the city of Vijayawada using GIS. In addition, 150 socioeconomic sample surveys were collected from five different pilot case slums to comprehend and ascertain the socioeconomic status and the travel pattern of slum dwellers. Further, the relationship between the workplace and place of residence has been established through the origin-density (OD) survey. Spatial dynamics of land-use typology and their relation with slum location through destination attraction survey points are simulated.

3 Case Area

Vijayawada is the second largest city in the newly formed state of Andhra Pradesh, India. Vijayawada is surrounded by the Krishna River on the east and west and the Budameru River on the north. The northern, northwestern, and southwestern parts of the city are covered by a low range of hills, while the central, southwestern, and northwestern parts are covered by rich and fertile agricultural lands with three major irrigation canals [12]. Vijayawada Municipal Corporation (VMC) population as per the 2001 and 2011 census was 8, 45,217, and 10,48,000, respectively. The biggest contributors to the city's population growth are mainly the natural increase and immigration from the surrounding villages. Vijayawada also serves as a business hub in terms of wholesale agricultural products, particularly specializing in vegetables, fruits, and spices [13]. Further, Vijayawada is known for automobile body building, garment, iron, and hardware small scale industries. The Jawahar Lal Nehru Auto Nagar Industrial Estate in Vijayawada, is one of the largest hubs of the automobile industry in Asia. There are about 3000 automobile accessory units in and around the city and about 100,000 workers are employed in these industries. Apart from workers of the city, people from other parts of the state come for work in the industrial estate [12]. The slum population of VMC as per the 2001 and 2011 census was 1,69,043 and 2,87,983, respectively. The total area of the city is 61.88 Km^2 and the slum area is estimated to be 9.27 Km^2 which is 14.98% of the total area as shown in Fig. 1. A

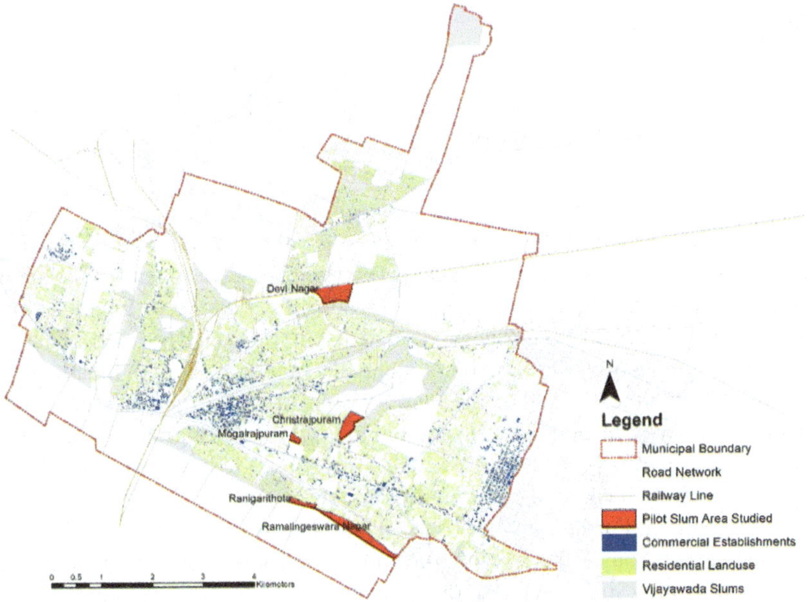

Fig. 1 Pilot slum case study location along with residential and commercial land-uses for VMC. *Source* Map generated by Authors, based on secondary sources, 2020

Table 1 Demographic details of the five pilot case study slums

S. No.	Name of the slum	Ward No.	Population	No. of households	Area (in ha)	Slum density
1	Christurajapuram	3	4458	1276	17.30	258
2	Moghalrajapuram Hill Area East	7	757	208	5.80	131
3	Ramaligeswara Nagar	14	2541	818	4.54	560
4	Ranigarithota	16	7642	2102	9.40	813
5	Devi Nagar	53	3693	1101	16.64	222

total of 111 slums spread over 59 electoral wards are found in VMC, of which 20 are hazardous and 91 are nonhazardous slum category. Out of 111 slums, 58 slums are on state government land, 22 on private land, 27 on lands owned by the local body, and 4 are on railway land. Further, it is observed that 81 slums are in Residential areas, 22 in Commercial areas, and 8 are in Institutional areas. Eighteen slums are spread in an area of more than 15 Ha, with Prakash Nagar Slum being the largest occupying a maximum area of 35 Ha. It is observed that 81.32% of the slums are located in tenable locations in the city. It is also observed that 81% of the slums have an area greater than 1.2 Ha [6]. In this pilot study, Ramalingeswara Nagar, Ranigarithotta, Christurajapuram, Moghalrajapram, and Devi Nagar slums as shown in Fig. 1 are studied. Basic demographic details of the identified five pilot case study slums are shown in Table 1.

4 Analysis and Results

A primary survey was carried out between 16–21 February, 2020 at Moghalrajapuram , Devi Nagar, Christurajapuram , Ranigarithota , and Ramalingeswara Nagar, the five pilot slum case areas. Thirty survey samples were taken from each of these slums. Three main characteristics of the slum dwellers were mapped during the survey, i.e., their travel expenses, mode, and distance traveled for work. These outputs of the survey questions were spatially mapped to understand slum locations and their workplace dynamics. Based on the survey results, it is observed that most of the slum dwellers live with meager income and the majority have migrated to the city around twenty years ago from the nearby villages for jobs. Most of them are illiterate and are from socially backward communities and the majority of them work as unskilled daily wage labor. Further, it was observed that many of the slum dwellers are tenants and do not own houses. As a result, they are settled across the city in an unorganized manner illegally. Since, these settlements are in existences for over a few decades now, the municipal authorities have facilitated access to water and electricity services to these slums, however, despite the lack of land tenure records.

4.1 Landuse Intensity and Slum Locations in VMC Area: Multiple Regression

Based on landuse intensity analysis and multiple regression, it is observed that slum landuse is very intense (2.8–7.5) in about ten wards. Of these wards, the highest residential landuse is found in ward no. 23, 36, 45, 54, 57. Interestingly, about 80% of the slums are located in residential landuse across the city. These wards are in the vicinity of commercial zones with a commercial landuse intensity of 4.2–8.6 units. Slum landuse intensity is observed to be very low (up to 1.6) in ward no. 8,9, 19, 21,28, 39, 40,41,42, 47, and 50. Coincidently, the commercial intensity is observed to be very high ranging from 1.6–8.6 units. Therefore, confirming the slum location proximity and nearness to the workplace. Further, multiple regression for each of the slum with a buffer of 0.5, 1, and 2 km were analyzed and within each buffer multiple regression analysis are carried out, taking the population as the dependent variable and residential, commercial, mixed use, and public-semipublic land use intensity as the independent variable.

From the multi-regression analysis, it is found that the R-square value within the range of 0.5 km, 1 km, and 2 km is 0.18465, 0.210851, 0.154301, respectively. The results signify that the landuse intensity (residential, commercial, public-semipublic, and mixed land use) accounts for 18, 21, and 15% for the same range. The overall regression results obtained are found to be significant and for specific distance following R^2 values are obtained,

1. within 0.5 km buffer, $F\ (4{,}93) = 5.27$, $p < 0.001$ (0.000723009), $R^2 = 0.18465$
2. within 1 km buffer, $F\ (4{,}93) = 6.21$, $p < 0.001$ (0.000178442, $R^2 = 0.210851$
3. within 2 km buffer, $F\ (4{,}93) = 4.24$, $p < 0.00$ (10.003373), $R^2 = 0.003373$

Hence, from the analysis, it can be said that the slum population is strongly affected by the variations in landuse intensity which demands the attention of physical planners to consider the same while planning cities or to retrofit the existing master plans.

4.2 Case Area Buffer Analysis with Landuse Intensities

Within the buffer of 500 m commercial landuse intensity (CLUI), mixed landuse intensity (MLUI), and public-semipublic landuse intensity (PSPLUI) is 0.7, 1, and 0.9, respectively in Moghalrajapuram , it is evident from the survey results that 30% of the respondents travel less than 1 km for their job. 10% of people travel more than 5 km. About 35% of the people rely on two-wheelers for commuting, while 80% of the females travel less than 1 km for work. In Christurajapuram, 50% of the samples surveyed walk/cycle up to 3 km for job purposes, as CLUI, MLUI, and PSP LUI is 1.03, 1.15, and 1, respectively. However, the settlement is on a steep slope and vehicular access is not feasible, as a reason, only 7% of the surveyed respondents own

Table 2 R-Square values for different buffer zones

S. No.	Item	Buffer of 0.5 km	Buffer of 1 km	Buffer of 2 km
1	R-Square	0.18465	0.210851	0.154301
2	Significance F	0.000723009	0.000178442	0.003373

two-wheelers. While about 43% rely on auto and public mode of transport. MLUI, PSP LUI is 0.65 and 0.74, respectively, within the 500 m range in Ranigharithota and about 31% of people go to work by walk, while 31% prefer two-wheelers. Further, CLUI, MLUI, PSP LUI is 1.49, 1.59, and 1.35, respectively, with 50% of the people commute to work within the 2 km range. In Ramaligeswara, about 45% of people who go to work travel by two-wheelers and about 20% rely on a public mode of transport, while 50% of the people travel more than 5 kms for work. In Devi Nagar slum, 17% of people prefer walking and 41% rely on two-wheelers, for commuting (about 60% of people travel more than 5 kms for work) to the workplace. Further, it may be noted that MLUI of 0.61 within 500 m signifies the people going to work by walk. Table 2 shows landuse intensity for all the slums with a buffer range from 500 to 2000 m (Table 3).

4.3 Origin-Destination Matrix

Based on the primary survey conducted and the response obtained from slum dwellers on the place of work, origin-destination (OD matrix) sightlines were generated using GIS. Figure 2 shows all the five pilot slum case locations and responses obtained on travel destinations for work. It is observed from the results, that Auto Nagar, Besant Road, Sidhartha Nagar are the primary workplace destinations of the pilot case slums. It is observed from the analysis that; furthest slums have higher and longer travel sightlines indicating an increased number of trips from slum location to the workplace. However, most of the slum dwellers as shown in the Fig. 2 travel within a 1 km radius from the place of residence indicating the natural location choices of slums. Further, almost 90% of the travel is within a 3 km radius as indicated from the survey results. Thus, a strong relationship and spatial dynamics exist between the slum location and workplace. In addition, it is observed from the study that, over a period of time, most of the slums have organically evolved around the commercial activities of the city. This also reiterated from the OD matrix as shown in Fig. 2, where most of the slum dwellings travel within a 3 km radius to save travel cost. However, slums which are located 8—0 km away from the city, such as Ramalingeswara Nagar show higher and longer sightlines, indicating high volumes of daily travel pattern among the slum dwellers. Thus, the workplace and slum location spatial dynamics are clearly visible through the pilot study. Such a relationship is well established in major literature as well. However, such influx of travel from the informal settlements must be absorbed and contained through spatial policy and deliberate urban planning

Table 3 Land use intensities in the case area

Name of the slum	Land use intensity	Christurajapuram	Mogalrajapuram hill area east	Ramalingeswara nagar (Canal Hutting)	Ranigarithota, (Bhaskara Rao Pet)	Devinagar
Population		4458	757	2541	7642	3693
Buffer-500 m	Commercial	0.18	0.70	0.20	0.53	0.36
	Mixed	0.28	1.00	0.38	0.65	0.61
	PSP	0.46	0.92	0.36	0.74	0.17
	Residential	0.71	1.23	0.57	0.93	1.01
Buffer-1000 m	Commercial	0.25	0.80	0.76	0.64	0.25
	Mixed	0.48	1.10	0.89	0.96	0.47
	PSP	0.47	0.67	0.58	1.36	0.29
	Residential	0.81	1.19	0.66	0.78	0.94
Buffer-2000 m	Commercial	1.03	1.16	0.81	1.49	0.46
	Mixed	1.15	1.40	0.83	1.59	0.81
	PSP	1.00	1.16	0.94	1.35	0.69
	Residential	1.16	1.20	0.73	0.81	0.85

Source Extracted by authors from Landuse intensity analysis using GIS

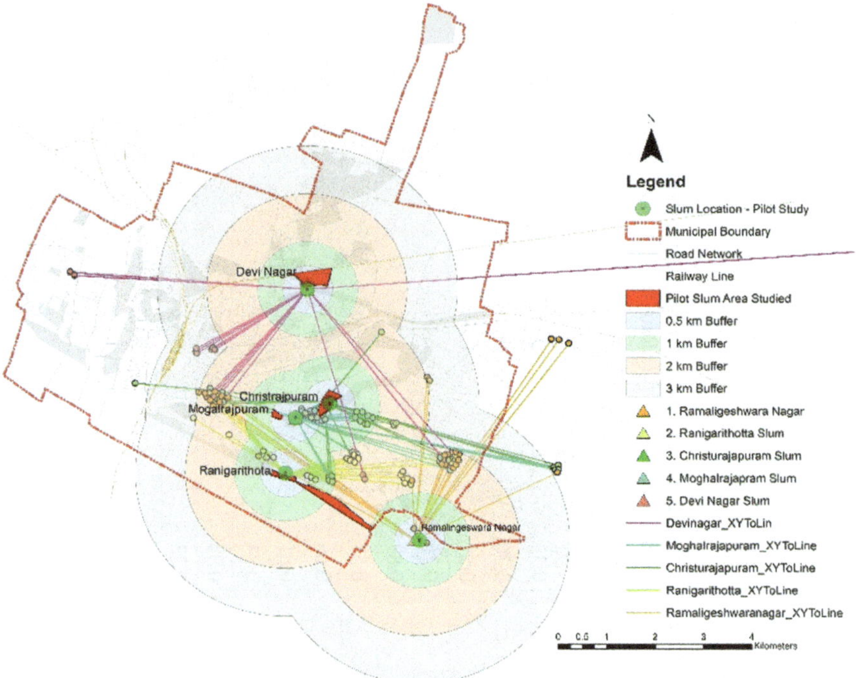

Fig. 2 OD Matrix of pilot case slum locations and their workplace, 2020

interventions. Master plans prepared in this regard must include the allocation of lands for informal housing within the vicinity of commercial activities. Such thoughtful inclusion of informal housing settlements would pay the way for just and inclusive city planning. Amendments in the existing city planning policy and regulation are required to be taken up for inclusive growth.

4.4 *Discussion*

The results from the pilot case studies indicate that many of the slum dwellers prefer living close to their workplaces. Similar observations are made in many pieces of literature . These results also highlight and reconfirm that socioeconomic conditions of people are one of the decisive factors in slum location determinants. Further, it may be noted that 80% of the slums are located within residential land-use typology and the reach of commercial zones (major workplace spots) as shown in Fig. 1, confirming the slum location and workplace nearness dynamics.

Further, many of the slum dwellers prefer to live as close as possible to their workplaces, so to avoid travel-related expenses. However, such observations are rarely adhered while preparing master plans and development plans. Location specific guidelines must be developed for handling slums and informal settlements in the city. Master plans must be dynamic in nature to observe such fluxes. New policy and planning measures must be developed to include informal settlements within the planning norms of city planning. In the case of existing cities, required amendments in the development plan must be carried out. Integrating informality into the existing land-use typology must be experimented. Further, workplace and slum location dynamics must carefully be studied, understood documented across the planning area, while preparing development plans. The influx of migrants for economic opportunities shall continue in major cities, therefore, inclusive sustainable urban planning and policy measures are required to be institutionalized at the earliest.

5 Conclusion

In this paper, the spatial relation between slum location and their workplace dynamics is presented with an intent to draw directions for developing inclusive informal settlement planning guidelines. Five identified informal settlements in Vijayawada, India, are studied as pilot cases. Random surveys from pilot cases were interviewed for establishing socioeconomic status and their travel pattern. Overlay analysis and origin-destination matrix for pilot cases were simulated using GIS to understand the spatial dynamics. The results indicate and reconfirm that socioeconomic conditions of people in informal settlements force them to live as close as possible to their workplace. Land-use planning and policy measures must take into consideration to allow the existence of slums within the central business districts of the city and commercial zones in a planned manner. Thus, there is a need for inclusive planning for the urban poor within the accessible range of the workplace along with affordable housing options. Further, the study draws the attention of researchers, planners, and policy makers to carry out further research in this direction to define the accessibility range of travel for the urban poor, housing location, and choices. Further, the use of scientific and advanced spatial planning support tools must be encouraged and used for attaining sustainable and inclusive planning of cities. As established and proved in the paper that slums are found near the work locations. The majority of the workplaces fall within the commercial, public-semipublic, and mixed landuse. People who come and settle down in slums are economically weak. The intent behind this study is to emphasize the need for affordable residences for the migrants and low-income people while planning commercial areas. Public-semipublic and mixed landuse areas are the potential workplace for low-income slum dwellers. Therefore, the provision of affordable residential near the workplace, on the one hand, shall fulfill the housing requirements of the low-income people and also aid in the formation of new slum clusters. Further, if "X is the commercial, mixed or Public-semipublic landuse intensity" then it would result in "Y number of slums population" in and around such landuse typologies. So, while population projection, this "Y number of slum population" must be considered. Further, research to integrate the same into urban planning processes may be carried out.

References

1. Badami, M., Tiwari, G., & Mohan, D. (2004). Access and mobility for the urban poor in India: Bridging the gap between policy and needs. In *Forum on Urban Infrastructure and Public Service Delivery for the Urban Poor*. National Institute of Urban Affairs, Delhi, India.
2. Balchin, P., Kieve, J., & Bull, G. (1995). *Urban land economics*. London: Macmillan.
3. Baviskar, A. (2003). Between violence and desire: space, power, and identity in the making of metropolitan Delhi. *International Social Science Journal, 55*(175), 89e98. http://dx.doi.org/10.1111/1468-2451.5501009.
4. Bhide, A., Crenshaw, K., Shaban, A., De Neve, G., Donner, H., HLRN, Banerjee-Guha, S., et al. (2019). Housing poverty in urban India: The failures of past and current strategies and the need for a new blueprint. *Economic and Political Weekly, 10*. https://doi.org/10.2307/40277859.
5. Harvey, J. (1996). *Urban land economics* (4th ed.). Macmillan Press Ltd: London.
6. IIHS-UNDP. (2017). *Urban resilience baseline study*. Vijayawada.
7. Jordaan, A. C., Drost, B. E., & Makgata, M. A. (2004). Land Value as a function of distance from the CBD: The case of the eastern suburbs of Pretoria, *7*(3), 532–541.
8. Lall, S., Lundberg, M., & Shalizi, Z. (2008). Implicationsof alternative policies on welfare of slum dwellers: Evidence from Pune, India. *Journal of Urban Economics, 63*, 56–73.
9. Mohupa, G. O. I. (2013). Compendium, a statistical. Slums_in_India_Compendium_English_Version.
10. Narayan, Laxmi. (2014). Slum free cities: Strategies and solutions. *International Journal of Research, 1*(7), 577–588.
11. Phe, H. H., & Wakely, P. (2000). Status, quality and the other trade-off: Towards a new theory of urban residential location *37*(1), 7–35.
12. Rajeswari, M. V. (2014). *Growth and structure of Indian Railways*. Sri Venkateswara University, Department of Economics. Retrieved from http://hdl.handle.net/10603/106644.
13. Srikonda, R., Razak, A., Kaja, N., Daketi, S., & Tarafdar, A. (2016). *Building inclusive urban communities city resume Vijayawada*.
14. Tiwari, Geetam. (2002). Urban transport priorities—Meeting the challenge of socioeconomic diversity in cities; A case study of Delhi, India. *Cities, 19*(2), 95–103.
15. United Nations. (2018). World urbanization prospects. *Demographic Research, 12*. https://doi.org/10.4054/demres.2005.12.9.

The Value of a Life: Potentials and Challenges for Road Safety of Non-motorized Transport Users in Ahmedabad

Aparna Ramesh and Pratiksha Surpuriya

Abstract Despite its high modal share, Ahmedabad suffers from poor quality, uncomfortable and unsafe infrastructure for walking and cycling. As a result, pedestrians and cyclists have become the most vulnerable road users, accounting for 43% of the fatalities despite their significantly shorter trip lengths (JP Research India (2017) Accident research status report Ahmedabad January to July 2017). This paper aims to identify the barriers to providing safe, secure and comfortable facilities for pedestrians and cyclists in Ahmedabad. In November 2018, as the first phase of this study, a road design and safety audit was undertaken at three of the city's accident hotspots so as to examine the situation on ground. The results of the survey revealed the appalling conditions of the footpaths and the complete absence of segregated cycle tracks in the selected accident hotspots. To understand the possible reasons for this gap, the second phase of the study mapped the roles of stakeholders involved in the allocation of road space for walking and cycling and interviewed five diverse stakeholders among them. Discussions with the stakeholders revealed three critical, interrelated issues that affected the implementation of NMT infrastructure in the city—conflicts in the technical versus socio-political decision-making on road space allocation, lack of a coordinated, multidimensional approach by the government for urban transport and skewed cultural perceptions about walking and cycling.

Keywords Urban mobility · Non-motorized transport · Road safety · Pedestrian infrastructure · Road safety audit

1 Introduction

Non-motorized transport (NMT), such as walking and cycling, is truly a low carbon mode of mobility. It is not dependent on the use of fossil fuels, produces no tailpipe emissions and is associated with several positive externalities like improved health,

---

A. Ramesh (✉)
New Delhi, India

P. Surpuriya
Pune, India

A. Jana and P. Banerji (eds.), *Urban Science and Engineering*, Lecture Notes in Civil Engineering 121, https://doi.org/10.1007/978-981-33-4114-2_17

reduced traffic congestion and lower levels of air pollution [1]. Comprehensive studies show that walking and cycling account for more than a third of all trips in Indian cities [2]. Apart from being the preferred mode for short trips, first- and last-mile connections to public transport, several low-income households are often entirely dependent on these modes to access essential services like education and employment.

Despite the high modal share, Indian cities often have very poor quality, uncomfortable and unsafe infrastructure for walking and cycling. As a result, pedestrians and cyclists have become the most vulnerable road users, accounting for nearly half of all road accident fatalities [3]. Road traffic accidents always have severe and irreversible impacts on the lives of their victims. Apart from the immediate pain, grief and emotional suffering, the victim's families inevitably endure the inimical ripple effects of the accident for many generations to come. In November 2015, India signed Brasilia Declaration on Road Safety, thereby officially committing to United Nation's Sustainable Development Goal 3.6—to halve the total number of fatal road accidents by 2020. If we are to achieve this goal, it becomes imperative to re-examine the issue of road safety keeping in mind the needs of the most vulnerable users.

2 Aim

Using the city of Ahmedabad as a reference case, this paper aims to identify the barriers to providing safe, secure and comfortable facilities for both pedestrians and cyclists in Indian cities. By examining the current situation on ground and then, analytically mapping the conflicts and contestations among the various stakeholders, this paper highlights the three most critical challenges that need to be overcome in order to ensure the provision of essential NMT infrastructure.

3 Ahmedabad—Past Plans and Projects

Ahmedabad, home to 5.5 million people, is the commercial capital of the state of Gujarat [4]. The Ahmedabad Municipal Corporation (AMC) is the local self-government in the city and covers an area of 466 km^2 [5]. Similar to other Indian cities, the modal share of pedestrians and cyclists is significantly high at 37% and 9%, respectively [6]. They account for nearly 36 lakh trips every day with an average trip length of 2 km by walk and 3 km by cycle. In stark comparison, cars account for only 4% of trips in the city with average trip lengths of 5.5 km.

Since 2005, the city, with the support from the state and national governments, has invested heavily in building mass rapid transit infrastructure—nearly 100 km of Bus Rapid Transit System (BRTS) and approximately 40 km of metro rail [7, 8]. In order to integrate these transport infrastructure projects with the land use planning, a regional-level transport plan, "Integrated Mobility Plan for Greater Ahmedabad Region" (IMP), was prepared alongside Development Plan 2021.

Several attempts have been made to plan for a well-designed and safe NMT infrastructure in the city. The IMP proposed the preparation of Local Area Access Plans to improve footpaths and provide cycle tracks around mass transit stations. Similarly, the street redesign schemes for the BRTS project also proposed wide footpaths and over 20 km of segregated cycle tracks. The most recent plan, Smart City Plan for Ahmedabad, proposed pedestrian-friendly development with plazas, wide footpaths and vehicle-free zones as a key feature of its area-based development project [9]. However, the service-level benchmarking exercises reveal a huge gap between what is planned and what exists on ground. In 2015, only 16% of Ahmedabad's major roads had footpaths; a meager 3% of its total road network had cycle tracks out of which 50% was encroached upon at the time of survey [10].

Studies of road accidents in the city paint a sordid picture. From the accidents examined, pedestrians and cyclists together account for 43% of the fatalities despite their significantly shorter trip lengths [11]. The same study also identifies the lack of road infrastructure, particularly poor road marking/signage, lack of pedestrian infrastructure for crossing and substandard intersection design, as one of the three main factors causing accidents. As a result of increasing motorization, deteriorating NMT infrastructure and high risks to NMT users, the share of NMT users has declined rapidly over the last four decades in Ahmedabad [3]. This is a distressing trend and, therefore, must be addressed.

4 Road Design and Safety Audit

As the first phase of this study, a road design and safety audit was undertaken at three of the city's accident hotspots. These hotspots were identified based on the locations of road accident fatalities from Urban Road Accident Study conducted by JP Research India in 2016–17. The selected hotspots were

1. Patiya Circle (Narol-Naroda Road)
2. S.G. Highway (between Thaltej and Ambli Road)
3. Naya Vadaj Circle (toward Sabarmati railway station)

They were located on major arterial roads in different parts of the city and exhibited varying road design and traffic characteristics (Table 1).

The parameters used to audit the NMT facilities were identified based on guidelines issued by Indian Road Congress (IRC) and Urban Road Safety Audit Toolkit published by Ministry of Urban Development [12, 13]. The audit was divided into two parts—facilities for walking and facilities for cycling. Separate parameters for walking and cycling examined the quality and level of service of footpaths, crosswalks and cycle tracks present in the city. The combined list of parameters for walking and cycling were

Table 1 Characteristics of road design and safety audit locations. Average network speed for four-wheelers was sourced from IMP [6]

Area/Road name	Right of way width	Avg. speed for four-wheelers	Character	Length surveyed
Patia Circle (Narol-Naroda Road)	60 m	Partly >30 kmph and partly 15–25 kmph	Major arterial road with small commercial retail outlets. This location serves as a stop for intercity services by GSRTC (Gujarat State Road Transport Corporation). BRTS route runs along this road	1.16 km
SG Highway (between Thaltej and Ambli Road junctions)	75 m	>30 kmph	Major arterial Road with commercial retail outlets and large shopping malls. Service Road provided on both sides	1.36 km
Naya Vadaj Circle (toward Sabarmati railway station)	30 m	>30 kmph	Major arterial road with small commercial outlets. BRTS route runs on this road. The proposed metro rail is also being constructed along this stretch	1.25 km

1. Footpath/Cycle track width
2. Footpath/Cycle track surface
3. Obstructions
4. Encroachment
5. Potential of Vehicular conflict
6. Continuity
7. Security
8. Comfort
9. Walking environment

While the first six pertained to the physical characteristics of the footpath facility, the latter three were user-based factors (Annexure 1). A survey form was prepared to conduct the road design and safety audit and the available facilities were graded on a 3 point scale, from poor to good (Annexure 2). To collect information and grade the user-based parameters, the authors themselves walked and cycled at the selected locations during peak and off-peak traffic hours.

4.1 Results of Road Design and Safety Audit

Facilities for walking. At all three locations, the available footpaths were less than 1.5 m wide. The footpaths were paved with concrete paver blocks and had a raised edge. However, most often these blocks were damaged or uprooted from their locations leaving behind irregular and unevenly paved surfaces. The walking facilities were not continuous as significant stretches of the footpath were either absent, severely damaged or encroached by a variety of objects (electric poles, street lights, electricity meter boxes, advertisement boards, parked vehicles, bus stops and public toilets) and people (street vendors, auto-rickshaw drivers, homeless beggars, people waiting at the bus stop, etc.). At Patiya Circle, ironically the traffic police themselves had set up their monitoring station on the available footpath and, thus, had obstructed it entirely. Lack of continuous, segregated walking facilities resulted in pedestrians walking at the edge of the carriageway for vehicles. The high network speeds (>30 kmph) increased the potential of human-vehicular conflicts at all three locations. Crosswalks were spaced at an average distance of 500–700 m at Patiya Circle and Naya Vadaj Circle but on SG highway, the crosswalks were on an average more than 700 m apart. The crosswalks are uncomfortable and unsafe to use as the medians are higher than 450 mm with no refuge islands, and there are no pelican signals for cars to stop. As a result, crossing the road at these locations took more than 60 s, and the authors observed most other pedestrians choosing to jaywalk across the road wherever there was a break or damage in the median. The walking facilities were not shaded and did not have any street furniture for pedestrians. While they remained sufficiently well-lit at night with light poles at regular intervals, the lack of security arose from the poor quality of walking surfaces (Table 2).

Facilities for cycling. Segregated cycle tracks were planned, and space was allocated along the Narol-Naroda Road as a part of the BRTS project. However, at the time of the survey, the road space allocated to cycling was fully encroached by parked cars and trucks. The cyclists used the edge of the carriageway and wove dangerously through the pedestrians and motor vehicles. The authors' own experience of cycling at the survey locations was terrifying and daunting. With every automobile whizzing past on the undivided carriageway, their self-awareness heightened and every reflected light, every honk and every screech of car breaks metamorphosed into danger signals. Inevitably, cyclists were pushed to the tattered edges of the road, but the multitude of obstructions forced them to swerve back into the oncoming fast-moving traffic. The bicycles that were meant to be an aid to mobility turned into an Achilles heel, inhibiting freedom of movement (Table 3 and Fig. 1).

Table 2 Results of road design and safety audit for pedestrian facilities

	Parameters	Findings					
		Site (1)		Site (2)		Site (3)	
		Narol-Naroda Road		S. G. Highway		Naya Vadaj Circle	
		D1	D2	D1	D2	D1	D2
1	Footpath width	Absent	Absent	Absent	Good	Good	Good
2	Footpath surface	Absent	Absent	Absent	Fair	Fair	Fair
3	Footpath height	Absent	Absent	Absent	Bad	Good	Good
4	Obstruction	Bad	Bad	Bad	Bad	Bad	Bad
5	Encroachment	Bad	Bad	Bad	Bad	Bad	Bad
6	Continuity	Bad	Bad	Bad	Bad	Bad	Bad
7	Security	Fair	Bad	Bad	Good	Fair	Fair
8	Disable friendly infrastructure	Bad	Bad	Bad	Bad	Bad	Bad
9	Comfort	Bad	Bad	Bad	Bad	Bad	Bad
10	Walking Environment	Bad	Bad	Bad	Good	Bad	Fair
11	Availability of crossings	Bad	Bad	Fair	Fair	Fair	Fair
12	Time taken for crossing	Bad	Bad	Bad	Bad	Bad	Bad

Colour	Legend
Grey	Absent
Yellow-green	Good
Orange	Fair
Red	Bad

Table 3 Results of Road design and safety audit for cyclist facilities

	Parameters	Findings					
		Site (1)		Site (2)		Site (3)	
		Narol-Naroda Road		S. G. Highway		Naya Vadaj Circle	
		D1	D2	D1	D2	D1	D2
1	Cycle Track width	Absent	Absent	Absent	Absent	Absent	Absent
2	Cycle Track surface	Absent	Absent	Absent	Absent	Absent	Absent
3	Shade	Absent	Absent	Absent	Absent	Absent	Absent
4	Parking facilities	Absent	Absent	Absent	Absent	Absent	Absent

5 Conflicts and Contestations on Urban Roads

5.1 Mapping Stakeholders

The results of the survey revealed the appalling conditions of the footpaths and the complete absence of segregated cycle tracks in the selected accident hotspots. To understand the possible reasons for this gap, the second phase of the present study

Fig. 1 Typical road sections and photographs at survey locations

Table 4 Key and other stakeholders involved in the planning and provision of NMT infrastructure in Ahmedabad

Service	Key stakeholders	Other stakeholders
Laying out of norms, regulations and legislations with regard to NMT	Ahmedabad Municipal Corporation (AMC)	Roads and Buildings Department, Government of Gujarat; Ministry of Housing and Urban Affairs
Planning of footpaths, cycle tracks and pedestrian-priority zones	AMC, Private urban planning consultants contracted by AMC	Traffic Police, Ahmedabad Urban Development Authority, Academia, Research Cells such as Center of Excellence for Urban Transport and CEPT University
Undertaking investment and construction of planned infrastructure	AMC	Ahmedabad Urban Development Authority and National Highways Authority of India
Monitoring and data collection	AMC, Traffic Police	Academia, Research Cells and Civil society organizations
Operation and maintenance of infrastructure	AMC	Private construction companies and utility providers hired by AMC
Enforcement of regulations on the roads	AMC, Traffic police	Civil society

mapped the roles of stakeholders involved in the allocation of road space for walking and cycling and interviewed five diverse stakeholders among them. The stakeholders were mapped based on their service mandates and accordingly, "key" stakeholders who were empowered to make decisions about the outcomes of the projects as well as "other" stakeholders who were involved in the projects but had limited influence on the outcomes were identified (Table 4).

The stakeholders interviewed were primarily from three sectors—the city and state government, private consultants to the government and civil society organizations (Table 5). With the help of a structured interview (Annexure 3), the authors recorded the stakeholder's viewpoints on the barriers to the implementation of essential NMT infrastructure in the city.

5.2 *Analysis, Inferences and Recommendations*

Discussions with the stakeholders revealed three critical, interrelated issues that affected the implementation of NMT infrastructure in the city, namely

Table 5 Barriers to NMT implementation from the stakeholders' perspectives

Sector	Role	Barriers to NMT
Government	Senior road engineer, AMC	Rising number of private motorized vehicles require more space on the roads
		Poor maintenance of existing road infrastructure by numerous private contractors who are hired to lay utility pipelines below the roads
		Increased availability of public transport caused a modal shift from cycling
		Personal discomfort during walking and cycling in high outdoor air temperatures
		Walking and cycling are viewed as the poor man's mode of transport
Government	Senior executive officer, AMC	Different road functions—street vending, parking and NMT are not coordinated
		Limited use of existing cycling facilities
		Encroachment of footpaths by street vendors
Private	Urban designer and planner	Popular misconception of AMC that wider footpaths lead to nuisance with the influx of hawkers and footpath dwellers
		Lack of holistic city-level policy or plan that covers pedestrians and cyclists and also providing universal access to the differently abled
		Unclear traffic rules and signs and hence, lack of awareness among motorists and cyclists on how to drive in the presence of each other
Private	Transport planner and street design expert	Those in charge of executing NMT infrastructure projects are primarily road engineers trained to prioritize the smooth flow of vehicular traffic
		Misconceptions about the need for wider roads for smooth traffic flow result in existing footpaths being narrowed to accommodate more traffic lanes

(continued)

Table 5 (continued)

Sector	Role	Barriers to NMT
		Misconceptions about hawkers as a menace result in AMC breaking down and narrowing footpaths
		AMC's piecemeal and uncoordinated efforts provide only short-term solutions. Evicting hawkers from footpaths does not solve the core issue of inequitable allocation of road space
Civil Society	CEO, NGO promoting road safety and green mobility	Lack of legal accountability among road infrastructure providers—both public and private
		Limited opportunities for engagement from civil society members at the planning stage

1. Technical versus socio-political decision-making on road space allocation,
2. Lack of a coordinated, multidimensional approach by the government and
3. Skewed cultural perceptions about walking and cycling.

The politics of road space allocation. The results of the road safety audit revealed the extreme inequities in road space for non-motorized modes such as walking and cycling. These inequities are not merely the result of purely rational, technical decisions but the result of a socio-political division of available road space.

Urban sociologist Lefebvre theorized that space is not only that which is physically built but also that which is socially interpreted, produced and reproduced many times over through social relations [14]. Post liberalization in the 1990s, the steadily rising incomes of the burgeoning middle-class population increased access to private motor vehicles. As a result, the ideal Indian city came to be popularly imagined as global and "world-class", where wide, smoothly paved, unobstructed roads became primary conduits for "free-flowing traffic" [15]. Since then, much effort has gone into the production and reproduction of this particular idea of the road. It is now so deeply ingrained in people's minds that it continues to dominate despite the cost paid by all other alternative modes (cycle, walk and pushcarts) and activities (street vending, socializing, resting, etc.).

During the interviews with the stakeholders, the city engineer cited the rising number of private vehicles to be the rationale for widening carriageways at the cost of narrowing footpaths. Street vendors were seen as a hindrance to traffic flow and hence were being evicted forcefully in a series of anti-encroachment drives across the city [16, 17]. Additionally, the private consultants also highlighted instances where innovative proposals for footpaths wider than specified in the national guidelines were rejected as they would lead to the loss of precious road space for motorists. While national-level documents such as National Urban Transport Policy have already put forth alternate, more equitable visions for urban road space, these ideas have not percolated to the city-level actions in the last decade.

Lack of a coordinated, multidimensional approach by the government. Ahmedabad lacks a holistic policy document that considers the needs of pedestrians, cyclists and ensures universal access to the differently abled in the city. As a result, AMC's efforts over the years have been contradictory and piecemeal in nature. The senior engineer himself highlighted the troublesome issue of "Road Opening (RO)" permits issued by the AMC to nearly 90 odd private contractors. He complained that these contractors had dug up over 800 km of road edges to lay electricity and gas supply pipelines in the city but did not replace the paving after completion of their task. Similarly, one consultant cited the case of AMC haphazardly removing cycle tracks and narrowing footpaths over 2 m wide along stretches of the BRTS corridor to prevent street vendors from using these spaces.

By preparing a comprehensive NMT policy, a comprehensive network plan for cycling along with statutory city-specific street design guidelines for footpaths, cycle tracks and vending, the AMC along with other stakeholders can coordinate their efforts toward achieving well-defined goals and measurable targets.

Cultural bias against walking and cycling. Discussions with the stakeholders revealed that the city's wealthier residents viewed walking and cycling as the poor man's mode of transport. The city's senior engineer cited this bias as the primary cause behind the failure to attract cyclists to the few cycle tracks in the city. In stark contrast, data on cycle ownership reveals a different picture—48.8% of Ahmedabad's households own one or more cycles [4]. With increasing investment in mass rapid transit systems, the demand for safer, more efficient and comfortable first- and last-mile connections on foot is going to increase. Therefore, it is essential to reimagine the societal ideas of poor and rich man's transport. After all, in different situations, the pedestrian is the motorist and vice versa.

Learning from successful cities. Several cities have overcome similar issues and have successfully integrated non-motorized transport into their daily lives by adopting a multi-pronged approach—a drastic modification of the existing physical infrastructure, technology integration, radical changes in transport policy and, most importantly, reimagining, as a society, the act of cycling and walking. Cities like Chicago and Copenhagen have tackled the issue by preparing comprehensive pedestrian and cycling policies and plans. In Chicago, the Plan sets out policy actions that stretch beyond the purview of Department of Transportation, and involve various city agencies and community organizations in the development and maintenance of vibrant, livable streets [18, 19]. In the Copenhagen city government, an integrated planning organization encourages knowledge exchange between urban, transport and bicycle planners and thus, creates an environment of understanding for different professional views on planning [20]. In Germany, Bike Academy funded by National Cycling Scheme was specially set up to provide continuing education to city government officials on the benefits of cycling [21].

Closer home, Pune, has taken decisive steps by institutionalizing its best practices for NMT. In 2016, with the help of vocal civil society organizations and a visionary government, the city published its own city-specific street design guidelines that prioritized people not vehicles to recreate a vibrant public realm on Pune's roads [22]. Backed by a comprehensive transport policy for Pune Metropolitan Region, Pune Municipal Corporation introduced a special Non-Motorized Transport cell to implement the comprehensive bicycling plan for the city [23]. A technical committee on NMT in the city government ensured that various stakeholders from within the government and outside voiced their concerns.

Along with such systemic changes, cities across the world are consciously promoting the creation of cycling and walking culture. Take, for example, Copenhagen's Cycle Chic campaigns that were able to transform the meaning of cycling in the Danish imagination, associating it with positive values such as freedom, health and personal energy. Today, Copenhagen has nearly 265,000 bicycles with even the Danish Prime Minister cycling to work. Similarly, in Sweden, Ministry of Transport through their Vision Zero Initiative has changed the way people think about road safety. By setting the target of bringing down the road traffic accident fatalities down to zero, both the government and the public shifted their vision of road safety from more transport engineering-oriented goals to a more radical humanist approach—acknowledging that every road traffic accident is a human catastrophe with physical and mental health implications for all those involved.

6 Conclusion

Cities are in dire need of new paradigms for envisioning and implementing sustainable urban transport. In this scenario, non-motorized transport provides an invaluable opportunity to achieve the Sustainable Development Goal of improved road safety while also reducing greenhouse gas emissions and saving on rising import costs of fossil fuels.

The present state of NMT infrastructure in Indian cities is abysmal. In Ahmedabad's most accident-prone roads, footpaths are severely damaged or encroached, crosswalks are unsafe and poorly designed, and cycle tracks are absent altogether. Relegated to the tattered edges of the roads, pedestrians and cyclists have become the most vulnerable to road accidents.

From discussions with key stakeholders, three critical interrelated issues emerged as the primary barriers to implementing essential NMT infrastructure, namely (i) technical versus socio-political factors affecting the government's decisions of road space allocation, (ii) lack of a coordinated, multidimensional approach by the government and (iii) the skewed cultural perceptions about walking and cycling in the society at large. Cities across the world have overcome similar challenges by adopting a multi-pronged approach—a drastic modification of the existing physical infrastructure, technology integration, radical shifts toward a more humanist transport policy and, most importantly, reimagining, as a society, the act of cycling and walking.

Appendix 1: Road Design and Safety Audit Parameters

	Parameters	Reasoning/Description
1	Footpath/cycle track width	Minimum obstacle-free walkway width adjacent to Residential/Mixed use land uses: 1.8 m In case of commercial land use: 2.5 m This is based on the minimum width required for two people to cross paths comfortably
2	Footpath/cycle track surface	A firm, even paved and slip-resistant surface Vertical deviation not more than 5 mm Raised edge of footpath Tactile pavers laid continuously
3	Obstruction	Obstructions in the form of electric poles, trees, bins, etc. should not be within the minimum 1.8 m width of the walkway
4	Encroachment	The extent of encroachment should not rise above a level that the footpath facility becomes inaccessible/non-usable by the pedestrian
5	Potential of vehicular conflict	Footpaths to be distinctly separate from roads by raising footpaths and providing guardrails along roads where there is fast vehicle movement
6	Continuity	Raised crossings Change in level made clear using bright contrasting colors and tactile pavers Provision of curb ramps for maintaining the continuity
7	Security	Street lighting at 20–30 m, lighting fixtures of not more than 4 m height Separation of pedestrian and vehicular traffic
8	Comfort	Carefully planned provision of trees to protect from inclement weather Providing seats and benches and rain shelters at regular intervals Provision of disable friendly facilities
9	Walking environment	A clean footpath free of stink This can also be subjective and pertains to the individuals' experience

Appendix 2: Road Design and Safety Audit Survey Questionnaire for Walking and Cycling Facilities

Part 1—Facilities for walking

Indicators	(A)	(B)			(C)	
	Absent: 0	Good	Fair	Poor	Total	Remark
	Present: 1	(1)	(0.5)	(0.2)	(A) × (B)	
Width of the footpath		1.8–5 m	1.5–1.8 m	<1.5 m		
Footpath Surface		Concrete/Paver blocks/Tar and anti-slip, tactile	Tiles	Unpaved/non-metaled surface		
Height of footpath		Max <100 mm	100–300 mm	>300 mm		
Obstructions (such as trees, electric poles, and signage)		No obstructions	Pedestrian has to slow down sometimes	Pedestrian has to slow down most of the time		
Encroachments (by hawkers, vehicles, etc.)		No encroachment	Pedestrian has to slow down sometimes	Pedestrian has to slow down most of the time		
Continuity (curb ramps, distinctive change in level)		Presence of both	Provided in some places	No continuity		
Security (light poles at 20–30 m, height max 4 m)		Light poles at regular interval	Light poles less frequent	Mostly no light poles		
Provision of disable friendly infrastructure		Present	Some infrastructure available	Mostly absent		
Comfort (shading devices, benches and seats)		Present	Some infrastructure available	Mostly absent		
Walking Environment (cleanliness and maintenance)		Well maintained	Need better maintenance	Not maintained		
Availability of crossings (frequency)		Average spacing <500 m	Average spacing 500–700 m	Average spacing >700 m		

(continued)

(continued)

Indicators	(A)	(B)			(C)	
Time taken for road crossing		10–30 s	30–60 s	>60 s		

Part 2—Facilities for cycling

Indicators	(A)	(B)			(C)	
	Absent: 0	Good	Fair	Poor	Total	Remark
	Present: 1	(1)	(0.5)	(0.2)	(A) × (B)	
Width of the cycle track		1.8–5 m	1.5–1.8 m	<1.5 m		
Cycle track surface		Concrete/Tar/Asphalt	Interlocking blocks	Unpaved/non-metaled surface		
Height of track		Max <100 mm	100–300 mm	>300 mm		
Shade		Complete	Mostly shaded	Mostly not shaded		
Parking facilities		Within 250 m of the station	250–500 m of the station	Informal parking within 500 m of the station		

Appendix 3: Interview Questions

To initiate the discussion on barriers to the implementation of essential NMT infrastructure, the following questions were asked to the various stakeholders.

1. What are the road planning, construction and maintenance projects you have worked on in the city?
2. Why do many of the city's roads lack essential infrastructure such as footpaths?
3. While several street design guidelines from the central government are readily available, they remain unimplemented on the city's roads. What were the barriers that prevented their implementation?
4. In your opinion, what are the ways to overcome these challenges?

References

1. Hook, W. (2003). *Preserving and expanding the role of non-motorised transport.* Institute for Transportation and Development Policy, GTZ Transport and Mobility Group.
2. Wilbur Smith Associates. (2008). *Study on traffic and transportation policies and strategies in urban areas in India.* Ministry of Urban Development, Government of India.
3. Tiwari, G., & Jain, D. (2013). *NMT infrastructure in India: Investment, policy and design.* Indian Institute of Technology, Delhi. UNEP Risø Centre on Energy, Climate and Sustainable Development Technical University of Denmark.
4. Census 2011. (n.d.). *District census handbook.* Retrieved August 2018, from http://www.censusindia.gov.in/2011census/dchb/DCHB.html.
5. Ahmedabad Municipal Corporation. (n.d.). *About Ahmedabad.* Retrieved August 2018, from https://ahmedabadcity.gov.in/portal/jsp/Static_pages/introduction_of_amdavad.jsp.
6. Ahmedabad Urban Development Authority. (2013). *Integrated mobility plan for greater Ahmedabad Region.*
7. Ahmedabad Janmarg Limited. (n.d.). *About us.* Retrieved August 2018, from http://www.ahmedabadbrts.org/.
8. Metro-Link Express for Gandhinagar and Ahmedabad (MEGA) Company Ltd. (n.d.). *Project profile.* Retrieved August 2018, from http://www.gujaratmetrorail.com/project/project-profile/.
9. Ahmedabad Municipal Corporation. (2015). *Smart City Ahmedabad.*
10. Ministry of Urban Development, Government of India. (n.d.). Retrieved August 2018, from Service Level Benchmark Urban Transport: www.utbenchmark.in.
11. JP Research India. (2017). *Accident research status report Ahmedabad January to July 2017.*
12. Indian Roads Congress. (2012). *Guidelines for pedestrian facilities IRC 103-2012.* Indian Roads Congress.
13. Transportation Research and Injury Prevention Programme. (2013). *Urban road safety audit.*
14. Lefebvre, H. (1991). *The production of space.* Oxford: Blackwell Publishing.
15. Joshi, R., & Joseph, Y. (2015). Invisible cyclists and disappearing cycles. *Transfers, 5*(3), 23–40.
16. John, P., & Sharma, S. (2018, August 10). *Rights of street vendors robbed in Gujarat.* Retrieved August 11, 2018, from The Times of India: https://timesofindia.indiatimes.com/city/ahmedabad/rights-of-street-vendors-robbed-in-gujarat/articleshow/65345313.cms.
17. Ahmedabad Mirror. (2018, July 14). *AMC to clear roads of encroachments.* Retrieved August 11, 2018, from Ahmedabad Mirror: https://ahmedabadmirror.indiatimes.com/ahmedabad/others/amc-to-clear-roads-of-encroachments/articleshow/64981984.cms.
18. Department of Transportation. (2015). *Chicago streets for cycling plan 2020.*
19. Department of Transportation. (2012). *Chicago pedestrian plan.*
20. Koglin, T. (2015). Organisation does matter—Planning for cycling in Stockholm and Copenhagen. *Transport Policy, 39,* 55–62.
21. German Institute for Urban Studies gGmbH. (2018). *About the portal.* Retrieved August 11, 2018, from Fahrradportal: https://nationaler-radverkehrsplan.de/de/ueber-das-portal.
22. Pune Municipal Corporation. (2016). *Urban street design guidelines.* Pune Municipal Corporation.
23. Pune Municipal Corporation. (2017). *Comprehensive bicycle plan for Pune.* Pune Municipal Corporation.
24. National Transport Development Policy Committee (NTDPC). (2013). *India transport report: Moving India to 2032.* Planning Commission, Government of India.
25. TRIPP, IITD. (2013). *Urban road safety audit.* New Delhi: Ministry of Urban Development, GoI.
26. Chicago Department of Transportation. (2013). *Chicogo streets for cycling plan 2020.* Chicago: City of Chicago.
27. Pune Municipal Corporation. (2017). *Comprehensive bicycle plan for Pune.* Pune.
28. GTZ, & Hook, W. (2003). *Preserving and expanding the role of non-motorized transport.* GTZ.

Urban Envelope: Representing Audio and Visual Features Extracted from Deep Learning Techniques

Deepank Verma and Arnab Jana

Abstract The study focuses on automated classification and representation of environmental features present in the surroundings. The aim of the study is twofold. Firstly, it compiles a set of Deep Learning (DL) algorithms that are capable of extracting features from street collected images and audio clips. Secondly, it proposes a graphical method to represent such features. The method explains the extracted audio and visual features in a simple illustration that can be automatically created and displayed by analyzing a pair of photographs and audio clips for a particular location. To achieve the objectives, camera and audio recorders are used to collect the street view photographs and sound clips at several locations. These clips are then processed by algorithms to obtain detailed information from such collection. At a given location, auditory and visual features are combined to represent the envelope of sensory information.

Keywords Urban surroundings · Deep learning · Sound classification · Urban envelope · Street view images

1 Introduction

Quantifying encountered visual and acoustic environments is essential in understanding and documenting urban areas. The information regarding the presence of sounds of birds, vehicles, and crowds and count of pedestrians and vehicles can be used for justifying site planning, development regulations, and other local bylaws [1]. These measurements may assist in scientific research related to urban health, sociocultural aspects, urban design, and planning. Studying these sensory features is especially relevant to urban perception research, which deals with human preferences in urban surroundings [2]. Such studies frame relationships between a variety of these features with human thoughts and emotions.

D. Verma (✉) · A. Jana
Centre for Urban Science and Engineering, Indian Institute of Technology Bombay, Mumbai 400076, India
e-mail: deepank@iitb.ac.in

A. Jana and P. Banerji (eds.), *Urban Science and Engineering*, Lecture Notes in Civil Engineering 121, https://doi.org/10.1007/978-981-33-4114-2_18

The current methods of recording environmental characteristics require rigorous manual surveying, which consistently loses its significance when dealing with large urban areas due to biases in logging information. Recent advancements in data analysis fueled by Artificial Intelligence (AI) techniques and the availability of large datasets have made such investigations possible at the city-wide scale [3]. Studies have widely utilized large databases such as Google and Tencent street view imagery to calculate greenery [4], sky view factor [5], perception of safety and liveliness [6], and health prospects [7].

The quantification or counting of the features is a two-step process (a) data collection, and (b) data analysis. The data is actively collected and analyzed in situ by sensory organs such as eyes, nose, and ears. Alternatively, devices can also be used to record audio and video, which can help gather inferences passively with the help of data analysis algorithms. The devices such as camera and audio recorder emulate the visual and auditory stream closely to that of humans; however, tools to record smells are relatively uncommon [8]. Describing smells has been linguistically challenging and much dependent upon the inherent cultural and linguistic experience of the observer [9]. Hence, the research involving assessment of the olfactory landscapes has mainly comprised of subjective evaluation. The researchers have relied on interviews and personal descriptions of the encountered smells to gauge the perception of smell and to create smell maps at the local and regional scale [10, 11].

2 Machine-Based Quantification

While the human brain is trained to identify, count, and perceive elements present in the sensory realms, intentionally indulging in such an endeavor is time taking and mentally and physically exhausting. With the recent push in large-scale data collection and analysis, mainly using AI, gathering inputs on the presence of features in the surroundings can be made fully automated. However, not all the sensory streams could be processed by machine-based systems. As discussed, visual and auditory datasets such as photographs and sound clips are easier to collect and interpret, while olfactory details are not. This study focuses on street-level images and audio datasets to capture environmental attributes.

The data processing algorithms are capable of extracting High and Low-level features from the datasets. The algorithms detect Low-level Features (LLFs) through variation in intensities in values of local fundamental units in the data such as pixel information in images or audio signals in the recorded audio clip. The values of pixels can be used to generate an estimation of colors, edges, shapes, and blobs in the image. Similarly, the values in the waveform in sound clips can be processed to provide Sound Pressure Levels (SPL) measured in decibels and other psychoacoustical features. The High-level features (HLFs) offer details that are more in tune with how we perceive the audio and visual data with our own senses. The algorithms to capture LLFs are simpler to implement, whereas HLFs mostly require the use of relatively complex algorithms such as Deep Learning (DL) models. Two of the extensively used and proven DL algorithms, Convolutional Neural Networks (CNN) and Recurrent Neural Networks (RNN), are used in the study.

CNNs are the category of Deep Neural Networks that specializes in image processing. It utilizes Convolution layers, which act as filters that capture spatial dependencies between parts of the images. The variations of the regular CNN models are used to perform tasks such as Image Classification, Object Detection (OD), and Semantic Segmentation (SS) [12] (Fig. 1). Similarly, Recurrent Neural Networks (RNN) specializes in learning features from the data that contain temporal dimensions such as time series analysis, language translation, stock prediction, and audio classification. RNNs store variable-length sequences of information in memory layers, which are utilized to make predictions [13]. The regular CNNs are modified to achieve various tasks such as (a) detecting and counting features. (b) determining the composition of the image, and (c) gathering contextual information from the pictures, such as information regarding the presence of slums, plazas, shops, and markets. Similarly, RNNs are utilized to classify sources of sounds and detecting sound events in the sound clips collected in urban streets.

On providing these algorithms with a chunk of datasets along with annotated labels, the model learns a mapping function from inputs to the outputs labels, and after sufficient training, provides inferences on similar sets of new data. The trained model can then be used as a standalone file to process the datasets for which the training datasets are not required. The classes used for training the models vary according to the specifics of the usage. In this study, Faster RCNN NAS-based

Fig. 1 Outputs from object detection and semantic segmentation tasks

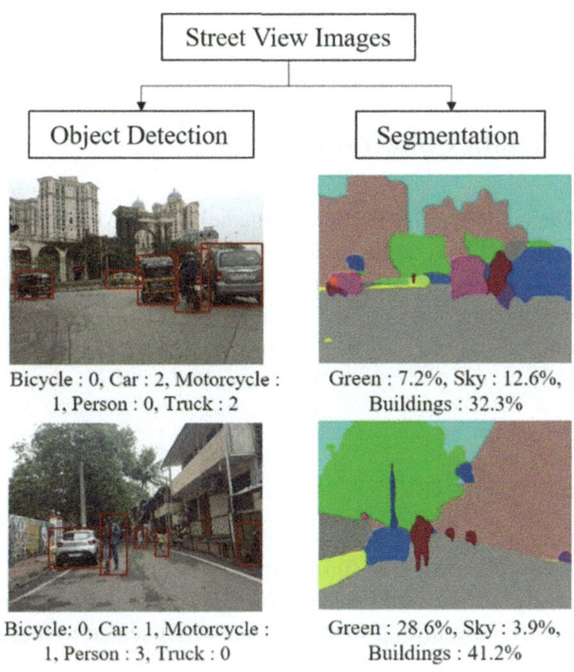

OD model trained on the COCO [14] dataset is used, which provides information on detected transient features of the scene, such as vehicles and pedestrians in the streets. Similarly, Pyramid Scene Parsing Network (PSPNet) [15] model trained on ADE20K [16] is used for SS, which estimates the proportion of different classes in the image.

Such pretrained models are not widely available for sound classification tasks. Most studies in this domain have developed solutions for particular use cases utilizing comparatively smaller datasets; therefore, most of these results are not generalizable for the task discussed in this study. This study follows the AudioSet data-based street-level sound classification [17].

We used three DL-based algorithms to analyze images and audio clips. The OD task counts the number of people and different transportation modes such as cars, trucks, motorcycles, and bicycles. We aggregated the results into two classes: (a) people and (b) vehicles. The detector identifies the exposed human body as a person; hence bike riders are counted towards the person class, while bikes, cars, and trucks are counted as vehicles. The SS task classifies each pixel in the image into classes such as sky, greenery, buildings, roads, vehicles, and persons. However, the categories, such as persons and vehicles, are not considered through this task as they are already accounted for in the OD. As the vehicles and person class pixels occlude the streets in most cases, the road pixels are removed from the SS output. The Sound Classification task takes a 10 s sound clip and provides the probability of the presence of each of the three classes such as (a) Anthropogenic, which consists of sounds of vehicle and engines, (b) Biotic, which includes sounds of bird calls and silence, and (c) Human-based, which comprises of sounds of people conversing, children playing and the crowd.

3 Data Collection

Data collection is done with the help of 16 Megapixel (MP) point-and-shoot camera and H4NPro handheld audio recorder with binaural microphones. With the help of algorithms, the information regarding the presence of audio and visual features is extracted. For this study, such image and audio pairs are collected for a few locations once each hour (0700–1900 h) during the day to capture the temporal characteristics [18].

4 Results

Each collected pair is introduced in the discussed DL algorithms to obtain information on eight audio and visual features. While the data gathered from one pair of the clip is not overwhelming, such information, if scaled to include multiple locations coupled

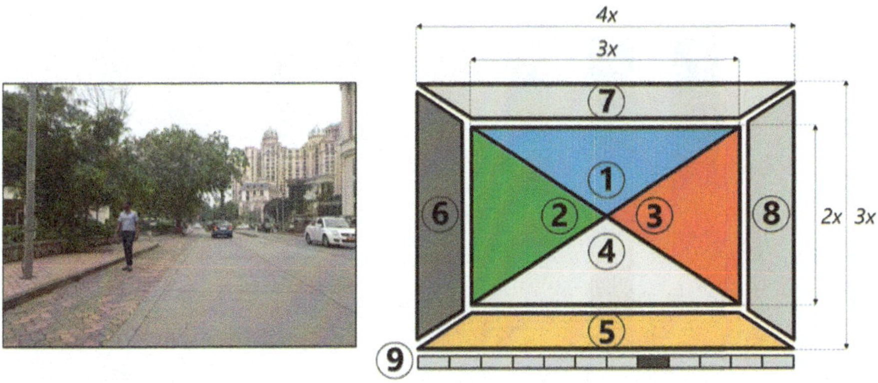

Fig. 2 [Left] Street view image collected at a random location *i*. [Right] Vector graphic for feature representation

with different time intervals, would prove challenging to study and compare via pie charts, bar graphs, and histograms.

The envelope-shaped vector graphic is designed to cater to such constraints that can represent the feature values with the help of colors and color gradients (Fig. 2). The graphic is created using basic shapes such as triangles and rectangles and saved as an SVG file format. Each shape object in the graphic corresponds to the value obtained from the analysis of audio or visual clip. These values are represented by color maps (https://matplotlib.org/3.1.0/tutorials/colors/colormaps.html) given by the Matplotlib python library.

The inner shapes (1, 2, 3, and 4) represent street view image mapped as one-point perspective drawing with the vanishing point on the horizon. Shape 1 corresponds to the percentage of visible sky represented by Blues sequential color map (Table 1). Shape 2 and 3 correspond to the mix of street's built and green character on the left and right side of the road, respectively. The output of the SS task is divided into right and left halves to calculate these values. The color in each of these shapes is determined by the proportion of greenery to that of built. In other words, for each half of the image, the higher value of greenery (>0.5) corresponds to the greenish hue and vice versa, while the color yellow represents the same value (0.5) of both the classes. The number of vehicles is represented by shape 4, while the number of pedestrians, by shape 5. Similarly, the results of audio classification are represented by shapes 6, 7, and 8 with the probability values of anthropogenic, biotic, and human-based sounds, respectively. Shape 9 is subdivided into 12 shapes, indicating the time interval at which the data is collected.

As evident from Table 1, all the values except the count of vehicles and persons are proportions with defined intervals. The counts, on the other hand, are tricky to map with the selected color maps without predefined upper bounds. From the results gathered after analysis, it is found that not more than 20 persons and 15 vehicles

Table 1 Description of features, data types and selected colormaps

S. no	Features	DL tasks	Data type	Range	Name of color maps	Color map/Color
1	Sky	Semantic segmentation	Float (%age)	0.0–1.0	Blues (Sequential)	
2 and 3	Greenery + buildings	Semantic segmentation	Float (%age)	0.0–1.0	RdYIGn (Diverging)	
4	Vehicles	Object detection	Integer (count)	0–15	Greys (Sequential)	
5	Persons	Object detection	Integer (count)	0–20	Wistia (Sequential)	
6	Anthropogenic	Sound classification	Float (%age)	0.0–1.0	Greys (Sequential)	
7	Biotic	Sound classification	Float (%age)	0.0–1.0	Greys (Sequential)	
8	Human-based	Sound classification	Float (%age)	0.0–1.0	Greys (Sequential)	
9	Time interval	NA	Binary	0, 1	Grey (Off), black (On),	

are detected in one street view image captured from the eye view level. Hence, we defined the upper bounds of these classes to 20 and 15, respectively.

Figure 3 shows the 12-interval representation of audio and visual features for a randomly selected location. The percentage of the sky, built, and green character is the fixed characteristics at *location i*, which do not change with time. On the other hand, the gradual increase and decrease of intensities in transient features such as sounds and counts of vehicles and people are seen over the peak hours in the morning and evening. Sounds of birds are audible during morning hours (t_1 and t_2) and during evening hours (t_7 and t_8) due to less amount of vehicle sounds at that time. A gradual increase in the number of vehicles is evident peaking at t_4 in the morning hours

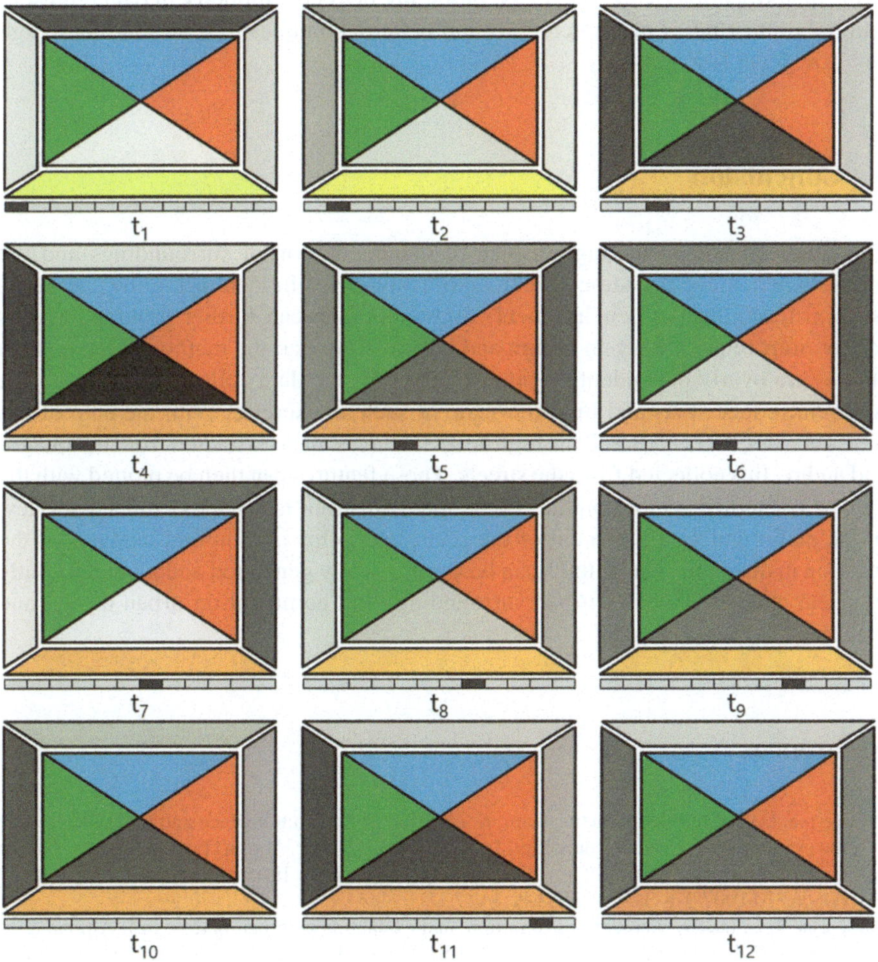

Fig. 3 Temporal representation of audio and visual features collected at random location *i* from t_1 to t_{12}

and t_{10} and t_{11} in the evening hours. Further, the number of pedestrians remains relatively constant during the day, except during morning hours. It can also be seen that the counts of vehicles and pedestrians are correlated with anthropogenic and human-based sounds.

As the created envelope shape is provided as an SVG file, it can be automatically generated and rendered with color with simple programming scripts. While this method offers a new view on data analysis and representation to map urban surroundings as experienced by the bystander, one of the main limitations is important to discuss. This method selectively focuses on presenting HLFs only. Although previous studies have extensively worked on understanding the relationship between the presence of LLFs such as colors, edges, noise levels, which are equally crucial in the quantification of surroundings, this study favored simplicity in data representation. Future studies may attempt to find better techniques to represent both High and Low-level features.

5 Conclusion

This study discussed the significance of documentation of surroundings and the methods to achieve the same. The quantification helps in identifying the characteristics of the immediate environment, studying which can further provide answers to questions related to urban design and health. However, the methods used in such studies are overly dependent on human surveyors for data collection and analysis, which introduces bias and limit the scale of such an exercise. With the help of DL tools such as CNN and RNN, the high-level features can be extracted from the images and audio clips collected from the streets. These features can then be plotted with the help of a designed vector graphic. The graphic illustration provides a quick overview of the audio and visual content of the scene and helps in temporal analysis of the location in question. The outputs can be automatically generated and plotted to study transient characteristics of urban surroundings and comment on urban design and planning.

References

1. Verma, D., Jana, A., & Ramamritham, K. (2019). Artificial intelligence and human senses for the evaluation of urban surroundings. In W. Karwowski & T. Ahram (Eds.),*Intelligent human systems integration* (Vol. 722, pp. 852–857). Cham: Springer International Publishing. https://doi.org/10.1007/978-3-319-73888-8.
2. Nasar, J. L. (1989). Perception, cognition, and evaluation of urban places. In I. Altman & E. H. Zube (Eds.), *Public places and spaces* (pp. 31–56). Boston, MA: Springer US. https://doi.org/10.1007/978-1-4684-5601-1_3.
3. Dubey, A., Naik, N., Parikh, D., Raskar, R., & Hidalgo, C. A. (2016). Deep learning the city: Quantifying urban perception at a global scale. In *Lecture Notes in Computer Science (including*

subseries Lecture Notes in Artificial Intelligence and Lecture Notes in Bioinformatics). LNCS (Vol. 9905, pp. 196–212). https://arxiv.org/abs/1608.01769.
4. Larkin, A., & Hystad, P. (2018). Evaluating street view exposure measures of visible green space for health research. *Journal of Exposure Science and Environmental Epidemiology,* 1–10. https://doi.org/10.1038/s41370-018-0017-1.
5. Li, X., & Ratti, C. (2018). Mapping the spatial distribution of shade provision of street trees in Boston using Google street view panoramas. *Urban Forestry & Urban Greening, 31,* 109–119. https://doi.org/10.1016/j.ufug.2018.02.013.
6. Rossetti, T., & Hurtubia, R. (2019). Validity of immersive videos in stated preference surveys are immersive videos close enough to reality ? An assessment of the ecological validity of immersive videos in stated preference surveys.
7. Kelly, C. M., Wilson, J. S., Baker, E. A., Miller, D. K., & Schootman, M. (2013). Using Google street view to audit the built environment: Inter-rater reliability results. *Annals of Behavioral Medicine, 45*(SUPPL.1), 108–112. https://doi.org/10.1007/s12160-012-9419-9.
8. Henshaw, V. (2013). *Urban smellscapes: Understanding and designing city smell environments.* Routledge.
9. Majid, A. (2015). Cultural factors shape olfactory language. *Trends in Cognitive Sciences, 19*(11), 629–630. https://doi.org/10.1016/j.tics.2015.06.009.
10. McLean, K. (2017). Smellmap: Amsterdam—Olfactory art and smell visualization. *Leonardo, 50*(1), 92–93. https://doi.org/10.1162/LEON_a_01225.
11. Xiao, J., Tait, M., & Kang, J. (2018). A perceptual model of smellscape pleasantness. *Cities* (January), 0–1. https://doi.org/10.1016/j.cities.2018.01.013.
12. Lecun, Y., Bengio, Y., & Hinton, G. (2015). Deep learning. *Nature, 521*(7553), 436–444. https://doi.org/10.1038/nature14539.
13. Goodfellow, I., Bengio, Y., & Courville, A. (2016) *Deep learning.* MIT Press.
14. Lin, T. Y., et al. (2014). Microsoft COCO: Common objects in context. In *Lecture Notes in Computer Science (including Subseries. Lecture Notes Artificial Intelligence. Lecture Notes Bioinformatics).* LNCS (Vol. 8693, no. PART 5, pp. 740–755). https://doi.org/10.1007/978-3-319-10602-1_48.
15. Zhao, H., Shi, J., Qi, X., Wang, X., & Jia, J. (2017). Pyramid scene parsing network. In *Proceedings of 30th IEEE Conference on Computer Vision and Pattern Recognition, CVPR 2017* (Vol. 2017-January, pp. 6230–6239). https://doi.org/10.1109/CVPR.2017.660.
16. Zhou, B., Zhao, H., Puig, X., Fidler, S., Barriuso, A., & Torralba, A. (2016). Semantic understanding of scenes through the ADE20K dataset. https://arxiv.org/abs/1608.05442.
17. Verma, D., Jana, A., Ramamritham, K. (2019). Classification and mapping of sound sources in local urban streets through AudioSet data and Bayesian optimized Neural Networks. *Noise Mapping, 6*(1), 52–71. https://doi.org/10.1515/noise-2019-0005. https://www.degruyter.com/view/journals/noise/6/1/article-p52.xml.
18. Verma, D., Jana, A., & Ramamritham, K. (2019). Machine-based understanding of manually collected visual and auditory datasets for urban perception studies. *Landscape and Urban Planning, 190*(103604), 103604. https://doi.org/10.1016/j.landurbplan.2019.103604.

Desk-Farmer: Cultivation on the Office-Desk

Rishikesh Madan, Sourav Jena, Rathin Biswas, and Kavi Arya

Abstract In this study, we developed a framework to explore how cultivating food plants in an office-workplace affects its occupants and deployed a solution to test it. The solution was a low-cost, low-resource, low-maintenance device called Desk-Farmer, which consists of a small plant holding container, a nutrient solution tank, and an automated, self-contained lighting & ventilation module. It was implemented after a thorough review of multiple variables involved in indoor hydroponic growth, such as the growing medium, nutrient solutions, lighting systems, temperature, humidity, etc., while incorporating insights gained from industrial design to design a neat, ergonomic and convenient solution. In addition to food production, one of the core aims was also boosting psychological parameters such as concentration, comfort, etc. To measure these, we designed and conducted a survey ($n = 36$) to validate various psychological effects indicated in the existing literature. Our survey results indicate how the presence of plants in the workspace brings significant positive effects in certain psychological parameters viz. concentration, happiness, comfort, aesthetics, etc. A perception score was derived from aggregating responses to a structured questionnaire targeting the aforementioned psychological parameters. The presence of plants brings a 15% better perception score than the absence, that is, pre- and post-deployment of the system. This study is beneficial for designers, planners, and occupants of indoor spaces. Further studies can be carried out to experiment with the differences in these parameters with food versus ornamental plants.

Keywords Office · Hydroponics · Cultivation · Indoor · Desk · Farming

---

R. Madan (✉) · S. Jena · R. Biswas · K. Arya
Indian Institute of Technology Bombay, Mumbai 400076, India
e-mail: rishi.madan@iitb.ac.in

S. Jena
e-mail: souravjena@cse.iitb.ac.in

R. Biswas
e-mail: r.biswas@iitb.ac.in

K. Arya
e-mail: kavi@cse.iitb.ac.in

A. Jana and P. Banerji (eds.), *Urban Science and Engineering*, Lecture Notes in Civil Engineering 121, https://doi.org/10.1007/978-981-33-4114-2_19

1 Introduction

1.1 Background

There is a consensus on increasing food production to meet projected future demand, with current methods being altered in some ways to do so [1–3]. Solutions proposed include increasing efficiency of present land under cultivation, managing inputs such as water, soil and fertilizers more effectively to do so, and altering 'demand-side' variables such as consumption patterns, while addressing problems such as food wastage that occur throughout the supply chain. There is also a need to build resilience against supply chain disruptions, particularly in the light of events such as the COVD-19 pandemic raging at the time of this writing. Growing food locally is a solution to current and anticipated future disruptions in food supply [4, 5]. In urban areas, urban farming has the potential of providing a pathway for sustainable communities [6] while increasing food security [7].

On the other hand, various studies demonstrated the achievement of positive psychological and physical effects due to the presence of plants in their workplace [8, 9], and inside built environments in general [10].

In this study, we aimed to combine these two aspects, and measure whether producing plants as food in front of users' eye on their own office-desk, led to similar outcomes or not. This was done with the intention of introducing people to the parameters involved in food production, viz. plant choice, plant nutrition and environmental factors in an accessible and familiar setting.

As proposed by Hekkert et al. [11], the design seeks to merge novelty and typicality in the usage of the platform, retaining typical features and aesthetics of desk-lamps and indoor plants, with the novel aspect of the lighting being controlled automatically & wirelessly along with the potential to monitor plant health. This along with a focus on being easily implementable, low-cost and low-resource, were the main drivers of the design.

1.2 Recent Trends

Soil free farming, urban farming.

1.3 Purpose

The purpose of this study is to explore and develop a low-cost information and communications technology-enabled compact urban farming system that can be placed even at the smallest available space such as workspace, along with a pre and post-deployment analysis of human perception about the developed system.

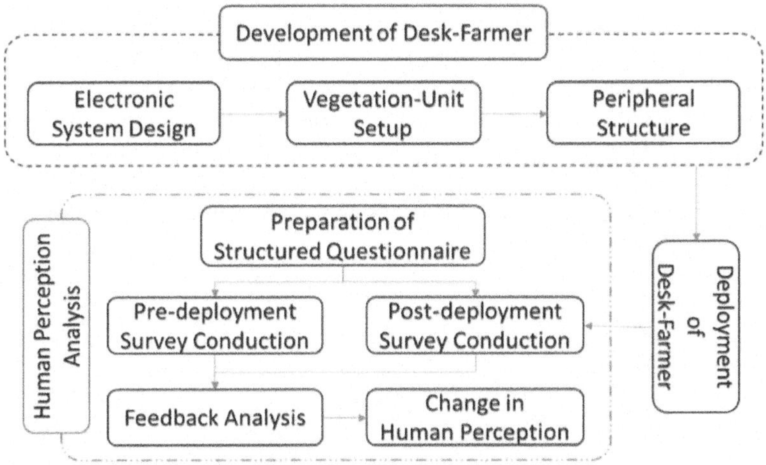

Fig. 1 Methodology

2 Method

The overall methodology is shown in Fig. 1, where there are two major segments in this study- development of a compact urban farming system and post-deployment analysis related to human or users' perception. For the development of compact systems suitable for farming in the smallest possible urban area, such as workspaces, a 2 × 3 feet conventional office desk or table with a desktop-computer (DC) is chosen for our study. Half of the remaining vacant space on the tabletop is considered to place our urban-farming system or Desk-Farmer (i.e., 1 × 2 feet approx.). The system consists of an electronic unit, a vegetation unit, and a peripheral structural unit. The detailed design and implementation are discussed in the subsequent sections. After the development, the system is deployed on the office-desks in an office. A survey was designed to assess human perception. The survey was conducted twice among users. In order to get both the pre and post-deployment feedback related to human perception, a total of 36 responses were collected and analyzed, out of which 10 respondents were common during both the pre and post-deployment phases.

3 Desk-Farmer

To develop the Desk-Farmer, off the shelf components are used like 50 W Grow Light, ESP32 Devkit, BXTS13A—CPU Cooling Fan by Intel, AC to 5 V DC converter, AC to 12 V DC converter and a BT136 triode for alternating current (TRIAC) module. All these components are readily available; that's the reason behind our choice so that anyone who wants to replicate this can easily do it without having to search for uncommon components. The circuit is designed to enable the user to easily switch on and off the grow light via an App. For this setup, the nRF Connect App [12] is used, which is available on the Google Play Store. This App was chosen because of its perceived ease of use and intuitiveness compared to other similar apps. Bluetooth Low Energy (BLE) is used to establish one-way communication between the Desk-Farmer and the App. ESP32 is used to add Bluetooth capabilities to the Desk-Farmer. It turns on or off the grow light and the cooling fan when it receives commands from the App via a BLE link.

3.1 Circuitry

Regular wall outlet which gives AC supply is used to power the Desk-Farmer. Some of the components in the circuit (Fig. 2) require DC supply; therefore, two AC to DC converters are used. The first AC to DC converter is used to supply pure 5 V DC to ESP32 Devkit. The second AC to DC converter is used to supply 12 V DC to the cooling fan.

The 50 W Grow Light that is used in the Desk-Farmer requires AC supply. To switch on and off the AC power supply to the Grow Light and the AC to 12 V DC converter a BT136 TRIAC is used. The base of this TRIAC is connected to pin

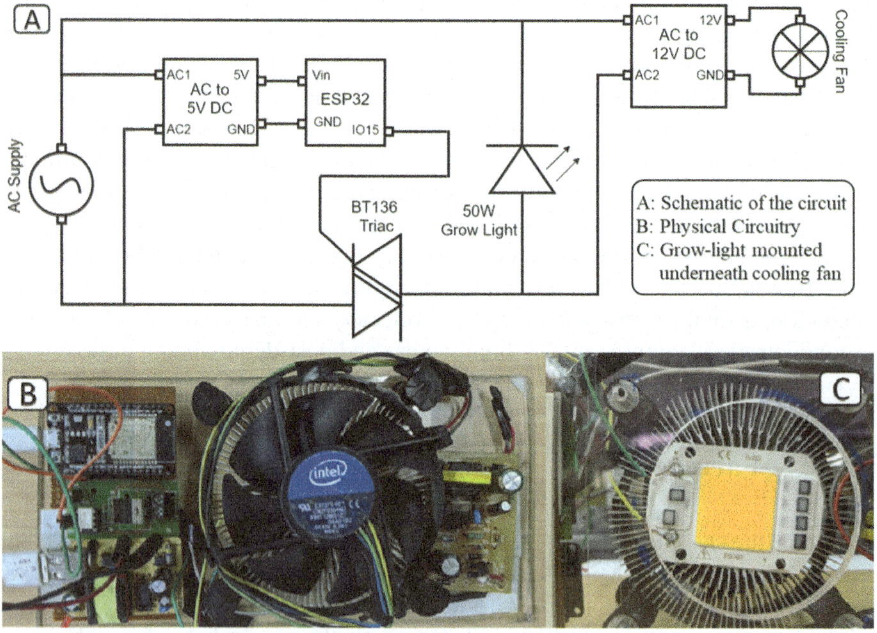

Fig. 2 Circuitry

GPIO15 of ESP32, which is a general-purpose input/output (GPIO) pin. When the ESP32 receives the command to turn on the Grow Light from the App via BLE link, then ESP32 makes GPIO15 high, thus turning on the Grow Light and the Cooling Fan together. When the ESP32 receives the command to turn it off, then it makes GPIO15 go low, which turns off the TRIAC, thus cutting off the power supply to the Grow Light and the AC to DC converter of the Cooling Fan.

3.2 Vegetation Unit and Peripheral Structure

Desk-Farmer can accommodate two vegetation units, as shown in Fig. 3. For a single vegetation unit, a small plant holding container is used to hold the stem of the plant straight to avoid the plant lean or sag while it is growing. This container is placed in a nutrient solution tank made up of a 2 L plastic soda bottle. This tank holds a 1.75 L of 30% fertilizer solution, which is enough for the lifecycle of a leaf cabbage plant. This growing method is derived from the suspended pot non-circulating hydroponic method [13] and does not require electricity or pumps for water circulation, thus reducing cost and maintenance requirements.

The skeleton of the Desk-Farmer is made using PVC pipes. Because PVC pipes are sturdy, lightweight and easily available, they are suitable for this purpose. To hold the lighting unit on top of the plant, zip ties are used to fix it with the PVC pipe structure. To make Desk-Farmer aesthetically pleasing, the PVC pipes are painted in black and a thin cardboard sheet is used to cover the vegetation units. Four faux leather fabric sheets are fixed at the top, in case the user wishes to cover the side(s) of the unit. These sheets can be rolled and unrolled, making them act as blinds for the user.

Fig. 3 Vegetation unit and peripheral structure

4 User-Perception Analysis

4.1 *Survey Design*

In order to measure human or users' perception, a survey questionnaire was designed. It consists of a total of 14 statements or seven attributes represented in two ways, seven with positive sentiments and seven with negative sentiments. These statements are used to assess human or users' perception on various attributes such as- comfortability, attractiveness, performance, concentration, experience and feeling on indoor air quality, impacted by the presence of plants in the workspace. Users were asked to evaluate these statements on a Likert Scale (LS) of "1" to "5" where "1" represents "Strongly Disagree" and "5" represents "Strongly Agree." The survey was conducted twice- once before the system was deployed and once after the deployment, to draw the comparison illustrating the change in human perception.

4.2 *Assessment*

The Likert Scale responses (LS_response) of various Perception-Attributes (PA) are recorded to form an Individual Score (IS) for every individual participant. This scoring is partially inspired by the mechanism of usability scoring stated by Brooke [14]. For individual participants, each attribute's score contribution ranges from 0 to 4, corresponding to the LS response from 1 to 5.

- For attributes with positive statements, the score contribution towards IS is (1 − LS_response).
- For attributes with negative statements, the score contribution towards IS is (5 − LS_response).

After adding all the score contributions for both types of statements, the sum value is normalized and further divided by 56 to get the normalized IS, where 56 is the highest possible score since there are a total of 14 statements and each of them can carry a maximum-score of 4. The calculated and normalized IS further converted into the percentage. The IS from all the participants are averaged out to get the final Composite Score (CS). The whole assessment is carried out twice to get the two CS, one for pre-deployment and another for post-deployment of Desk-Farmer, which helps to draw a clear comparison on the change in participants' or users' perception.

4.3 *Impact*

The average Likert Scale (LS) responses of various Perception-Attributes (PA) are demonstrated through Figs. 4 and 5. For positive statements, the higher LS response denotes the positive change in human perception. For negative statements, a lower

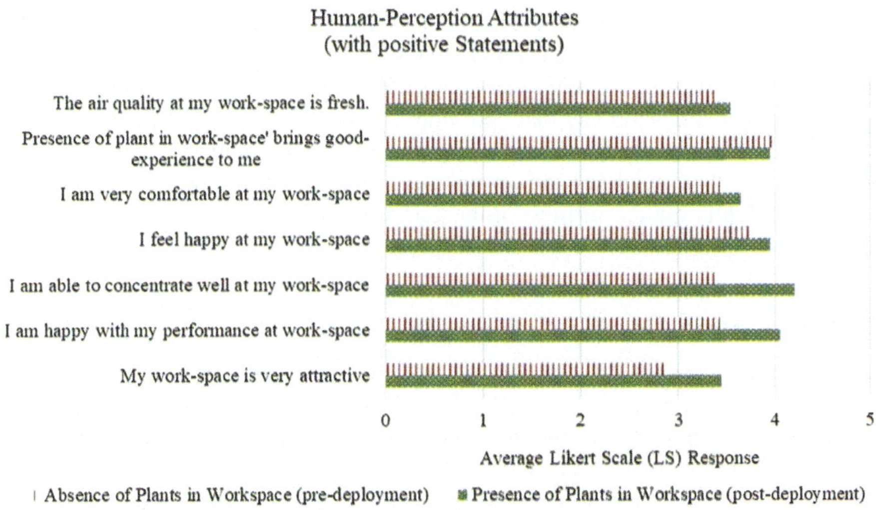

Fig. 4 Human-perception attributes with respect to positive statements

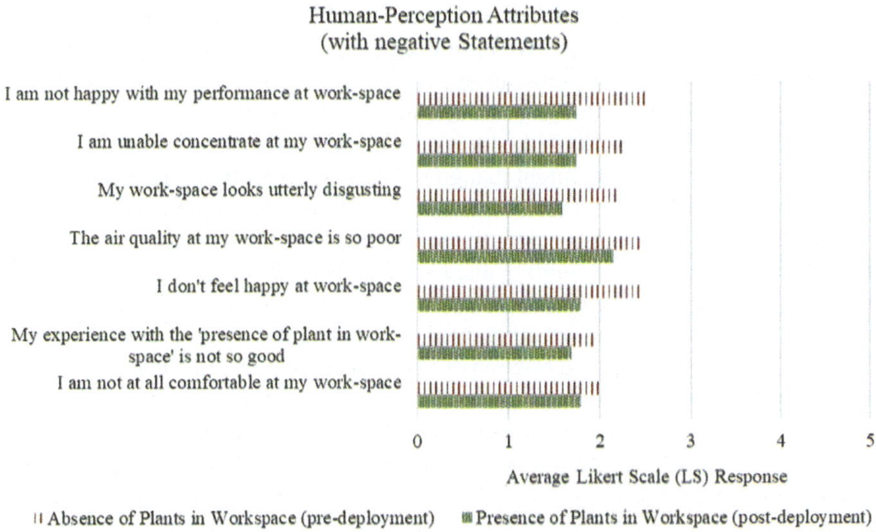

Fig. 5 Human-perception attributes with respect to negative statements

LS response denotes the position change in perception. It is clear from both the figures that "plant in work-space" can bring significant positive changes in human perception related to performance, concentration, aesthetics, comfort, happiness and feelings over air quality. Only the "work experience" is slightly not following its peers in case of human-perception attributes with a negative attitude.

The Composite Score (CS) is calculated by averaging the Individual Score (IS). There are two CS, one is before deploying Desk-Farmer and the other one is after the deployment. The CS for pre-deployment was 65% and was 75% for post-deployment, which clearly justifies the fact that "presence of plants in work-space" significantly advances (~15%) the positivity in human perception. A comparison of Individual Score (IS) distribution among the common-respondents shows the post-deployment scores are slightly higher for most of the participants.

5 Discussion

By 2050 the world will need to feed an additional two or three billion people [15] living in cities. Not only will the demand for food increase, but space available for agriculture in urban areas is expected to reduce. As found from literature, "soil-free vertical farming" [16] can be the way to deal with such pressing issues. Deriving from this literature, we developed a conceptual framework to explore how cultivating food plants in the workplace affects its occupants and deployed a solution to test it. We implemented a low cost, low resource (capital and labor) "Desk-Farmer" approach focusing on the maximum utilization of indoor space (such as office desks) for sustainable food production. We optimized the system to require minimal maintenance. We have measured how our system affects users' perceptions on various psychological parameters such as concentration, comfort, etc., through a survey. This solution aims to bring benefits in the form of cultivated products for offices and other indoor environments with minimal manpower. Desk-Farmer was implemented after a thorough review of multiple variables involved in indoor hydroponic growing, such as the growing medium, nutrient solutions, lighting systems, temperature, humidity, etc., while incorporating insights gained from industrial design to design a neat, ergonomic and convenient solution. It consists of a small plant holding container, a nutrient solution tank and an automated, self-contained lighting and ventilation module. The system occupies a fraction of a typical office-desk. In our study, we used Desk-Farmer to grow leaf cabbage in an office space in the IIT Bombay campus. Leaf cabbage was chosen for multiple reasons: (I) The amount of water consumed by a leaf cabbage plant in its lifecycle can be accommodated in the aforementioned nutrient solution tank, which is present below the plant holding container, thus circumventing the need for regular watering, leading to a set-it-and-forget-it system. (II) Due to its large green coverage, in order to experiment with the psychological effects of plants on the workspace occupants. We further designed and conducted a survey to validate various psychological effects. In conclusion, we designed and deployed Desk-Farmer, a low-cost, modular, low-maintenance system for growing food plants by efficiently using existing office-desk space. Our survey results indicate how the presence of plants in the workspace brings significant positive effects in certain psychological parameters viz. concentration, happiness, comfort, aesthetics, etc. A perception score was derived from aggregating responses to a structured questionnaire targeting the aforementioned psychological parameters. The presence of plants brings a 15%

better perception score than the absence, that is, pre- and post-deployment of the system.

6 Conclusion and Future Scope

In this study, "Desk-Farmer" is introduced, designed and deployed, which is a low-cost, modular, low-maintenance system for growing food plants by efficiently using existing office-desk space. The core aims of developing Desk-Farmer were food production and boosting psychological parameters such as concentration, comfort, etc. In order to measure the latter, a questionnaire was designed, and a survey was conducted among a focused-group of users two times, before and after the implementation of Desk-Farmer on users' table/desk inside the office. The outcome of the survey reveals "plant in workspace" can bring significant positive changes in human perception related to performance, concentration, aesthetics, comfort, happiness and feelings over indoor air quality.

For the next version of the Desk-Farmer, we wish to try hydroponic and aeroponic farming techniques. We want to compare the output of these techniques with the current technique which we are using. We also want to compare the feasibility and ease of setting up these versions of Desk-Farmer. We further wish to experiment with the difference in environmental and psychological parameters with food-growing and ornamental plants.

Acknowledgements We would like to thank all the participants who voluntarily participated in our survey and provided their valuable insights.

Funding Source The study presented here is a part of e-Yantra Project, funded by "Ministry of Human Resource Development" (MHRD), Govt. of India, under the National Mission on Education through Information and Communication Technology (NMEICT).

References

1. Kendall, H. W., & Pimentel, D. (1994). Constraints on the expansion of the global food supply. *Ambio*, 198–205.
2. Smith, P., & Gregory, P. J. (2013). Climate change and sustainable food production. *The Proceedings of the Nutrition Society, 72*(1), 21–28.
3. Davis, K. F., Gephart, J. A., Emery, K. A., Leach, A. M., Galloway, J. N., & D'Odorico, P. (2016). Meeting future food demand with current agricultural resources. *Global Environmental Change, 39,* 125–132.
4. Rothwell, A., Ridoutt, B., Page, G., & Bellotti, W. (2016). Environmental performance of local food: Trade-offs and implications for climate resilience in a developed city. *Journal of Cleaner Production, 114,* 420–430.
5. Baum, S. D., Denkenberger, D. C., Pearce, J. M., Robock, A., & Winkler, R. (2015). Resilience to global food supply catastrophes. *Environment Systems & Decisions, 35*(2), 301–313.
6. Amundsen, R. N. (2013). Urban farming: Victory gardens for sustainable communities. In *2013 Energy and Sustainability Conference* (pp. 1–3).

7. Endres, A. B., & Endres, J. M. (2009). Homeland security planning: What victory gardens and Fidel Castro can teach us in preparing for food crises in the United States. *Food and Drug Law Journal, 64,* 405.
8. Larsen, L., Adams, J., Deal, B., Kweon, B. S., & Tyler, E. (1998). Plants in the workplace: The effects of plant density on productivity, attitudes, and perceptions. *Environment and Behavior, 30*(3), 261–281.
9. Chang, C.-Y., & Chen, P.-K. (2005). Human response to window views and indoor plants in the workplace. *HortScience, 40*(5), 1354–1359.
10. Wolverton, B. C., Johnson, A., & Bounds, K. (1989). Interior landscape plants for indoor air pollution abatement.
11. Hekkert, P., Snelders, D., & Van Wieringen, P. C. W. (2003). 'Most advanced, yet acceptable': Typicality and novelty as joint predictors of aesthetic preference in industrial design. *British Journal of Psychology, 94*(1), 111–124.
12. Nordic Semiconductor. (2019). nRF connect for mobile. Nordic Semiconductor ASA via Google Play Store.
13. Kratky, B. A. (2003). A suspended pot, non-circulating hydroponic method. In *South Pacific Soilless Culture Conference-SPSCC 648* (pp. 83–89).
14. Brooke, J. (1996). SUS—A quick and dirty usability scale. *Usability evaluation in industry* (Vol. 189, no. 194, pp. 4–7).
15. Foley, J. A. (2011). Can we feed the world sustain the planet? *Scientific American, 305*(5), 60–65.
16. Benke, K., & Tomkins, B. (2017). Future food-production systems: Vertical farming and controlled-environment agriculture. *Sustainability: Science, Practice and Policy, 13*(1), 13–26.

Analysing the Impact of Interior Design on Indoor Ventilation in Low-Income Housing of Mumbai

Ahana Sarkar and Ronita Bardhan

Abstract The deteriorated indoor environment not only creates thermal discomfort but also coerce the inhabitants to utilize energy-intensive cooling measures. This calls for a design outline to improve the indoor environment in order to sustain good health. Interior design, despite a subject of individual choice and socio-cultural practices, has a spin-off influence on the indoor environment. The objective of this study is to identify and explore the association between interior design parameters and indoor ventilation perception. The study was initiated with a target-group primary survey of 80 low-income households. Information pertaining to the tenement's room characteristics, fenestration design aspects, occupant behaviour and furniture details were reckoned. Logistic Principal Component Analysis (LPCA) coupled with an ordered probit regression model, were performed to analyse the impact of indoor built-environment on indoor ventilation perception. Results explained that the households with smaller room area, attached to the double-loaded corridor, located at lower floors and with higher household size had a higher proclivity towards experiencing poor ventilation. A strong and statistically significant correlation was also observed between window-related variables, such as number, location, opening schedule with ventilation perception. This study establishes that interior design parameters might strongly impact the indoor ventilation, and hence should be rationally designed for a better future and also should be incorporated in future low-income habitat design guidelines.

Keywords Natural ventilation · Interior design · Low-income housing · Regression analysis · Logistic PCA

A. Sarkar (✉) · R. Bardhan
Centre for Urban Science and Engineering, Indian Institute of Technology Bombay, Powai 400076, India
e-mail: ahana.sarkar@iitb.ac.in

R. Bardhan
Department of Architecture, University of Cambridge, Cambridge, UK

A. Jana and P. Banerji (eds.), *Urban Science and Engineering*, Lecture Notes in Civil Engineering 121, https://doi.org/10.1007/978-981-33-4114-2_20

1 Introduction

With people spending 90% of their time indoors, indoor built environment becomes integrally crucial to human health and well-being [1]. Creating healthy indoor environments should be a priority for building designers, and evidence-based design decisions should be used to ensure that the built-environment delivers healthful benefits to the occupants [2]. The natural environment has a positive impact on psychological well-being. Therefore, connecting natural ventilation with the built-environment may improve occupant well-being.

Built-environment characteristics significantly impact the indoor environment and human health by controlling indoor air quality and thermal comfort. Building characteristics like building age, floor level had a strong and positively significant correlation with the incidence of sick building syndromes (SBS) in residential apartments of Hong Kong [3]. Enhanced ventilation can effectively improve occupants' health and also reduces health expenditure by 18% [4]. Built-environment design at neighbourhood [5] and building envelope level [6] considerably control indoor ventilation characteristics.

But it is yet to investigate whether interior layout affects the indoor ventilation perception of the occupants. Studies have examined the impact of interior material finishes and surface material like wall and floor materials on indoor air quality and human health. Indoor surface materials act as sorbents or sources of Volatile Organic Compounds (VOCs) emission, advertently affecting human health. Partition wall and furniture layout have been observed in a few pieces of research to affect the indoor electro-mechanical ventilation characteristics within offices. A recent study identified building typology, dwelling area, layout of the unit, window type, window operating schedule and ventilation system as major built-environment parameters that affect indoor air quality in low-income housing [7].

To date, studies dealing with the association between built-environment and ventilation have focussed either on macro-scale involving neighbourhood and envelope level built-environment design parameters or at micro level parameters like surface materials. Furthermore, research of the field has investigated single parametric based analysis. Relatively little is known about the aggregated impact of different interior level built-environment design parameters on indoor ventilation. The specific interior design parameters which need to be explored while designing healthy habitats still remain a blind-spot in built-environment and health literature.

In this study, we aim to investigate the impacts of different interior level built-environment design parameters on indoor ventilation perception at natural ventilation condition in slum rehabilitation low-income housing of Mumbai with the following objectives:

1. To identify the specific interior design parameters that impact indoor ventilation perception in natural ventilation conditions.
2. To investigate the strength of the impact of interior design parameters on indoor natural ventilation perception and incidence of occupant health risk.

2 Data and Methods

The proposed mixed-mode methodological framework is illustrated in Fig. 1. A primary survey of 121 households was undertaken in March–April 2019 with the intention of identifying the interior-level built-environment determinants of indoor natural ventilation performance evaluation criteria. A survey questionnaire was utilised here to quantify the interior built-environment determinants and their relationship with ventilation perception issues in the slum rehabilitated colonies of Mumbai. A self-administered questionnaire was developed based on the parameters obtained from literature as well as reconnaissance field visits.

Data were collected from specific slum rehabilitated sites of (1) Lallubhai compound, (2) Ramabai colony, and (3) Kanjurmarg colony in Mumbai. The questionnaire, divided into three parts, initiated with the collection of respondents' general information like age, household size, etc. However, personal information representing the respondent's education level and employment status were not enquired. While the second section involved an examination of four variables pertaining to 'Room Characteristics' of a tenement unit, including (i) Room Area, (ii) Corridor Type, (iii) Floor level and (iv) Room Density (Table 1). The respondents were asked to deliver information about the tenement unit regarding its total area, the floor number of the dwelling and whether the tenement was abutting to a single-loaded or double-loaded corridor. Another parameter crucial to indoor ventilation level is the 'Opening Characteristics'. The variables under this category included information regarding the fenestrations in the tenement like number and type of windows, their locations, presence of air-outlet (ventilator). These variables were selected, keeping in mind the evidence-based effectiveness of cross-ventilation strategy and better indoor airflow

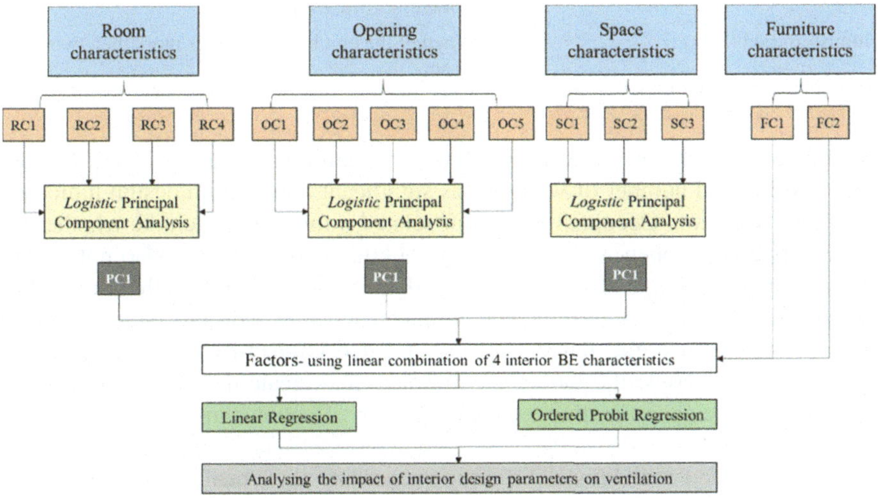

Fig. 1 Methodology adopted

Table 1 Indoor built-environment variables

Interior built-environment characteristics	Description (1 if yes, 0 if no)
Room area (RC1)	Whether the area of the tenement unit is less than 250 sq.ft or not
Corridor type (RC2)	Whether the internal corridor is double-loaded or not
Floor level (RC3)	Whether the floor level is between the 1st and 3rd floor or not
Room density (RC4)	Whether the room density is more than 0.0160, that is, number of household members staying in 225 sq.ft area is more than 4 or not
Number of windows (OC1)	Whether the total number of windows in living space is more than or equal to two or not
Type of window (OC2)	Whether the window is fully-openable or not
Position of window (OC3)	Whether the windows are placed on the opposite side or not. (Assumption: opposite placed windows enable cross-ventilation)
Presence of ventilator (OC4)	Presence of high-level air outlet (ventilator) on the opposite side of the window or not. (Assumption: Air-outlet enhances cross-ventilation)
Window opening schedule (OC5)	Whether occupants open window during cooking or not
Kitchen type (SC1)	Whether the kitchen is segregated or not
Bedroom type (SC2)	Whether the unit has a separate bedroom or not
Storage system (SC3)	Whether the tenement unit has a bunk or storage or vertically segregated sleeping space or not
Number of wardrobes (FC1)	Whether the unit has equal to or more than two storage wardrobes or not
Number of bed (FC1)	Whether the unit has equal to or more than two beds or not

characteristics in the presence of windows. Literature has majorly identified size, type, position and number of windows as significant fenestration design indicators affecting ventilation. Occupant behaviour related information suchlike whether the inhabitants used to open windows during cooking activities, were also noted. The questions in the subsequent sections included information about whether the kitchen was segregated and whether the tenement had a separate bedroom. Studies in low-income housing have elucidated that unsegregated kitchen-living zone within the tenement units degrade the indoor environment and create thermal discomfort by increasing high temperature trapped zones, thereby affecting human health condition [1, 2].

It also recorded individual information regarding their health status, in particular, the respiratory diseases related symptoms. In health-related studies, symptoms included discomfort or problems regarding breathlessness, frequent coughing and other allergies. First, the respondents were required to answer whether any of their household members were affected with any health risks in 'yes' or 'no'. Next, they were asked specific questions regarding diseases like tuberculosis, upper respiratory diseases, asthma, coughing problem, skin allergies and others. Respondent's likelihood of indoor ventilation and perception of overall indoor air quality was also noted.

2.1 Data Description

Out of 121 surveyed data, a final set of 80 responses were screened, with Lallubhai Compound (29), Kanjurmarg SRA (21), and Ramabai SRA colony (30). The exploratory analysis elucidated that only 10% of the respondents were satisfied with indoor ventilation levels, while 25% of the households reported health risks occurrences. Of the room specific characteristics, 81.3% of the studied houses were connected by a double-loaded corridor. Around 90% of the surveyed households were located at lower floor levels (i.e. from floor level 1 to 3), while the buildings were six to eight floors high (see Table 2). 76.3% of households had high room density, indicating that maximum households had more than four members residing in a single multi-purpose tenement unit. The tenement units' size in the surveyed buildings varied from 180 to 270 sq.ft; with 93.8% of the households having room area less than 250 sq.ft. Figure 2 illustrates the double-loaded corridors and single-

Table 2 Data description showing the percentage of surveyed households with built-environment details

Interior built-environment characteristics	Percentage of households ($N = 80$) (%)
Room area (RC1)	93.8
Corridor type (RC2)	81.3
Floor level (RC3)	90
Room density (RC4)	76.3
Number of windows (OC1)	65
Type of window (OC2)	5
Position of window (OC3)	22.5
Presence of ventilator (OC4)	11.3
Window opening schedule (OC5)	86.3
Kitchen type (SC1)	86.3
Bedroom type (SC2)	42.5
Storage system (SC3)	7.5
Number of wardrobes (FC1)	93.8
Number of beds (FC2)	90

Fig. 2 Double loaded corridor, Windows closed, Partition wall acting as space separators, Bunks used for storage and sleeping areas in different households

multi-purpose tenement units in typical low-income slum rehabilitation housing of Mumbai.

When 'opening characteristics' concerning variables were explored, it was reckoned that although 65% of the units had equal to or more than two windows, only 22.5% of the households had possibilities of cross-ventilation. This can be attributed to the positioning of both the windows on the single side due to design constraints, thus inhibiting cross-ventilation. Furthermore, only 5% of the households had fully openable windows. Figure 2 also illustrates that most of the occupants in SRA units tend to close their windows during daytime, especially during cooking times, to maintain privacy, safety and security. While other occupants obstruct the windows by utilising window space as storage areas. While 86.3% of the households had segregated kitchen space, 13.8% of households were exposed to open cooking-living area. It was reckoned from the field visits that the majority of the households used taller items of furniture such as wardrobes or curtains as a space-separator between cooking and living zones. However, despite being smaller in size (one multi-purpose unit with an average area 225 sq.ft), nearly 90% of the households were found equipped with furniture such as wardrobes and beds. Due to larger household size, around 25% of households had equal to or more than two beds and wardrobes, indicating extreme space-constraint. Figure 2 also shows the occupant behaviour of vertical and horizontal space segregation using varying types of bunks and partitions.

2.2 *Statistical Analysis*

Despite the rehabilitation scheme deliver similar size and layout designs, the indoor built environment characteristics turn heterogeneous because interior design, space arrangement and furniture layout is a predilection of individual choice, occupant behaviour and socio-cultural context. The coupled method of binary dimensionality reduction, Logistic Principal Component Analysis (LPCA) and ordered probit regression was utilised here to analyse the impact of interior design parameters on ventilation perception (see Fig. 1).

Of the various aspects studied, 12 criteria were selected for the LPCA. The variables were initially clustered under homogenous groups of 'room', 'opening', 'space' and 'furniture' characteristics. Logistic Principal Component Analysis (LPCA), an extension of Pearson's PCA is a dimension reduction technique used to construct a unidimensional measure out of many variables to capture variability in the indoor built-environment characteristics, especially when the captured variables are binary. It is motivated as an extension of ordinary PCA employing a matrix factorisation, akin to the singular value decomposition, that maximises the Bernoulli log-likelihood and minimises Binomial deviance [8]. 'LogisticPCA' package developed by Andrew J. Landgraf in 2016 in R software was employed here to perform the LPCA model analysis. Three individual logistic PCAs were executed for the three clusters of 'Room Characteristics', 'Opening Characteristics', and 'Space Characteristics' as shown in Fig. 1.

Nevertheless, the two specific design variables under 'Furniture characteristics' were individually accounted for here. The output from the logistic PCAs delivered a component score coefficient matrix, which was used to generate the three final components. Next, the first principal component (PC1) component factors were utilised through an ordered probit regression analysis to understand the aggregated impact of interior built-environment characteristics on indoor ventilation perception levels.

3 Results and Discussion

The first linear regression model, refer to Table 3, was undertaken to understand the impact of all built-environment characteristics accounted individually on the indoor ventilation perception of the occupants. The R-square value of the model is 0.325. No variables were removed from the model based on their low significance level since all variables are of interest and are expected to have effects on ventilation. As Table 3 denotes, the space-constrained single multi-purpose households with a higher number of tall and bulky items of furniture like wardrobe had a lesser likelihood towards improved ventilation perception. Due to the extreme space constraints, occupants often tend to locate the furniture near the windows, advertently hindering the incoming air path by acting as an obstruction.

Model 1

The linear regression results also showcased that window-type affected ventilation effectiveness. The households with fully openable windows had exhibited improved indoor ventilation rates. A fully openable window delivers a higher open area for air exchange, which increases the ventilation levels than that of the sliding windows, where 50% of the window area remains blocked. Model 1 also estimated that the households with high room density, that is larger household size and lesser room area, were found dissatisfied with indoor ventilation levels. Space constraint in the tenement units of slum rehabilitation colonies in addition to larger household size leads to

Table 3 Linear regression model of ventilation perception and individual parameters

R-square: 0.325	Beta	t-stat
Dependent variable: Ventilation perception levels (1 = very satisfied, 2 = satisfied, 3 = neutral, 4- dissatisfied, 5 = very dissatisfied)		
Intercept	2.80	0.17
Double loaded corridor	0.18	0.58
Low floor level	−1.39**	2.11
High room density	−0.76**	−2.13
Low room area	−1.87**	−2.21
Bed more than equal to 1	0.28	0.63
Wardrobe more than or equal to 1	1.03*	1.89
Windows more than or equal to 2	0.14	0.50
Windows with cross ventilation	−0.23	−0.60
Fully openable windows	0.27**	2.51
Open windows during cooking time	0.28	0.72
Ventilator present	0.11	0.25
Presence of bunk for sleeping or storage	2.11**	2.50
Segregated Kitchen	0.38	0.98
Segregated Bedroom	0.22	0.69

Note * significant at 90%, ** significant at 95%, *** significant at 99% CI

overcrowding, which in turn deteriorates air freshness and ventilation effectiveness. The survey identified that in 76.3% of the households, more than four members reside in a single multi-purpose tenement unit of less than 250 sq.ft area. Furthermore, the households staying at lower floor levels were reckoned to be satisfied with indoor ventilation rates.

In Model 1, a few built-environment related parameters turned non-significant. Hence, the binary variable dimensionality reduction technique was next applied to explore the association between the clustered built-environment characteristics and ventilation perception. When the parameter of 'Room characteristics' was taken into consideration, most of the correlations were found significant at alpha < 0.05. Among the four 'Room Characteristics' related design variables considered, a high correlation was found between the variables like Floor Level (RC3) and Room Area (RC1), and medium correlation was observed between Room Density (RC4) and Room Area (RC1). Similarly, while accounting 'Opening Characteristics', four out of five parameters were found significant at alpha < 0.5. A medium correlation was observed between the variables of the Presence of Ventilator (OC4) and Number of Windows (OC1) and Position of Windows (OC2). For 'Space Characteristics', a medium correlation was observed between Storage System (SC3), Kitchen Type (SC1) and Bedroom Type (SC2). No two variables in any of the clustering were strongly correlated, and therefore, all variables were considered for PCA and regression process. The results of three logistic PCAs showed KMO values above 0.5.

In the LPCA model, a matrix (x) was initially formed with all binary variables. The k value, referring to the number of principal components to return from the model was considered equal to 1. The value (m) to approximate the saturated model was taken 4 based on literature [8]. The first logistic PCA for 'Room Characteristics' resulted in the formation of one single principal component which was above the Eigenvalue of 1. The data used to derive the principal component (PC) from composite interior built-environment parameters specifically indicates the influence on ventilation performance of the tenement unit. The first PC is the linear combination of exploratory variables that explains the maximum variation in the data. From logistic PCA results, it was noted that the first principal component of 'Room characteristics' explains the maximum variance in the data (32.4%). Similarly, the first component for the logistic PCAs of 'Opening', and 'Space' characteristics had 26.9%, and 41.5% variance.

In order to relax the restrictive functional assumption, this study re-estimated the data using an ordered probit model with a multinomial distribution function which is demonstrated below in Eq. 1. The ordered probit can be estimated via several commercially available software packages and is theoretically superior to other models for the ordinal data analysed in this work. The following specification was used here:

$$T_n^* = \beta Z_n + \varepsilon_n(1) \tag{1}$$

where T^*_n is the latent and continuous measure of ventilation satisfaction reported by the households n, Z_n is a vector of explanatory variables describing the built-environment design variables, β is a vector of parameters to be estimated, and ε_n is a random error term (assumed to follow a standard normal distribution).

To analyse the effects of the ordinal responses, the observed and coded satisfaction variable, T_n, is determined through the 'threshold concept' from the model as follows [9]:

Tn = {1 if $-\infty \le T^*_n \le \mu_1$(very dissatisfied),
2 if $\mu_1 \le T^*_n \le \mu_2$(dissatisfied),
3 if $\mu_2 \le T^*_n \le \mu_3$(neutral),
4 if $\mu_3 \le T^*_n \le \mu_4$(satisfied),
5 if $\mu_4 \le T^*_n \le \infty$ (very satisfied),

where the μ_i represent thresholds to be estimated (along with parameter vector β).

Model 2 in Table 4 showed a high significance level of the ventilation perception *very dissatisfied, dissatisfied and neutral* with respect to *very satisfied* ventilation perception, thereby indicating the importance of employing *ordered* probit regression. Due to the increasing nature of the ordered classes, the interpretation of the model's primary parameter was set as follows: positive sign indicates higher satisfaction level as the associated variables increase, while negative signs suggested the converse [9].

Table 4 Ordered Probit Regression demonstrating the relationship between the LPCA derived constructed design variables and the perception of indoor ventilation

Dependent variable: ventilation perception levels (1 = very dissatisfied, 2 = dissatisfied, 3 = neutral, 4 = satisfied, 5 = very satisfied)	Beta	t-stat
Thresholds		
μ_1	−3.09***	−4.60
μ_2	−1.96***	−3.06
μ_3	−1.43***	−2.27
μ_4	−0.20	−0.32
Space characteristics	0.43	1.28
Opening characteristics	−0.67*	−1.81
Room characteristics	−0.84*	−1.87
Furniture characteristics		
Bed more than or equal to 2	−0.15	−0.33
Wardrobe more than or equal to 2	−0.99*	−1.87
−2 Log-likelihood	221.63	

Note * significant at 90%, ** significant at 95%, *** significant at 99% CI
Chi-square: 294.960, df: 303, Sig: 0.627, R-square: 0.055.

Model 2

Model 2 estimated that the clustered built-environment design parameter of 'Room characteristics' and the number of furniture in the tenement unit exhibited a statistically significant and negative relationship with the ventilation perception. This indicates that the households with lower room area, higher room density, located at lower floors and attached to double-loaded corridor had a higher probability of being dissatisfied with indoor ventilation with respect to the very satisfied households. Also, the households with a higher number of wardrobes were observed to be less satisfied with the indoor ventilation than those satisfied households. The constructed parameter of 'Opening characteristics' had a negative and statistically significant relation with indoor ventilation perception, indicating that the households with fully openable cross-ventilated windows, ventilators or air-outlets as well as those who used to keep windows open during cooking were surprisingly dissatisfied with ventilation rates. Despite this result is counter-intuitive, one possible reason behind this phenomenon can be attributed to the floor-level of the surveyed tenements. Most of the SRA colonies are located in close proximity to the garbage dump-yards, incinerators, chemical and petroleum factories, etc. Moreover, the recent SRA housing designs are characterised by tall and bulky buildings adjacent to 3 m wide alleys. These unutilised alleys, ultimately turn into public refuse and garbage dumps due to lack of adequate maintenance. Hence, the degraded environmental condition in the tenements, especially in the lower floors, are forced to keep their windows closed due to extreme outdoor garbage-induced foul smell. Since 90% of the surveyed respondents dwell on lower floors, they tend to keep their windows shut in order to avoid extreme adverse outdoor conditions and perceive better ventilation by not opening windows.

This result shows that outdoor maintenance should be given increased priority, as degraded outdoor atmosphere and setting results in higher dissatisfaction from ventilation, even if a cross-ventilated higher number of windows, outlets are present in the tenements. Therefore, even with the same window-to-wall ratio (WWR), the tenements at lower floors are forced to under-utilise the ventilation benefits of openings due to degraded outdoor conditions (foul smell, the breeding ground of mosquitoes and flies).

Lastly, a check was made of the robustness of the statistical results concerning the health hazards and risk occurrences. In the questionnaire, respondents were asked to state whether they had been suffering from any health hazards, particularly respiratory problems. The responses showcased that 25% suffered from respiratory problems. It was observed that 52 out of 69 respondents who had segregated kitchen did not suffer from health hazards. 67.65% of the tenement units with separate bedrooms and cooking zones did not suffer from any health hazards.

4 Conclusion

This section attempted to identify an association between the interior-level built-environment design parameters and indoor ventilation perception levels. The results showcased that built environment parameters like crowding factor, floor-level of a tenement; the unit area affected the indoor ventilation perception in the slum rehabilitated colonies. Due to extreme space constraints, adequate items of furniture within a single multi-purpose room hindered the indoor airflow characteristics and deteriorated the ventilation rates. Hence a negative correlation was observed between a higher number of bulky items of furniture and indoor ventilation perception (see *Model 1 and 2*). Other crucial design parameters included the number and design typology of windows. Future extension of this study would cover the analysis of increased sample size and built-environment variables to strengthen the association. These results shed light on the fact that indoor air quality and ventilation efficiency is a subject of interior built-environment design. This study would aid in the revision of forthcoming sustainable and liveable low-income habitat design guidelines.

References

1. Sarkar, A., & Bardhan, R. (2019). Optimal interior design for naturally ventilated low-income housing: A design-route for environmental quality and cooling energy saving. *Advances in Building Energy Research, 1*, 33. https://doi.org/10.1080/17512549.2019.1626764.
2. Sarkar, A., & Bardhan, R. (2020). Improved indoor environment through optimised ventilator and furniture positioning: A case of slum rehabilitation. *Frontiers of Architectural Research*. https://doi.org/10.1016/j.foar.2019.12.001.

3. Wong, S. K., Wai-Chung Lai, L., Ho, D. C. W., Chau, K. W., Lo-Kuen Lam, C., & Hung-Fai Ng, C. (2009). Sick building syndrome and perceived indoor environmental quality: A survey of apartment buildings in Hong Kong. *Habitat International, 33*(4), 463–471. https://doi.org/10.1016/j.habitatint.2009.03.001.
4. Dutton, S. M., Banks, D., Brunswick, S. L., & Fisk, W. J. (2013). Health and economic implications of natural ventilation in California offices. *Building and Environment, 67,* 34–45. https://doi.org/10.1016/j.buildenv.2013.05.002.
5. Ramponi, R., Blocken, B., de Coo, L. B., & Janssen, W. D. (2015). CFD simulation of outdoor ventilation of generic urban configurations with different urban densities and equal and unequal street widths. *Building and Environment, 92,* 152–166.
6. Elshafei, G., Negm, A., Bady, M., Suzuki, M., & Ibrahim, M. G. (2017). Numerical and experimental investigations of the impacts of window parameters on indoor natural ventilation in a residential building. *Energy and Buildings, 141*, 321–332 (2017). https://doi.org/10.1016/j.enbuild.2017.02.055
7. Lueker, J., Bardhan, R., Sarkar, A., & Norford, L. (2020). Indoor air quality among Mumbai's resettled populations: Comparing Dharavi slum to nearby rehabilitation sites. *Building and Environment, 167*. https://doi.org/10.1016/j.buildenv.2019.106419.
8. Landgraf, A., & Lee, Y. (2015). Dimensionality Reduction for Binary Data through the Projection of Natural Parameters. *Technical Report No. 890,* Ohio State University.
9. Jana, A., & Harata, N. (2017). A framework to analyze variation of the satisfaction of patients for outpatient needs: A case of West Bengal, India. *Public Health and Welfare: Concepts, Methodologies, Tools, and Applications. IGI Global.* https://doi.org/10.4018/978-1-5225-1674-3.ch011.

Optimal Routing of Solid Waste Collection and Disposal in Nagpur City Using GIS

Smita

Abstract In the Indian context, Municipal Solid Waste Management (MSWM) has become a critical subject of concern because of lopsided urbanization and wavering consumption pattern and the serious health risk it poses due to the degrading environment. Although municipal authorities in India endorse the eminence of adequate solid waste collation and disposal as well as resource recovery and recycling, it is a pressing challenge to deal effectively with the growing amount of solid waste generated by the expanding cities. The present study analyzes MSW collection system and provides optimal routing for solid waste collection and disposal in Nagpur City. Based on the preliminary research, short/long term strategies were recommended with the aim of improving MSW collection system. Web-based applications such as Open Data Kit (ODK) and ARC GIS were employed for mapping waste generation and the collection as well as the optimization of routes for transporting wastes and compost plant setting. Maps generated using GIS builds the relationship between various factors responsible for the generation of solid waste and its management. This study could significantly enhance the planning, strategies and decision-making of municipalities to optimise their transport distance, time and cost for MSW.

Keywords Municipal solid waste Management (MSWM) · Nagpur · ARC GIS

1 Introduction

India, with a rapid increase in urbanization, industrialization, and population, has put a substantial challenge to its MSWM infrastructure and policies [1, 2]. Ministry of Housing and Urban Affairs, in its annual report for the year 2016–17, had estimated a total generation of solid waste of approximately 1,50,000 T/day. Although approximately 90% (1,35,000 MT/day) is collected but due to infrastructural, financial and institutional debilities, only 20% (27,000 MT/day) is processed and the remaining 80% (10,8000MT/day) goes to dump sites [3]. Mismanagement of MSW not only

Smita (✉)
Department of Architecture and Planning, Indian Institute of Technology Roorkee, Roorkee, Uttarakhand, India
e-mail: smita@ar.iitr.ac.in

A. Jana and P. Banerji (eds.), *Urban Science and Engineering*, Lecture Notes in Civil Engineering 121, https://doi.org/10.1007/978-981-33-4114-2_21

poses a serious health risk with a degrading environment but also can greatly impact the socio-economic wellbeing of the people. Discrete advents related to MSWM are important to study for the waste generation (domestic household wastes) in Indian city so that burden on depot site can be restricted. The right technological tool to support decision making in planning/policies and reliable collection of data can provide substantial help in MSWM. Web-based applications such as Open Data Kit (ODK) and ARC GIS [4–6] can help in multi-criteria evaluation of geographical sites for optimization of several competing variables for efficient MSWM [7].

Objective and Novelty: The objective of the present study is to analyze MSW collection system and provide optimal routing for solid waste collection and disposal in Nagpur City. Here Open Data Kit (ODK) and ARC GIS are employed for data collection and optimization of transport routes, including identifying the potential site for compost plant setting. Additionally, we also identify pertaining challenges to solid waste collection and disposal system and recommendations to address the same. Primary and secondary data is collected by employing a survey research method [8] in the study area to cover each ward of Nagpur, which is divided into ten zones. Data collected was analyzed and maps were developed according to the need in ARC GIS. The present study could significantly help in improving planning, strategies and decision making of municipalities to optimise their solid waste collection and disposal system, which in turn will optimise cost, time and infrastructural needs [9].

2 Materials and Methods

Overall, extensive literature as well field study was conducted on SWM in Nagpur city. Information regarding prevailing policies, administration and general information on solid waste management was collected from Nagpur Municipal Corporation (NMC) and various past studies. Information about waste quantities and characteristics were collected from the NMC. Field studies were conducted solely by the author to understand the dynamics of solid waste management of the city. Web-based application Open Data Kit (ODK) and ARC GIS were employed for carrying out various tasks such as data collection, generating maps, analysing mapped data, etc., throughout this study.

2.1 Study Area

Nagpur is the third-largest city in the state of Maharashtra, produces approximately 1100–1200 tons of waste per day. Despite being a progressive urban local body, with a population of 2.4 million (census 2011), the city faces several challenges related to waste management. According to the Swachh Sarvekshan Survey 2017 by Govt. of India, Nagpur was ranked 137 out of a total of 434 cities surveyed [10].

The study is limited to household wastes only. The survey research method is employed to collect data at the household level. For this, a web-based application; 'Open Data Kit' was used in the entire study to collect data. Nagpur city is divided into 10 zones for administrative purposes and is further sub-divided into 136 wards. Samples were collected from all 136 wards. A ward-wise list of households was obtained from the NMC and a simple random sampling technique was employed for the selection of 4700 households from all the wards. This sample size is sufficient and more than 10 times larger than the minimum sample size (margin of error 5%, confidence interval 95%) obtained from the Krejcie and Morgan [11] formula considering the total population of 24 lakhs for Nagpur.

3 Generation and Characteristics of Urban Solid Waste

Waste generation and its estimation is the key component of waste management. In 2017, a study related to solid waste management and its problem in detail was conducted by the National Environmental Engineering Research Institute (NEERI), Nagpur in order to provide NMC with an economically feasible solution for the implementation of proper waste management. The characterization of municipal solid waste was one of the main components of this study. Table 1 presents the details of waste characterization of 34 samples which were collected from all the 10 zones of the city.

3.1 Physical Characterization of Waste

The results for the zone-wise waste characterization (primarily residential waste from households) can be seen in Table 2 that indicates a very high percentage of the organic fraction in the waste (77%), followed by plastics (11.60%), and paper (7.66%). The balance (3.74%) constitutes inert, textile and cardboard.

Table 1 Details of samples collected for waste analysis

SI No	Particular	Number of samples	Zones (Z)
1	Residential area	10	All zones
2	Secondary collection point	14	All zones
3	Commercial area	3	Zone 5 and 7
4	Institutional area	3	Zone 1 and 4
5	Dumping site	4	2 old and 2 new

Table 2 Average Physical Characteristics of Solid Waste of all 10 zones (Z) in Nagpur

S. No	Item	Z-1	Z-2	Z-3	Z-4	Z-5	Z-6	Z-7	Z-8	Z-9	Z-10	Z Avg
1	Paper	8.9	7.99	4.0	ND	8.91	12.5	11.6	ND	12.5	10	7.66
2	Cardboard	8.9	ND	ND	ND	ND	ND	ND	ND	ND	ND	0.9
3	Plastic	9	7.86	11	9.6	11.8	15	21.6	6.5	13	10	11.6
4	Textile	9.0	1.33	ND	ND	ND	ND	ND	ND	ND	ND	1.04
5	Organic	55	79.8	8	90	79	72	66	93	74	80	77.2
6	Inert	8.3	2.93	4.4	ND	ND	ND	ND	ND	ND	ND	1.57
7	Wood	ND	ND	ND	ND	ND	ND	ND	ND	ND	ND	ND
8	Thermocol	ND	ND	ND	ND	ND	ND	ND	ND	ND	ND	ND
9	Metals	ND	ND	ND	ND	ND	ND	ND	ND	ND	ND	ND
10	Glass	ND	ND	ND	ND	ND	ND	ND	ND	ND	ND	ND

ND- Non-Detectable

3.2 *Existing Solid Waste Management in Nagpur*

In 1996, NMC started with door to door collection in which only 30% area was covered. Due to the absence of timely services, in 2003, NMC entrusted this responsibility of implementing the project to CDC on a tender basis and later to private contractors. Presently, NMC has hired a private company, Kanak resources, to manage the solid waste management collection and transportation services to the dumping site. Figure 1 describes the administrative setup for waste management of NMC.

Solid waste quantification: The data collected from all the 136 wards were run through ARCGIS to generate maps showing the estimation of the quantity of waste generated from all the wards. When quantified mathematically, it turned out to be approximately between 1100 and1200 tonnes of waste. The red areas in the map presented in Fig. 2a show the highest waste generation wards of the city that is higher than 15 tonnes of waste generation/day. Blue, Orange and light green wards show the frequency of 10–15 tonnes, 5–10 tonnes and 0–5 tonnes of waste generation/day, respectively. Zone 9 was found to generate maximum waste as it is characterized mostly by slums. Zone 6 and zone 7 are the core areas of the city, producing more waste as they are densely populated. Although the area of zone 7 and 4 is comparatively less but generates more waste because of mixed land-use.

Solid waste collection and segregation: Nagpur has a door to door collection system whose efficiency is found to be 97% from the survey. The GIS map in Fig. 2b gives ward wise data of waste collection. The area in the map marked with light green, dark green, orange, red shows the frequency of waste collection is every day, every alternate day, once in a week, once in 15 days, respectively.

The numbers of community bins have been significantly reduced in the city under the 'bin –free city' project. Now there are only 170 bins located overall in Nagpur city, which are also collection points for garbage through bigger vehicles. Some areas

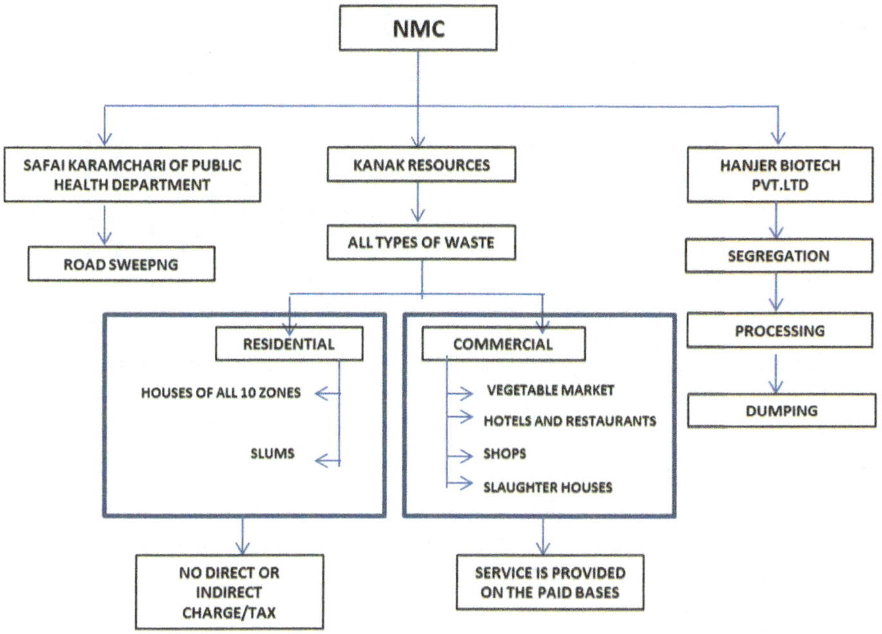

Fig. 1 Administrative setup of Nagpur Municipal Corporation in waste management

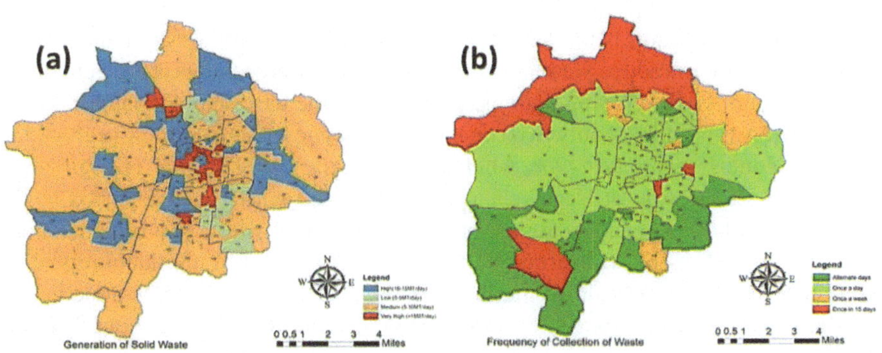

Fig. 2 GIS based map showing ward wise data of **a** waste generation **b** waste collection

of zone 10, 9, 6 and 5 experience open dumping at the points where previously bins were located. From Fig. 3a, it is clear that no zone is experiencing 100% collection. Some wards of zone 10 and zone 9 are facing issues in the waste collection as they are fringe areas of the Nagpur city and transportation vehicles do not come often for collection. Whereas, some fringe areas of zone 8 and zone 3 experience garbage collection once a week and in the rest of the zones, some wards experience the collection of waste on alternate days. We also explored whether the gap between waste generation and collection efficiency is influenced by income groups. Data

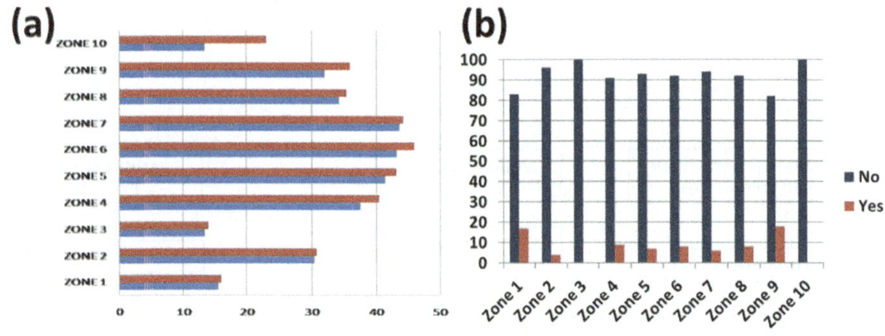

Fig. 3 **a** Bar chart showing comparison between waste generated and waste collected **b** Percentage of waste segregation at the zone level

comparison revealed that both factors are independent of each other. The core areas produce a high amount of wastes because they are densely populated, irrespective of income groups. These areas are predominant by Medium Income Group (MIG) though some are Low Income Group (LIG) as well.

3.3 *Solid Waste Segregation*

Solid waste segregation is the key step if we are planning for waste management. Developed nations practice waste segregation at point source that is at the generation stage itself, unlike developing nations. Similarly, is the case seen in Nagpur. When surveyed, it was found that only two zones out of 10 zones practice source segregation of waste. Zone 1 segregates 17%, whereas zone 9 does it 19% only. Though NMC does waste segregation at transfer stations before the waste is transported to the dumping site, but segregation at the source is negligible, which is a big challenge. Recyclables such as plastics, iron and papers are separated by the labourers at all the transfer stations and sold to the rag pickers. As the segregation is done before transportation to the dumping site, NMC is saving about Rs. 10 lakhs monthly on 60 TPD of garbage. Figure 3b presents a scenario of source segregation at zone level in Nagpur.

4 Problems and Perspectives

We did the SWOT Analysis after collecting and analysing all the data. To identify the problems at the smallest unit of the city that is at ward level, the data collected from the primary survey was divided into four stages of waste management, whereas data collected from the secondary survey was divided into two stages, mainly, transportation and processing and disposal. Figure 5 describes the identified problems after analyzing data.

- **Collection of waste**: The collection efficiency is 97%.
- **No source segregation**: The waste is collected from door to door without any segregation. About 50–60% of the waste collected from households is organic in nature. Hence it could be treated as a resource and can be used in the making of vermicomposting.
- **Transportation and collection efficiency**: Some wards generate a higher amount of wastes and collection frequency is low. The reason is Nagpur city has radial expansion and it has evolved with organic transformation. The core areas were also not planned; rather, they have evolved. Roads leading to the interior of the city make transportation difficult as vehicles could not enter those areas. These are the areas of high density with more waste generation. Secondly, due to less transfer stations, more vehicles are engaged in the collection and transporting of waste to the dumping yard, which results in higher fuel consumption and hence increased transportation cost.

Processing and disposal: Despite of being a million-plus city, Nagpur lacks the facilities and infrastructure for the efficient treatment of waste. From the institutional survey, it was found that by the end of 2020, Nagpur will start processing its waste. Presently, no sanitary landfilling is done, which is creating a nuisance to the people living in nearby areas of dumping yard.

5 Recommendations

As per the scenario in Nagpur, advertising the 3R concept could help in minimising waste generation. For this, awareness campaigns at the ward level should be organized by the NMC. For slum areas, NMC should organise an awareness program for the daily disposal of waste to the nearby temporary bins (provided by NMC). Segregation of waste at source should be made mandatory. NMC should provide separate bins to each household for the segregation of waste without any user charge. The waste collector should refuse to collect the waste if it is not segregated.

5.1 *Collection and Transportation*

NMC should use GPS technology for tracking and monitoring vehicles, whether the vehicle is collecting the waste effectively or not. This will likely enhance the efficiency of the collection, especially in the fringe areas of the city [12]. Every waste collector should be facilitated with separate bins for the collection of waste. Some user charges at a weekly/monthly basis should be charged. To strengthen the collection and transportation of waste in a decentralised way, seven transfer stations, as per accessibility to the landfill site, are proposed. The location of transfer stations is linked with the outer ring road of the city so that the vehicle can move timely without facing the congestions of the internal roads. From the transfer station to the dumping site, the separate vehicle should be used for biodegradable and non-biodegradable waste.

5.2 *Processing*

Processing of waste should be done in a decentralised way so that the waste send to the landfill site can be minimised. This will also reduce the transportation cost of solid waste from the transfer station to the dumping site. Construction of compost plant in the fringes of the city and biogas plant at the proposed transfer stations for processing of biodegradable waste within the city.

Biogas Plant: Since 50–60% of waste is organic, this can be treated in the decentralized biogas plant located in the city. The outcomes are the manure and methane gas, which can be used for generating electricity. The area required for installing a 5 MT/day biogas plant is 500 m^2 and the cost of the plant is approximately 40–50 lacs. The maintenance cost of the biogas plant is also low as only three to four people are required for efficient working of the plant. A 5 MT/day biogas plant can generate 300–400 units of electricity depending upon the quality of the waste.

Figure 6 demarks the position of all proposed biogas plants, the position of all the proposed transfer stations and the proposed route from all transfer stations to the dumping site. Nagpur city has an outer ring road. Transfer stations have been proposed on such locations so that the waste collection vehicles of higher volumes do not face the traffic problem. Hence connecting all the transfer stations, a route to the dumping site has been proposed, which is going from the outer ring road. Every zone is proposed with a biogas plant so that the organic fraction of the waste can be used as a resource.

Compost Plant Setting

Compost plant should be located at least 10–15 km radius outside of the airstrip aerodrome of township so as to avoid bird menace for aircraft safety. A detailed survey, soil and hydrological investigation should be carried out, including the status of the groundwater table so as to design civil structures properly. The site selected

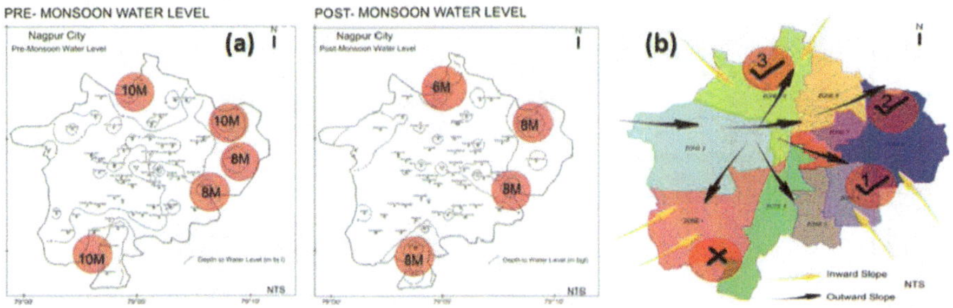

Fig. 4 **a** Pre-Monsoon and Post-Monsoon water level in Nagpur. **b** Site preference for compost plant as per water level in the city

should not be a waterlogged or marshy land as it would increase the civil work by more than 100%. Figure 4a shows the pre and post-monsoon water levels in Nagpur city. Based on the above criteria, site 1, site 2 and site 3 in zone 5, 6 and 10, respectively, are suited for compost plant. Figure 4b presents the criteria for site selection. Site 1 lies on the existing disposal path and also near the agricultural land where by-products can be readily used. Site 2 includes areas such as Pardi in Nagpur, which consists of undeveloped land where land cost is relatively low and better transport connectivity due to the outer ring road. At site 3, a small composting plant can be proposed to minimise open dumping in zone 9 and zone 10, as observed from the primary survey. 100 TPD composting site of area 1.1 hectare at site 1 whose production capacity is 30.76 TPD can be developed (Figs. 5 and 6).

6 Future Scope

The study was conducted at the ward level and is restricted to household waste only. Studies on Hospital waste, construction and demolition waste could also be conducted in order to analyse the proper composition and projection of wastes. Studies at a basic level such as ward level propose a good estimation for policy and decision making for Urban Local Bodies (ULB).

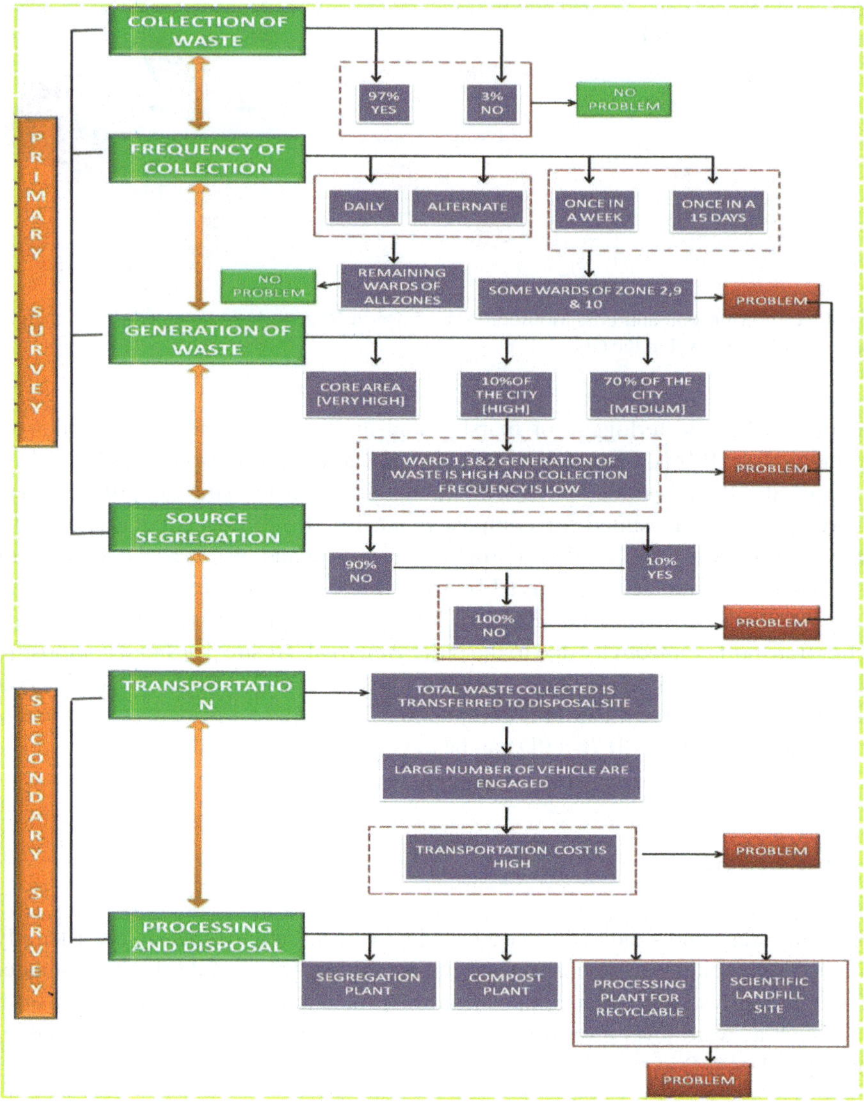

Fig. 5 Problems identified at ward level for waste management in Nagpur

Fig. 6 Proposed route from transfer stations to dumping site and biogas plant

References

1. Gupta, N., Yadav, K. K., & Kumar, V. (2015). A review on current status of municipal solid waste management in India. *Journal of Environmental Sciences, 37,* 206–217.
2. Kawai, K., & Tasaki, T. (2016). Revisiting estimates of municipal solid waste generation per capita and their reliability. *Journal of Material Cycles and Waste Management, 18*(1), 1–13.
3. https://www.indiaenvironmentportal.org.in/files/file/Generation%20of%20Waste.pdf.
4. Carver, S., Comber, A., McMorran, R., & Nutter, S. (2012). A GIS model for mapping spatial patterns and distribution of wild land in Scotland. *Landscape and Urban Planning, 104*(3), 395–409.

5. Grabaum, R., & Meyer, B. C. (1998). Multicriteria optimization of landscapes using GIS-based functional assessments. *Landscape and Urban Planning, 43*(1), 21–34.
6. Signore, A. (2016). Mapping and sharing agro-biodiversity using open data kit and google fusion tables. *Computers and Electronics in Agriculture, 127,* 87–91.
7. Vu, H. L., Ng, K. T. W., & Bolingbroke, D. (2018). Parameter interrelationships in a dual phase GIS-based municipal solid waste collection model. *Waste Management, 78,* 258–270.
8. Zia, H., & Devadas, V. (2008). Urban solid waste management in Kanpur: Opportunities and perspectives. *Habitat International, 32*(1), 58–73.
9. Sanjeevi, V., & Shahabudeen, P. (2015). Optimal routing for efficient municipal solid waste transportation by using ArcGIS application in Chennai India. *Waste Management & Research, 34*(1), 11–21.
10. Swachh Survekshan, GOI https://swachhsurvekshan2020.org/Images/SS_2017_Report.pdf.
11. Krejcie, R. V., & Morgan, D. W. (1970). Determining sample size for research activities. *Educational and Psychological Measurement, 30*(3), 607–610.
12. Kallel, A., Serbaji, M. M., & Zairi, M. (2016). Using GIS-based tools for the optimization of solid waste collection and transport: Case study of Sfax city Tunisia. *Journal of Engineering, 2016,* 4596849.

Need for Integrated Planning of Environmental Services for Small Town: A Case Study

Meghraj Garad, Megha Nikam, Bakul Rao, Kanchan Malkhede, and Hemlata Suryawanshi

Abstract India has been facing the challenge of overcoming the problem of open defecation, provisioning proper sanitation services, and supplying safe drinking water to the entire population. For bigger cities, a degree of systematic efforts have gone in designing model Water and Sanitation (WATSAN) systems, but similar efforts for smaller towns are rare. Several councils have tried to undertake the designing of WATSAN systems with local consultants, with a lesser degree of a systemic approach, resulting in poor performance as depicted in various assessments. The study to assess the need for integrated planning was conducted in Vadgaon Maval. A social survey of 100% households was conducted to understand the performance of the four systems and water quality testing was also done, wherein it was found that there was high inequality in water distribution. A total of 50 out of 53 samples were with *E. coli* signifying inefficient treatment or intrusion of sewage in the pipeline while supplying. The piped drainage system served for a limited part of the town but with inaccurate slopes and it carried both wastewater and storm-water. The solid waste collection was not 100%, resulting in people throwing waste in the open, which ultimately led to drain blockages. Thus, solid waste blocking the drains and drains contaminating the water supply lines crossing them, affecting the drinking water quality, is the current scenario. This necessitates the need for integrated planning and designing of four systems with consideration of the town's expansion while the implementation could be in phases.

Keywords Integrated planning · Water and wastewater · Small towns · Sustainable solutions

---

M. Garad · M. Nikam · B. Rao (✉) · K. Malkhede · H. Suryawanshi
Centre for Technology Alternatives for Rural Areas, Indian Institute of Technology Bombay, Mumbai, India
e-mail: bakulrao@iitb.ac.in

M. Garad
e-mail: meghrajgarad3713@iitbombay.org

A. Jana and P. Banerji (eds.), *Urban Science and Engineering*, Lecture Notes in Civil Engineering 121, https://doi.org/10.1007/978-981-33-4114-2_22

1 Background

In the development arena, India has been facing the challenge of overcoming the open defecation problem and provisioning of proper sanitation services in both urban and rural areas. Added to this is the problem of supplying safe drinking water to the entire population. Some of the issues related to water and sanitation systems commonly found in India are as follows.

Water supply issues	Sanitation issues
• Inadequate water supply • High UFW (Unaccounted for water) • Corrosion and scaling in pipes • Illegal connections • Intermittent water supply • Low pressures • Improper planning of natural and storm-water drainages • Inadequate storm-water drainages • High reliance on groundwater • Inadequate water treatment • Low Taxes and Inability and Unwillingness to pay taxes	• No sanitation systems • No wastewater collection systems • No/inadequate water quality monitoring • Mixed drainages between storm-water and wastewater • No treatment of wastewaters • On-site treatment of sewage–septic tanks options inadequate • Septic tanks with no soak pits, water left into drains • Inadequate FSM Technology available • Focus on conventional sewerage systems • Inadequate capacity of towns • Inadequate technical support from the centre to the state to the towns

Bigger cities and towns have systematic efforts in designing model WATSAN systems, but similar efforts for smaller towns are rare. Challenges for town administrations include a lack of capacity to undertake such projects on their own. Several councils have tried to undertake the designing of WATSAN systems with local consultants, with a lesser degree of a systemic approach, resulting in poor performance as depicted in various assessments. This study indicates the need to tackle the problems systematically while understanding the town in totality. The first phase of the ongoing research is to assess the on-ground reality of the four environmental services, that is, Water Supply, Waste Water Management, Solid Waste Management, and Storm Water Management, and the need for planning them by understanding the town.

2 Literature Review

In India, around 32% of HHs live in urban areas. Among these, about 62% of urban households receive water from a treated source through tap while 8% of urban HH receives it from an untreated source through tap and remaining HH receive the water from other sources such as open wells, borewells, and other sources without the tap connection [1]. Even households connected to the public supply system receive, on average, only three hours of drinking water supply a day and an average of 75 L per capita as opposed to the norm of 135lpcd [2]. Following the same, the wastewater collection scenario is completely different. According to Census 2011, nearly 45% of urban households are connected to closed drainage, and only 33% of the urban households are connected to a piped sewer system [1]. In numbers, only 300-odd cities in India are estimated to have a sewerage network in place [3]. Most of the treatment capacity is in Class I cities. There are 211 treatment plants in Class I cities as compared to 31 in Class II cities and 26 in the remaining cities [4]. In the case of Maharashtra, only 56% of HH receive treated water through tap connection, and 63% HHs have closed drainage out of these, only 38% are with piped sewer connections. Though the collection is there, most cities lag in treating the wastewater; however, the percentage of water treated is less than 10% in class B and class C cities [5]. Out of 230 Municipal councils, only 12 have underground sewage connections, and 9 have STPs [5]. The lag in the water cycle leads to the sanitation problem, and poor sanitation is a major cause of diseases in developing countries such as India, and the urban poor suffer disproportionately from this [6].

A statistical study conducted by the Indian Institute for Human Settlements (IIHS) shows that small and medium-sized towns have a more significant percentage of households with inadequate access to water and sanitation [7]. This study also emphasizes the policy-making considering a more balanced approach to cover a wide range of towns and essential to recognize that a differentiated approach might be needed for these towns. The study by Wankhade K. also highlights the need for focusing on access to water and the full cycle of sanitation for the urban poor, as fundamental to addressing the sanitation challenge [8]. Sewerage networks, where they exist, are poorly maintained: there are frequent blockages, siltation, and missing or broken manhole covers. There is hardly any preventive maintenance, and repairs are made only in case of crises [9]. Often, storm-water enters sewers, which are not designed to take these loads, leading to overflow onto the surrounding areas. Improper disposal of solid waste also tends to block sewer lines [8].

National Strategy for 'Water Supply and Sanitation 2014' by the Government of Bangladesh aims at the integration of water supply, sanitation and hygiene [10]. The study conducted by the World Bank shows the potential and importance of rapidly growing towns in Africa on integrated urban water management as an alternative to lack of natural resources and adoption towards urbanization [11]. In this approach, the African water facility came forward and started reforms through the Sustainable Water and Sanitation in Africa (SUWASA) program in sub-Saharan Africa, taking up IUWM concepts [12].

In relevance to this, the Government of India also recommended that water supply of 135lpcd can be supplied through piped water supply in the towns where sewerage system is existing/contemplated [13]. The Government of India is focusing on water and wastewater management through various schemes such as 'Atal Mission for Rejuvenation and Urban Transformation (AMRUT)' and 'Urban Infrastructure Development Scheme for Small and Medium Towns Scheme' (UIDSSMT) launched on 5 March 2012 which had funding pattern of 80:10:10, that is, 80% by GOI, 10% by state and balance 10% by Council. After completion of this program, the Maharashtra government launched the scheme. 'Maharashtra Suvarna Jayanti Nagarosthan Maha-Abhiyan' to provide funding support to small cities and towns for basic facilities such as water supply, sanitation and solid waste management and other such types of works. The funding support ranges from 90 to 70% based on types of Municipal Council, that is, C-type to A-Type, respectively. GOI has a program called 'Total Sanitation Campaign' (TSC), which was later renamed as 'Swaccha Bharat Mission' (SBM), to tackle the sanitation problem in India. All these programs run independently and there is a lack of interaction of water, wastewater and solid waste.

The program on Integrated Urban Water Management (IUWM) conducted by Arghyam, Banglore, showed similar field-level complexities in the water sector in small towns [14]. IUWM started eight pilot projects in different wards of four major cities, that is, Jaisalmer and Kishangarh in Rajsthan state and Ichalkaranji and Solapur in Maharashtra state. The results of this pilot project helped in developing a toolkit for planning and guideline about sharing the idea of IUWM to more than 30 cities in Maharashtra.

Another problem with the inefficient planning and operation of such large schemes is the inadequate capacity at the Municipal Council level. All the Municipal Councils come under the Directorate of Municipal Administration at the state level and each Municipal Council is headed by Chief Officer in Maharashtra. However, in terms of technical capacity, each Municipal Council has only one or two city engineers who look after all the works such as sanctioning the plans, preparation of estimates, operation of schemes, etc. Operators at the local level are the one who operates and run the water supply, wastewater and solid waste schemes and they lack in technical knowledge.

3 Study Area and Methodology

The town was selected for the study after considering parameters such as population, size of the town, density variation among the town, and accessibility. The chosen town Vadgaon Maval, with a population of 15,041 as per the Census 2011, is located near Pune and is a newly formed Nagar Panchayat. This town is the headquarters of the Maval block in the Pune district.

The methodology comprises the collection of secondary data from various secondary sources, the collection of primary data, and field observations related to the status of environmental services in the town. The primary data was collected through a social survey of 100% households (3927 HH) using the KoBoCollect app [15] (an open-source Android-based app) by structured interviews. The format for the social survey comprised of questions related to demographics, drinking water, wastewater, solid waste, and storm-water. Water quality was analyzed by testing water samples from source, water treatment plant, and household tap connections and compared with BIS 10500: drinking water quality parameters [16]. The existing water supply scheme was studied by mapping pipe networks with valve operators and measuring the flow and pressure at various endpoints of the network during water supply hours (Fig. 1).

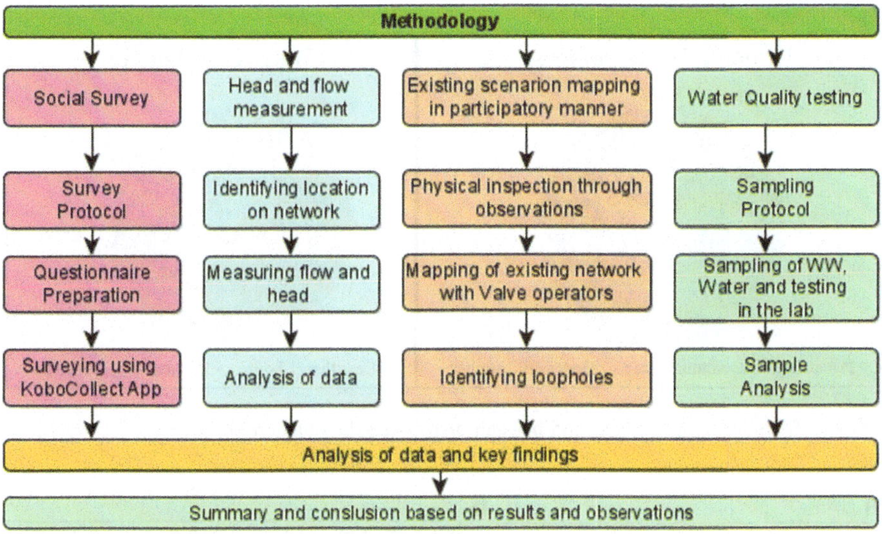

Fig. 1 Methodology for study

4 Results and Discussion

4.1 Water Supply

Amongst the 3,927 households surveyed, 77% HH had tap water connection, and 18% had community tap connection while the remaining 5% had other sources such as open well, bore well or packed drinking water, etc. Amongst the 77% private connections, 41% were receiving water for less than an hour, that is, they were receiving less than 80 L per capita per day (lpcd) with the average flow rate measured, and 47% received the water supply for 1 to 3 h while 12% received water for more than 3 h (Fig. 2).

Only 17% of households did not practice any household-level treatment while responding that they were satisfied with the quality of the water supplied. The remaining 87%, households (HH) practice household level treatment such as the use of filter or alum or chlorine, etc., before consumption, thus reflecting a negative perception towards the water quality despite spending on treatment by the Nagar Panchayat. To assess the quality of the water supplied, a total of 53 water samples were tested in the lab in four phases. It was found that 50 samples were contaminated with *E. coli*, while only three samples were tested negative for *E. coli* as per IS 10500 [16] and ASTM testing [17]. These three samples were taken immediately after the chlorination unit of one of the two water treatment plants, indicating a substantial scale of intrusion of sewage into the water supply during distribution (Fig. 3).

The unavailability of the water distribution network map at Nagar Panchayat leads to an unorganized expansion of the network, which further leads to unequal water

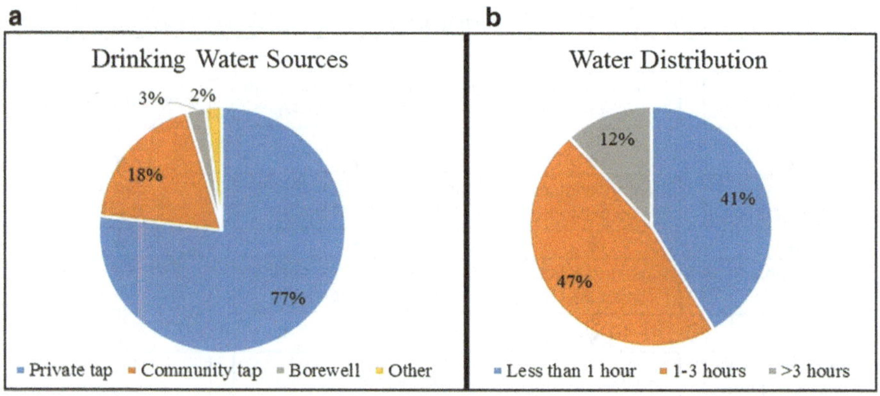

Fig. 2 **a** Drinking water sources and **b** Water distribution in Vadgaon Maval

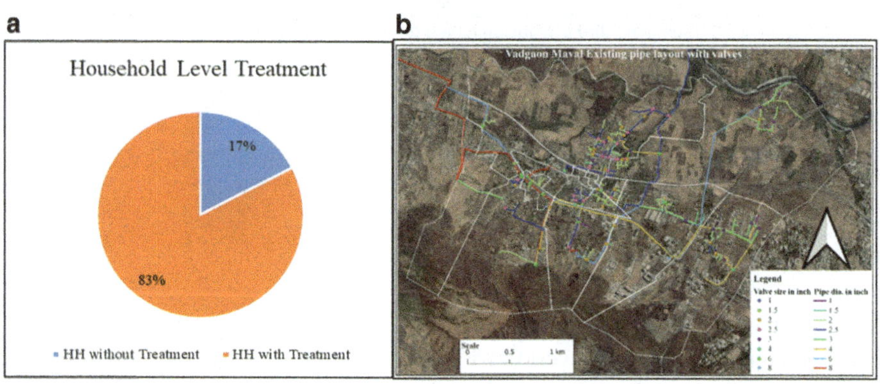

Fig. 3 **a** Household level treatment. **b** Map of water distribution network

distribution. During the mapping of the pipeline network, with the help of valve operators, it was found that at some places, larger diameters were connected ahead of smaller on stakeholders' request, which lead to a drop in the residual head thus leading to an inadequate water supply and ultimately forcing stakeholders to use motors to extract water from the pipeline. It was found that the town has two Water Treatment Plants (WTP). The old WTP was incapable of treating the water at the required efficiency due to a lack of regular maintenance of its components. The new WTP was not running at its full capacity because of low inflow due to the splitting of the rising main into two rising mains supplying water to two WTPs, that is, old and new WTP.

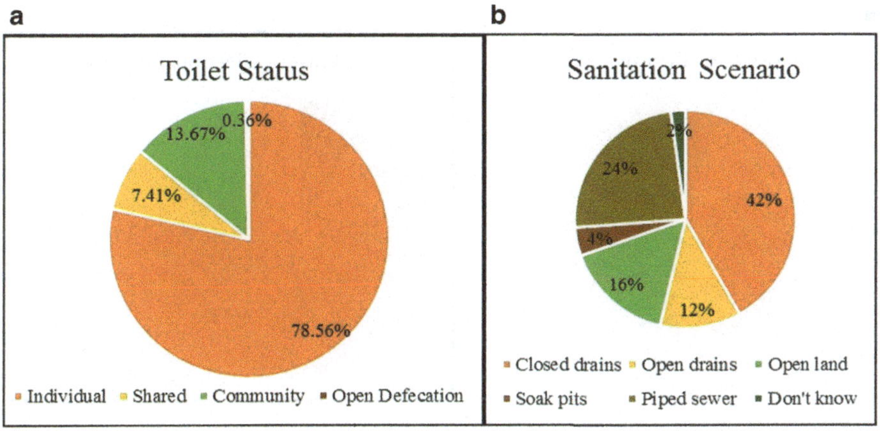

Fig. 4 **a** Toilet coverage. **b** Sanitation scenario in Vadgaon Maval

4.2 Wastewater and Sewage Treatment

According to Swachha Bharat Mission (SBM), the town is open defecation free [20]. However, 0.36% of households mentioned open defecation during household surveys, while the remaining 99.64% of households use either an individual or shared or community toilet.

As per census data 2011, only 32.7% of urban households have proper sewer connections in India, and as per the results of the HH survey in Vadgaon Maval, it was found that only 24% of HHs were connected to a piped sewer. This primarily includes the households in multi-storeyed buildings, while remaining HHs discharged the wastewater in a partially-closed/open drain or an open area. There was no wastewater treatment facility available in the town. Most households have septic tanks, and the septage coming from the septic tank was not treated at all. At several places, the piped drains were available, but the slope of the pipeline was not adequate, and storm-water also flows through the same pipe. There were blockages at several places and inadequate maintenance leading to non-efficient usage of the sewerage (Fig. 4).

4.3 Solid Waste Management

Vadgaon Maval was a Grampanchayat till 2018, and on 3 February 2018 got converted into an Urban Local Body as Nagar Panchayat. The entire solid waste is being dumped and burned in forest land without any treatment. Vadgaon Maval reported 92% collection of solid waste with door to door scenario and 6% with dust bin near the house while remaining 2% either deal with the waste on their own or throw it in the open spaces nearby their house. The frequency of solid waste collection was daily in 80% of households, while 10% of households replied with alternate day collection (Fig. 5).

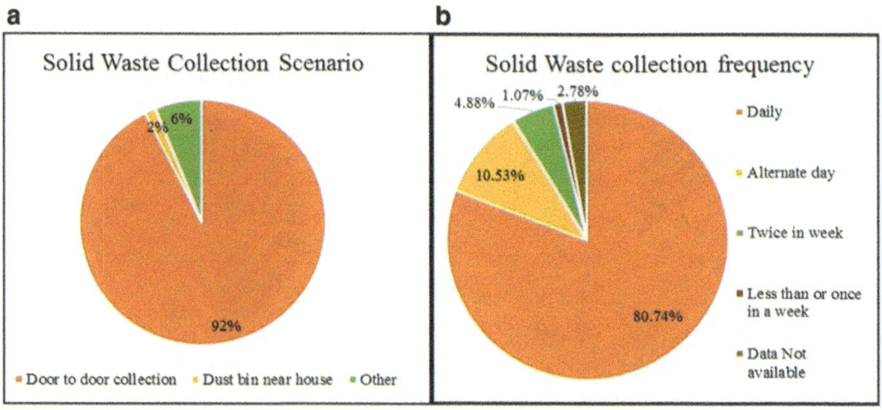

Fig. 5 **a** Solid waste collection status. **b** Solid waste collection frequency in Vadgaon Maval

4.4 Key Observations on the Field

The key observations based on the field survey are enlisted below:

1. The water supply network was not of adequate diameter throughout to serve the water to the population using intermittent supply leading to high inequality in distribution.
2. The water distribution network was a dead-end system leading to a high loss of water (Fig. 3b).
3. The three sources of water, with each having a different rising main connecting to two WTP or directly to the distribution reservoirs, led to the supply of untreated water to several parts of the town and treated water to the remaining parts.
4. Inadequate water treatment capacity and inefficient treatment by components resulted in degraded quality of water.
5. Few piped drainage lines were observed on the field, but due to improper slopes, the drainage is affected, leading to blockages.
6. At many places, water supply lines were interfering with the drains. Either these lines are passing through open drains/closed drains or crossing the drains at the same level of crossing (Fig. 6).
7. Most drains being open were clogged due to the dumping of solid waste into it (Fig. 6) or due to lack of regular cleaning of accumulated debris such as fallen leaves of trees, etc.
8. There were no separate drains to carry storm-water and sewage, so no treatment of sewage was possible, and thus, sewage and wastewater get discharged directly in natural drains and ultimately into the Indrayani River, which is one of the sources of drinking water for Vadgaon.
9. Whenever new pipelines for newly developed areas are required, the authority is permitted to extend the lines with approximate diameter from the nearest

Fig. 6 Photographs showing conflicts among four systems in the Vadgaon Maval

larger diameter without checking the capacity of connecting pipe and pipelines were laid along the roads and sometimes even through drains.

10. Many natural streams were blocked or diverted to open land or to the opposite direction of its natural slope during the construction of large buildings.
11. Storm-water gets accumulated at several places ranging from a few hours and days inside the dense locality due to improper drainage and gets mixed with sewage.
12. There was accumulated solid waste at several places, and the respondents nearby those areas responded to this that most of the factories/offices/labourers were not able to put their solid waste into collection vehicles due to a mismatch of timing, and they throw it on their way to the office.

The Fig. 6 shows some of the photographs of the field observation about conflicts among different environmental services.

5 Conclusion

In the current study, the assessment of all four environmental services viz. water supply, wastewater management, storm-water management, and solid waste management was conducted using a social survey of 100% households, measurement of flow and pressure, network mapping, lab testing of water samples, and on-field observations. The study found that drinking water supply was with high inequality having supply quantity of less than the norms of 135lpcd to more than half the population, and the quality was not up to the mark as per drinking water standards. Wastewater was not collected 100% and not treated at all. Observations found that the solid waste collection was not 100%, resulting in people throwing waste in the open, which ultimately led to drain blockages. Thus, solid waste blocking the drains and drains contaminating the water supply lines crossing them, affecting the drinking water quality, is the current scenario.

Three systems, that is, water supply, sewerage and wastewater treatment, and storm-water, are part of the water cycle and solid waste mismanagement as interference to them. Thus, it necessitates the need for planning and designing of four systems in an integrated manner with consideration of the town's expansion while the implementation of the system could be in phases.

Acknowledgements We acknowledge the Wheels India Niswarth (WIN) foundation for sponsoring the project, Directorate of Municipal Administration (DMA), Urban Development Department, Government of Maharashtra, and Nagar Panchayat, Vadgaon Maval for their support to conduct the study in the Vadgaon Maval.

References

1. The Census of India. https://censusindia.gov.in/2011census/Hlo-series/HH09.html. last accessed 2020/05/02.
2. MoUD, Service Levels. (2012). *in Urban Water and Sanitation Sector: Status Report (2010–2011)*. Ministry of Urban Development: New Delhi.
3. NIUA. (2005). Status of water supply, sanitation and solid waste management in urban areas. National Institute of Urban Affairs, New Delhi.
4. CPCB. (2013). Performance evaluation of sewage treatment plants under NRCD. Central Pollution Control Board, New Delhi.
5. Centre for Water and Sanitation. Performance Assessment System (PAS). https://www.pas.org.in/web/ceptpas/stateprofile. Last accessed 2020/05/15.
6. Andres, L. A., Briceño, B., Chase, C., & Echenique, J. A. (2014). Sanitation and externalities: Evidence from early childhood health in rural India. World Bank Policy Research Working Papers.
7. Wankhade, K., Balakrishnan, K., & Vishnu, J. (2014). Sustaining policy momentum, urban water supply & sanitation in India. IIHS RF Paper on Urban Water Supply and Sanitation in India, Banglaore.
8. Kavita, W. (2015). Urban sanitation in India: key shifts in the national policy frame. *Environment and Urbanization, 27*(2), 555–572.
9. WSP. (2008). *Technology options for urban sanitation in India.* Water and Sanitation Program, New Delhi.
10. Government of the People's Republic of Bangladesh. (2014). *National strategy for water supply and sanitation.* Ministry of Local Government: Rural Development and Cooperatives. Local Government Division.
11. Jacobsen, M., Webster, M, & Vairavamoorthy, K. (2012). The future of water in African cities: Why waste water? Integrated urban water management for Africa: background report.
12. United States Agency for International Development. (2015). Sustainable water and sanitation in Africa (SUWASA) final report September 30, 2009–September 29, 2015
13. CPHEEO. (1999). *Manual on water supply and treatment systems.* Ministry of Urban Development, New Delhi.
14. Sunita, N. (2012). *Approach to integrated urban water management (IUWM), the mulbagal experience.* Banglaore: Arghyam.

15. The KoboToolbox. https://kobo.humanitarianresponse.info/#/forms. Last accessed 2020/05/10
16. IS:10500:2012 (2012). Drinking water-specification, (Second Revision). Bureau of Indian Standard
17. ASTM (2018). *Standard methods for the examination for water and wastewater*, 23rd edn. American Public Health Association.
18. MoUD (2014). List of CSP cities. Available at https://moud.gov.in/sites/upload_files/moud/files/List_Of_CSP_Cities.pdf.
19. CPHEEO (2013). *Manual on sewerage and sewage treatment systems*. Ministry of Urban Development, New Delhi.
20. Swachha Bharat Mission (SWM). Government of India (GOI). https://sbm.gov.in/sbmReport/State.aspx#. Last accessed 2020/05/10

Understanding the Effect of Occupant Behaviour on Thermal Comfort and Energy Use in Slum Rehabilitation Housing

Jeetika Malik, Ronita Bardhan, and Pradipta Banerji

Abstract A transverse field study was conducted in 1216 slum rehabilitation housing units of Mumbai to understand the occupant behaviour towards thermal comfort and energy consumption. The data were analysed using a structural equation modelling approach to investigate the effects of occupant factors on thermal comfort perception within the low-income population. Three latent constructs *Household Income*, *Adaptive Behaviour* and *Built Environment Character* were developed. The causal model revealed that thermal comfort perception was significantly influenced by occupants' adaptive behaviour and their household income levels. The results also indicated the indifferent outlook of occupants towards their built environment. A direct effect of household income on adaptive occupant behaviour further informs that occupant behaviour can play an important aspect in diminishing the future rebound effects of energy consumption. The findings from this work would be helpful in integrating occupant behaviour into thermal comfort and energy consumption assessments to improve building performance and predicted energy gap within low-income housing.

Keywords Occupant behaviour · Thermal comfort · Perception · Energy use · Mumbai

J. Malik (✉) · R. Bardhan
Centre for Urban Science and Engineering, Indian Institute of Technology Bombay, Mumbai, India
e-mail: jeetika.malik@iitb.ac.in

R. Bardhan
Department of Architecture, University of Cambridge, Cambridge, UK

P. Banerji
Department of Civil Engineering, Indian Institute of Technology Bombay, Mumbai, India

A. Jana and P. Banerji (eds.), *Urban Science and Engineering*, Lecture Notes in Civil Engineering 121, https://doi.org/10.1007/978-981-33-4114-2_23

1 Introduction

Comfort is the outcome of interaction of physical exchanges, physiological, psychological, social and cultural rights and is thus considered as a complex subject [1]. The existing comfort standards often fail to predict comfortable conditions in other geographical regions and contexts, different from the ones that they were designed for [2]. This incorrect estimation of indoor comfort conditions results in overheating or overcooling leading to building performance gap. It is imperative to have a better understanding of thermal comfort; keeping in mind the socio-cultural and economic dimensions of comfort; to improve building sustainability and reduce energy consumption.

Occupants play a critical role in determining thermal comfort by interacting with their indoor environment to adapt to the thermal surroundings. Behavioural adjustment, that is, the use of adaptive actions; psychological effects such as habituation and expectations; and physiological influence related to age, gender, etc., collectively influence occupants' thermal adaptation and thus their comfort perception [3]. However, in most of the building design practices and operation processes, the impact of occupant behaviour on thermal comfort and energy use remains under-recognised [4]. This study emphases on occupant behaviour and related factors such as expectations, experience and perceptions within low-income housing of India. Housing policies for the low-income population often neglect the indoor comfort aspect and merely emphasize the cost and ease of construction of housing units. Nonetheless, providing comfort to the vulnerable low-income population is essential for improving productivity, health and well-being of the occupants. The key objectives of this study are- to investigate the effects of occupant behaviour on their thermal comfort perception and discuss how occupant behaviour can affect energy use in low-income residential buildings. The scope of this work is limited to the slum rehabilitation housing located in warm-humid climate of Mumbai, India.

2 Materials and Methods

Five slum rehabilitation housing (SRH) colonies of Mumbai were selected for data collection using a transverse survey method to capture occupant behaviour towards thermal comfort. SRH are low-cost tenement units provided to the slum inhabitants for free. These dwellings are constructed in public–private partnership under the 'Slum Rehabilitation Scheme' with an aim to improve the living conditions of slum dwellers. SRH units having an area of around 22 m^2 are stacked in six to eight storied vertical towers. Research within SRH suggests that these densely populated urban housing clusters have poor living conditions and inadequate indoor environment [5].

Data collection method comprised of cross-sectional household surveys administered in the summer months of Mumbai within five SRH location. Samples were randomly stratified according to the age of the respondent, floor location, and

Table 1 Questionnaire details

Section	Description
Socio demography	Age, gender, education and household size
Occupant behaviour	Adaptive actions, appliance setpoints and occupancy hours
Appliance ownership and usage	Appliances owned, daily usage and monthly electricity expenditure
Subjective votes	Comfort perception, indoor environment quality and housing satisfaction

condition of the housing unit. Surveys were conducted using a computer-aided personal interviewing system to improve the data quality and processing time. The questionnaire was carefully designed after a thorough literature review of relevant studies and consisted of six distinct sections. This study focuses on four sections: socio-demographic enquiry, occupant behaviour responses, appliance ownership and subjective comfort experience. Housing characteristics such as floor area, construction materials and spatial configuration were similar across the samples. Table 1 presents the structure of the survey questionnaire.

Data pre-processing, along with statistical analysis, was carried out using Statistical Package for Social Science (SPSS) version 25. Path analysis technique of structural equation modelling (SEM) was performed in AMOS-20 to understand the factors influencing thermal comfort perception and the causal relationships among them.

3 Conceptual Model

Based on the existing literature, a conceptual model is developed, as presented in Fig. 1, that links occupants' adaptive behaviour, household income and built environment to the perception of thermal comfort. The endogenous variable 'thermal comfort perception (TCP)' represents the thermal experience and was enquired on a seven-point thermal sensation scale ranging from cool (−3) to hot (+3) where 0 corresponds to thermal neutrality. Occupants voting within −1 to + 1 are considered as comfortable and the rest as uncomfortable.

The first exogenous variable, 'Household Income' (HI) represents the total income of the occupants' household and is based on the premise that higher income levels alter occupants' comfort expectations, and they desire thermal pleasure rather than thermal neutrality [6]. Thus, household income is expected to have a negative effect on thermal comfort perception. The next variable, 'Adaptive Behaviour' (AB) corresponds to commonly implemented adaptive measures by the occupants for improving their indoor thermal comfort. It is known that occupants adapt to a wider range of temperature or humidity through the use of adaptive measures, and thus thermal discomfort is expected to have a positive association with AB. The third exogenous

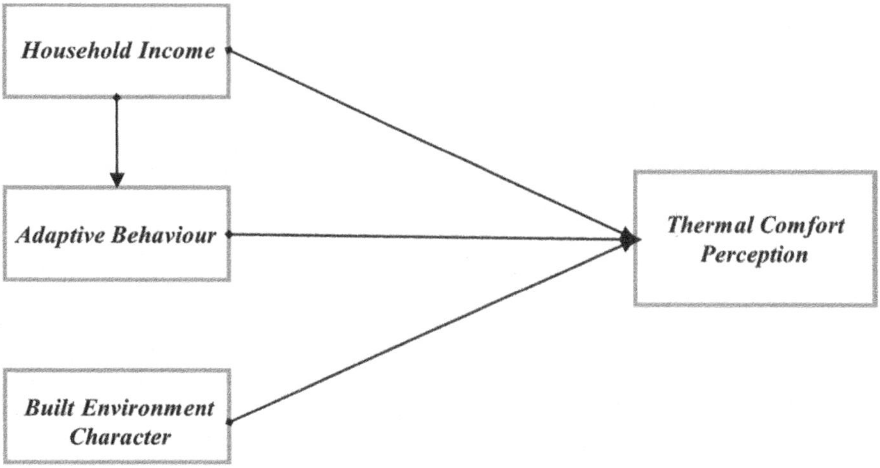

Fig. 1 Conceptual model

variable, 'Built Environment Character' (BEC) corresponds to the occupants' experience within their built environment and reflects characteristics of indoor environment. BEC is anticipated to have a positive effect on thermal comfort perception. Here, we propose that occupant factors- expectations (HI), behaviour (AB) and experience (BEC) shape their thermal comfort perception.

Figure 1 illustrates the conceptual model where the arrows represent proposed pathways and interdependencies among the exogenous and endogenous variables. Occupants' thermal comfort perception (TCP) is influenced by the occupant factors-household income, adaptive behaviour, and built environment experience. At the same time, occupants' household income also affects their adaptive behaviour towards comfort actions.

4 Results

The questionnaire survey yielded 1216 valid set of responses comprising of 23 variables from a total of 1267 sets. An item to item correlation was applied to remove multicollinearity and improve the reliability of the dataset leading to the removal of four variables. Next, the overall reliability of the survey instrument was examined using Cronbach' alpha test. A resultant value of 0.7 was observed, which was within the acceptable range [6], thus establishing the internal consistency of the dataset. The final dataset comprising of 19 variables, as presented in Table 2, was then subjected to two-step structural equation modelling analysis discussed in the subsequent sections.

Exploratory factor analysis (EFA) was performed to examine how well the response variables describe the latent constructs. Principal component analysis with varimax rotation was used to extract the factors. The adequacy of data was examined

Table 2 Description of the response variables

Code	Variable name	Variable type
Age	Age of the occupant	Ordinal (18–30 years = 1; 31–40 years = 2; 41–50 years = 3; 51–60 years = 4; Above 60 years = 5)
Gen	Gender of the occupant	Dichotomous (Female = 1; Male = 0)
Edu	Education level of the occupant	Ordinal (Below Primary = 1; Primary = 2; Secondary = 3; Graduation = 4; Above graduation = 5)
HCI	Housing Crowding Index (total number of co-residents per household divided by the dwelling size) [7]	Continuous
Comp	Appliance ownership: Computer	Dichotomous (Yes = 1; No = 0)
Ref	Appliance ownership: Refrigerator	Dichotomous (Yes = 1; No = 0)
AC	Appliance ownership: Air-conditioner	Dichotomous (Yes = 1; No = 0)
WM	Appliance ownership: Washing Machine	Dichotomous (Yes = 1; No = 0)
TV	Appliance ownership: Television	Dichotomous (Yes = 1; No = 0)
EC	Appliance ownership: Evaporative air-cooler	Dichotomous (Yes = 1; No = 0)
Windoor	Opening windows or door for comfort	Dichotomous (Yes = 1; No = 0)
Cur	Drawing curtains for comfort	Dichotomous (Yes = 1; No = 0)
Clo	Adjusting clothing levels for comfort	Dichotomous (Yes = 1; No = 0)
Rwet	Roof wetting or floor wetting for comfort	Dichotomous (Yes = 1; No = 0)
Pln	Using planters for comfort	Dichotomous (Yes = 1; No = 0)
EExp	Monthly electricity expenditure in INR	Continuous
IAQ	Indoor air quality	Likert scale (Very poor = 1; Poor = 2; Average = 3; Good = 4; Very good = 5)
HSAT	Housing satisfaction in terms of needs	Likert scale (Very much dissatisfied = 1; dissatisfied = 2; Average = 3; satisfied = 4; Very much satisfied = 5)
TCP	Thermal comfort perception measured as acceptable thermal conditions	Dichotomous (Comfortable = 0, Uncomfortable = 1)

through Kaiser–Meyer–Olkin (KMO) statistics and a value of 0.62 was found which is within the acceptable range. Next, Bartlett's test of sphericity was checked, which was highly significant ($p < 0.001$) and thus, factor analysis was considered as an appropriate technique for further analysis. EFA resulted in 14 variables under three distinct factors influencing thermal comfort perception. The first factor comprised of appliance ownership variables: *Comp, Ref, AC, WM* and *TV*. This factor serves a proxy to household income construct since income-related questions led to a response bias among the occupants. In the absence of income data, it is common to use household asset ownership as a surrogate. The next factor contains four variables related to adaptive occupant behaviour: *Windoor, Clo, Cur, Pln*. The third factor represents the built environment construct and includes *HCI, IAQ and HSAT*. Socio-demographic variables (*Age, Gen* and *Edu*), energy expenditure (*EExp*) as well as adaptive use of roof wetting (*Rwet*) were removed because of low factor loadings (<0.5). The reliability values of the selected factors were in the range of 0.76–0.87.

A confirmatory factor analysis (CFA) was carried out using maximum-likelihood estimation in AMOS 20. CFA examines the relationship between observed variables and underlying latent constructs [8]. The set of 14 variables derived from EFA were subject to CFA. The reliability and validity estimates of individual constructs were ascertained by the master validity tool [9]. Next, the model fit of the measurement model was improved through modification indices of covariance. The final model has all the model fit indices within the acceptable range (see Table 3), suggesting that the underlying constructs can be explained by the causal model to a great extent.

Figure 2 illustrates the structural equation model with standardized path coefficients (ß) and significant levels (p-value). The results indicate that adaptive behaviour (AB) and household income (HI) were statistically significant factors affecting thermal comfort perception of slum rehabilitation occupants, while built environment character (BEC) yielded non-significant results. AB has a positive effect on TCP with a path coefficient of 0.08 ($p < 0.01$). This association indicates that adaptation measures in the form of opening doors or windows, clothing adjustment, use of curtains and planters enhanced comfort levels among the respondents. Curtains and clothing adjustment had higher coefficients than the use of windows and doors or use of planters thereby suggesting greater influence on comfort perceptions. HI yielded a negative and significant effect on TCP, having a path coefficient of −0.10 ($p < 0.01$).

Table 3 Model fit indices of the resultant model

Model fit Indicators		Results	Acceptable values [10]
Absolute fit measures	Chi-square significance	0.00	<0.05
	Root mean square error of approximation	0.03	<0.06
	Adjusted goodness of fit index	0.97	>0.80
	Goodness of fit Index	0.98	>0.80
	Standardized root mean square residual	0.03	<0.08
Incremental Fit measures	Normed fit index	0.92	>0.90
	Comparative fit index	0.96	>0.90
	Incremental fir index	0.96	>0.90
	Tucker Lewis Index	0.94	>0.90
Parsimonious fit measures	Parsimonious ratio	0.70	>0.50
	Parsimonious normed fit index	0.65	>0.50
	Parsimonious comparative fit index	0.68	>0.50

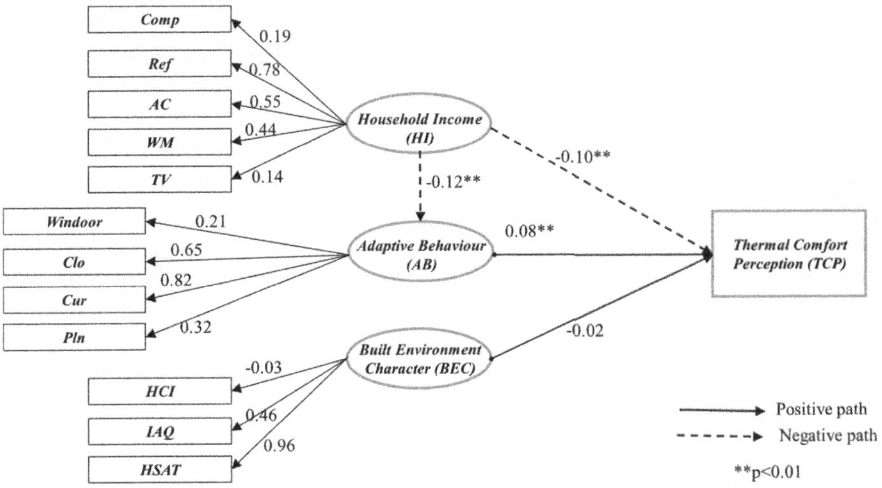

Fig. 2 Structural equation model for thermal comfort perception

Appliance ownership factors, which were treated as income proxy- refrigerators, air-conditioners and washing machines yielded standardized coefficients of 0.78, 0.55, 0.44, respectively ($p < 0.05$). This association could be attributed to the existence of 'thermal delight' among the occupants such that households having higher income levels exhibit higher comfort expectations and lower acceptability. Similar inferences have been drawn by researchers for other income categories within residential buildings [8]. BEC yielded insignificant results, indicating occupants' experience within their built environment does not influence in shaping comfort perceptions. These results seem plausible, given the peculiar characteristics of SRH. SRH are considered as 'vertical slums' with poor indoor environmental quality; however, studies have still reported occupants to be majorly satisfied with the number of rooms, housing needs or the indoor thermal environment [9, 11, 12]. These findings indicate that SRH occupant are indifferent about their built environment, and thus, their perception of comfort is not affected by the BEC factor. Further, the interrelationships among the latent factors reveal that AB is negatively affected by HI. The rationale for this relationship is that with an increase in income, occupants move towards energy-intensive comfort actions such as the use of air-conditioners or evaporative coolers, thereby not fully exercising the adaptive opportunity available to them. In other words, occupants give away the adaptive opportunity available in the form of passive measures.

The analysis presented in this study throws light on the thermal comfort perception of slum rehabilitation housing (SRH) dwellers, where the human aspect emerges to be a central factor. Next, we discuss the implications of electricity consumption, referred as energy use, on occupant behaviour within SRH. Monthly electricity expenditure was uncorrelated with thermal comfort acceptability in low-income households owing to the limited availability and restricted use of energy-intensive devices. However, it is expected that SRH will experience a rebound effect in energy use with the rise in household income levels and they would consume more electricity than anticipated [13]. A direct effect of household income on adaptive occupant behaviour ($ß = -0.12$, $p < 0.01$) further strengthens this supposition and informs that occupant behaviour can play an important aspect in diminishing the future rebound effects of energy consumption. It is therefore of utmost importance that housing schemes focusing on the low-income population such as Pradhan Mantri Awas Yojana, Slum Rehabilitation Scheme acknowledge and incorporate occupant behaviour and its related effects in estimating energy consumption. The study would be helpful in improving building sustainability within low-income housing apart from enhancing the comfort and well-being of the occupants.

5 Conclusion

Thermal comfort in slum rehabilitation housing is a multifaceted subject where occupants adapt to their thermal environment in myriad ways. A two-stage structural equation modelling approach was adopted to examine how occupant behaviour and related factors influence thermal comfort perception in slum rehabilitation housing of Mumbai, India. The results indicate that occupant behaviour and their household income are critical factors contributing to the behavioural and psychological thermal adaptation process and thus comfort perception. Adaptive comfort actions adopted by the occupants in the form of opening of windows or doors, using curtains to cut off glare, using planters for shade and adjusting their clothing levels witnessed a positive effect on thermal comfort. On the contrary, household income accounting for comfort expectations demonstrated a negative effect on thermal comfort levels in SRH. The study also highlights the indifferent attitude of SRH occupants towards built environment factors such as crowding index, housing satisfaction or air quality experience. The findings from this work would be helpful in integrating occupant behaviour within thermal comfort and energy consumption assessments to improve the building energy and performance gap within low-income housing. Additionally, the inferences would also aid in providing design recommendations for future housing units to be built under 'Slum Rehabilitation Scheme'. The broader goal is to improve occupant comfort and health of the marginalized population. The limitation of this study was that the data collection process did not include measurement of indoor environment variables, which could have helped in interlinking comfort conditions with the perception of comfort. Secondly, direct income data was not available due to reporting bias, which may have underestimated the results. Future efforts should include the investigation of physiological adaptation factors for a holistic perspective on comfort perception within low-income housing.

Acknowledgements Part of this study is funded by the Ministry of Human Resource Development (MHRD), the Government of India (GoI) project titled CoE-FAST, Grant No. 14MHRD005 and IRCC-IIT Bombay Fund, Grant No. 16IRCC561015. Any opinions, findings, and conclusions or recommendations expressed in this material are those of the authors and do not necessarily reflect the views of the MHRD, GoI and/or IRCC-IIT Bombay.

References

1. Fabbri, K. (2015). Indoor thermal comfort perception. In: *A questionnaire approach focusing on children* (pp. 1–302).
2. Rodriguez, C. M., Medina, J. M., Coronado, M. C., & Dálessandro, M. (2019). The development of data-collection methods for thermal comfort assessment in tropical countries. In: *IOP Conference Series: Materials Science and Engineering* (Vol. 603, No. 5).
3. De Dear, R., Brager, G., & Cooper, D. (1997). Developing an adaptive model of thermal comfort and preference-final report on RP-884. *ASHRAE Transactions, 104*(Part 1), 1–18.
4. Hong, T., Langevin, J., Luo, N., & Sun, K. (2020). *Developing quantitative insights on building occupant behaviour: Supporting modelling tools and datasets.* Elsevier Inc.
5. Nutkiewicz, A., Jain, R. K., & Bardhan, R. (2018). Energy modeling of urban informal settlement redevelopment: Exploring design parameters for optimal thermal comfort in Dharavi, Mumbai, India. *Applied Energy, 231,* 433–445.
6. George, P., & Mallery, D. (2003). *SPSS for Windows step by step: A simple guide and reference 11.0 update* (4th ed.). Boston, MA: Allyn & Bacon.
7. Burgos, S., Ruiz, P., & Koifman, R. (2013). Changes to indoor air quality as a result of relocating families from slums to public housing. *Atmospheric Environment, 70,* 179–185.
8. Indraganti, M., & Rao, K. D. (2010). Effect of age, gender, economic group and tenure on thermal comfort: A field study in residential buildings in hot and dry climate with seasonal variations. *Energy Buildings, 42*(3), 273–281.
9. Kshetrimayum, B., Bardhan, R., & Kubota, T. (2020). Factors affecting residential satisfaction in slum rehabilitation housing in Mumbai. *Sustain, 12*(6).
10. Fu, B., Kurisu, K., Hanaki, K., & Che, Y. (2018). Influential factors of public intention to improve the air quality in China influential factors of public intention to improve the air quality in China. *Journal of Cleaner Production, 209,* 595–607.
11. Malik, J., & Bardhan, R. (2020). Thermal comfort in affordable housing of Mumbai, India. In: *Energise-energy innovation for a sustainable economy* (pp. 66–74).
12. Zhang, Y. (2018). The credibility of slums: Informal housing and urban governance in India. *Land Use Policy, 79,* 876–890.
13. UKERC. (2007). *The Rebound Effect: an assessment of the evidence for economy-wide energy savings from improved energy efficiency* (Vol. 42, No. 4).

Governance of Domestic Water Supply in Small and Medium Towns of Karnataka

Aparna Ramesh and A. Mukherjee Basu

Abstract Following the liberalisation of the Indian economy in 1991, the framework of urban domestic water supply governance in Small and Medium Towns (SMTs) in Karnataka has been primarily impacted by the devolution of functions and finances through the 74th Constitutional Amendment Act 1992 and the conditional urban reforms led by multilateral financial institutions (MFIs). However, despite continued investments in institutional development, the urban local governments (ULGs) are constrained by poor financial resources and limited technical and managerial capacities. Therefore, this study attempts to understand how past reforms, particularly those in place as a result of conditional multilateral lending, helped in improving municipal capacities for domestic water supply provision in Karnataka's SMTs. Further, the study aims to identify the challenges and constraints still faced by ULGs in Karnataka's SMTs in the process of domestic water supply provision. The critical review of multilateral, national and state-level policies, programmes and associated reforms for urban water supply provision along with a comprehensive examination of municipal governance in two SMTs, Shivamogga Municipal Corporation and Chitradurga City Municipal Council revealed that coercive MFI-led reforms have resulted in the limited devolution of functions, centralised planning with a decentralisation of financial risks to ULGs, ad hoc institutional arrangements with state-level bureaucrats taking key decisions instead of democratically elected ULGs, private consultants and parastatal agencies steering the urban development agenda, key local level functions delegated to the district administration and finally, irrespective of varying size and capacities, all SMTs are prescribed uniform policies, legislations and recommendations.

Keywords Urban governance · Water supply · Small and medium towns · Multilateral financial institutions · Urban local government

---

A. Ramesh (✉) · A. M. Basu
CEPT University, Ahmedabad, India
e-mail: aparnaramesh237@gmail.com

A. M. Basu
e-mail: anurima.mb@cept.ac.in

A. Jana and P. Banerji (eds.), *Urban Science and Engineering*, Lecture Notes in Civil Engineering 121, https://doi.org/10.1007/978-981-33-4114-2_24

1 Introduction

Studies on urbanisation across the developing world have noted that Small and Medium Towns (SMTs) play an important role in the overall development of any region and are often seen as a viable alternative to the concentration of industrial productivity, high-income employment and economic development in the handful of million-plus cities [1–5]. Despite the important role of SMTs in India's urbanisation trajectory, their condition today is cause for much concern due to high levels of poverty, a large share of informal employment, lack of access to basic services, limited ability to raise their own revenues and limited technical capacities with the urban local governments [2, 6, 7].

The most recent definition provided in the Urban Infrastructure Development Scheme for Small and Medium Towns (UIDSSMT) programme guidelines terms all urban settlements with population ranging from 20,000 to 1 Million as SMTs [8]. Further, the Census of India noted that in 2011, 224 million people, accounting for 60% of India's urban population, lived in Class I towns (urban settlements with a population greater than 0.1 million) [9]. Moreover, 82% of the 502 Class I settlements had populations in the range of 0.1–0.5 million and 27% of India's urban population lived in such towns. Therefore, the present study was focussed solely on SMTs with a population between 0.1 million to 0.5 million.

2 Background: SMTs in Karnataka

With 23.6 million urban residents, the level of urbanisation in Karnataka is higher than the national average at 38.6% [9]. 22 SMTs (with a population between 0.1 and 0.5 Million) across Karnataka account for 22% (5.1 million) of the urban population. Further, the number of such SMTs has increased from 15 in 1991 to 22 in 2011.

Further, Karnataka has two legislations that deal with urban local governments in SMTs—the Karnataka Municipalities Act 1964 (KM Act 1964) and the Karnataka Municipal Corporations Act 1976 (KMC Act 1976) [10]. City Municipal Councils, Town Municipal Councils and Town Panchayats are governed by the former while Municipal Corporations are governed by the latter. Additionally, legislative provisions for the function of urban planning are provided under the Karnataka Town and Country Planning Act, 1961.

Despite the differences in sizes of each ULG, the functions devolved through the KM Act 1964 and the KMC Act 1976 are quite similar. Further, the scope of the function of 'Water supply for domestic, industrial and commercial purposes' has been devolved in a limited manner as an Obligatory function listed under Section 87(b) 'watering public streets and places' of KM Act 1964 and under Section 58(2) 'the watering and cleansing of all public streets and public places in the city' of KMC Act 1976.

In 1994, Amending Act 35 and Amending Act 36 amended the provisions of the KMC Act 1976 and the KM Act 1964 respectively to give effect to the provisions of 74th Constitutional Amendment Act. Soon after, the 11th Central Finance Commission as well as the Planning Commission's Working Group on Expenditure Norms, listed 'water supply' as a core function to be assigned to ULGs [11]. Yet, while several other functions listed under the 12th Schedule, such as 'maintaining schools for pre-primary education', 'slum improvements and up-gradation', 'urban forestry, protection of environment and promotion of ecological aspects' and 'urban poverty alleviation', were devolved through these Amending Acts 35 and 36, the scope of water supply provision by ULGs remained unchanged.

3 The Karnataka Model of Urban Development

Post liberalisation of the Indian economy in 1991, funding for development programmes from the central government declined and 'second generation reforms' for improvements in governance became an area of increased focus of both central government and multilateral financial institutions (MFIs), such as the World Bank and Asian Development Bank [12, 13]. Karnataka agreed to adopt rapid reforms in fiscal and public expenditure reform and administrative reform for better service delivery [12]. As a result, it was able to secure several sector-specific funding agreements from the Asian Development Bank (ADB) and the World Bank for various projects in urban development such as water supply, drainage, sewerage and public transport (Table 1).

4 Central Research Question, Aim and Objectives

The central research question for the present study is—how have past reforms, particularly those in place as a result of conditional multilateral lending, helped in improving municipal capacities for domestic water supply provision in Karnataka's SMTs. The scope of this study is limited to studying legislations, policies and programmes focused on urban domestic water supply provision and associated municipal governance reforms directed at SMTs in the State of Karnataka. It is supported by the preparation of in-depth case studies of two such SMTs.

5 Methodology

The objectives of the study were achieved in the two stages. The first stage involved a critical review of the policies, programmes and legislations of urban local government and urban water supply in the state of Karnataka. The findings from this initial study

Table 1 Focus areas of projects funded by multilateral financial institutions in Karnataka's SMTs [14–16]. Source: Compiled from various sources by author

Year	MFI	Name	Project components/focus areas
1995	ADB	Karnataka Urban Infrastructure Development Program (KUIDP)	Environmental sanitation Road improvement and truck and bus terminals Poverty reduction Development of industrial sites and services Implementation assistance and institutional strengthening Low-income housing finance
1999	ADB	Karnataka Urban Development and Coastal Environment Management Project (KUDCEMP)	Capacity building, Community participation, and poverty reduction Water supply rehabilitation and expansion Urban environmental improvements Street and bridge improvements Coastal environmental management, Implementation assistance
2001	WB	Karnataka Economic Reforms Loan (KERL)-1	Fiscal and public expenditure management reforms; Administrative reforms; Reforms to promote private sector development Poverty and human development monitoring
2002	WB	KERL-2	
2002	WB	Karnataka Municipal Reforms Project (KMRP)	Support for implementation of State-wide reforms and provision of technical assistance Support for upgrading urban services Bangalore Capacity Building Support and Bangalore Investment Support Project management support to KUIDFC and DMA
2004	WB	Karnataka Urban Water Sector Improvement Project (KUWASIP)	The state-wide urban water policy framework local-level reform Preparation of a business model and private sector participation process for service provision in the three participating ULGs
2006	ADB	North Karnataka Urban Sector Improvement Program (NKUSIP)	Sanitation infrastructure, Water supply infrastructure, Slum improvement, Nonmunicipal infrastructure, Urban transport infrastructure, Institutional development, and Investment Program assistance facility

(continued)

Table 1 (continued)

Year	MFI	Name	Project components/focus areas
2015	ADB	Karnataka Integrated Urban Water Management Investment Program (KIUWMIP)	Strengthen UWSS infrastructure in 3 towns support reforms such as the establishment of an urban local body incentive fund Support capacity development, including project management and administrative capacity
2016	WB	Karnataka Urban Water Supply Modernisation Project (KUWSMP)	City-wide access to a continuous piped water supply in three selected cities Strengthening service delivery arrangements at the city level

were mapped to create a timeline connecting policies, programmes and associated legislative reforms. In the second stage, two SMTs—Shivamogga and Chitradurga, within the population range of 0.1–0.5 million were chosen as case studies to examine municipal governance in practice in greater depth.

Semi-structured interviews were held with local government officials, both elected and appointed, at these two SMTs. Additionally, discussions with stakeholders, such as the local offices of KUWSDB, were also conducted. The budgets of the ULGs available online (2011–2016) were also scrutinised to understand the flow of funds required to carry out projects under various state government and national government programmes.

Two methods were adopted to critically examine municipal governance in practice. Firstly, an accountability framework was prepared to analyse the relationships between the various stakeholders in the recent urban water supply infrastructure projects in the two towns. Next, based on the patterns identified in the distribution of functions and the flow of finances, the various functions involved in the water supply were unbundled and an Activity Map was prepared. On the basis of analysis from these exercises, the challenges to realising an effective and empowered local government in the chosen SMTs were identified.

6 Karnataka: A Critical Review of Policies, Programmes and Legislations

Karnataka's history of urban reforms can be divided into three distinct phases—the first phase between 1995 and 2004 which saw the initial wave of lending from MFIs, the second phase between 2005 and 2013 which saw the convergence of national and state-level agendas for urban development and finally, the third phase from 2014 till the present day where water and sanitation in SMTs have emerged as areas of continued interest for both national and state governments, as well as MFIs (Table 1).

7 Shivamogga and Chitradurga: Examining Municipal Governance in Practice

A brief overview of the selected case study SMTs and their location in the state of Karnataka are provided in Table 2 and Fig. 1, respectively.

7.1 Projects and Processes: Analysing Implementation of Urban Water Supply Projects Through Accountability Frameworks

Interviews with public officials and an analysis of the two ULG's annual budget reports revealed that, in the last decade alone, funds from four different programmes have been used to invest in urban water supply projects in the two towns—(i) Chief Minister's Small and Medium Town Development Programme (CMSMTDP), (ii) Mukhya Mantri Nagarothana Yojana—Special Grant of Rs. 100 crore Program (100 cr. Program) (iii) Karnataka Municipal Reforms Project—Municipal Investment Component (KMRP) and (iv) Atal Mission for Rejuvenation and Urban Transformation (AMRUT) (Table 3).

While AMRUT is a national urban development scheme of the central government, CMSMTDP and the 100 cr. programmes are state government programmes

Table 2 Overview of the selected towns. Compiled by author from various sources [9]

Parameter	Shivamogga	Chitradurga
ULG	Shivamogga Municipal Corporation	Chitradurga City Municipal Council
Governing legislation	Karnataka Municipal Corporations Act 1976	Karnataka Municipalities Act 1964
Population (Census 2011)	322,650	145,853
Decadal growth rate	18% (2001–2011)	15.87% (2001–2011)
Area	77.10 km^2	29.5 km^2
Geographical location	Eastern foothills of Western Ghats	Rocky, drought-prone peninsular plateau
Annual rainfall	1813 mm	680 mm
Major water sources	Major rivers—Tunga and Bhadra Smaller rivers—Kali, Gangavathi, Sharavathi and Tadadi Nine major water reservoirs in the district, including the Linganamakki, Ambligola and Tunga reservoir	Lies in the Vedavathi river basin, which is a tributary of the Krishna river Existing sources of groundwater are subject to over-exploitation and contamination from nitrates and fluorides

Fig. 1 Location of Shivamogga and Chitradurga in the state of Karnataka

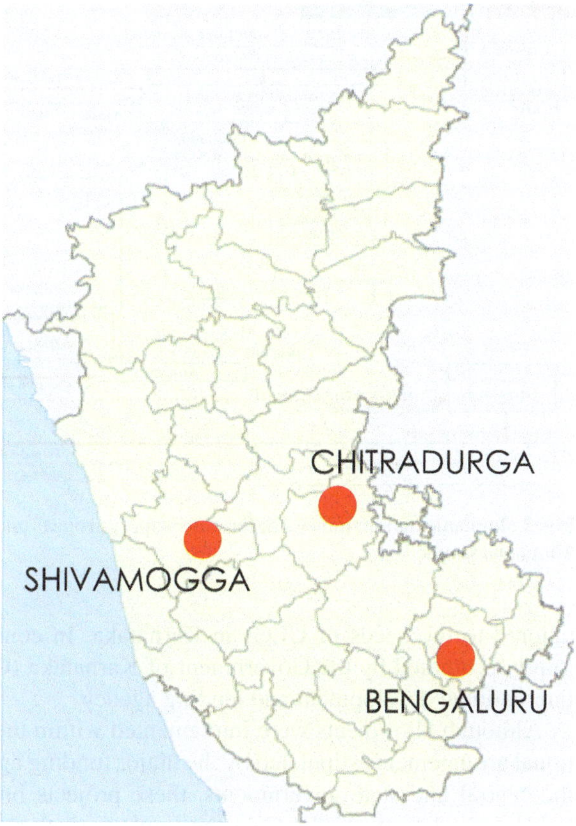

Table 3 Recent investments in urban water supply infrastructure in Shivamogga and Chitradurga

Programme	ULG	Funding pattern
Chief Minister's Small and Medium Town Development Programme (CMSMTDP)	CCMC	50% grant from state government 50% loan from KUIDFC
Karnataka Municipal Reforms Project—Municipal Investment Component (KMRP)	CCMC	40% Loans from KUIDFC 50% grants from state government 10% from ULG revenue
Mukhya Mantri Nagarothana Yojana—Special Grant of Rs. 100 crore Program (100 cr. Program)	SMC	50% grant from state government 50% loan from KUIDFC
Atal Mission for Rejuvenation and Urban Transformation (AMRUT)	SMC and CCMC	50% grants from central government 20% grants state government 30% from ULG revenue

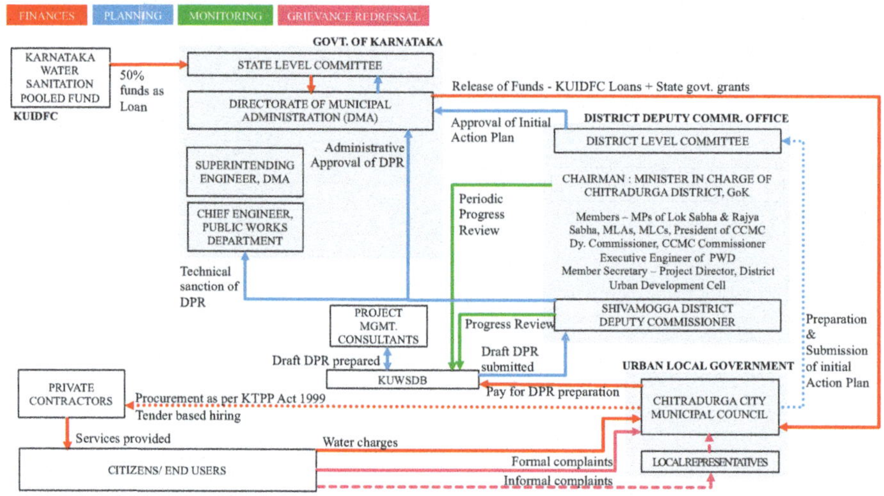

Fig. 2 Implementation framework for water supply projects under CMSMTDP in Chitradurga City Municipal Council

tailored to the needs of ULGs in Karnataka. In contrast, KMRP is a programme jointly promoted by the Government of Karnataka (GoK) and the World Bank, an international development and lending agency.

Although all projects were implemented within the municipal limits, the institutional arrangements stipulated by the major funding agents—the World Bank as well the central and state governments, these projects brought together various stakeholders apart from the ULG in the implementation process. Each of these stakeholders shared varying levels of accountability to the ULG, the democratically elected government of the town (Figs. 2, 3, 4 and 5). Details of the implementation process were gathered from published guidelines for each of the projects as well as from interviews with local and state government officials. Accountability frameworks detailing the project planning and implementation processes along with agreements for operations and maintenance were prepared for urban water supply projects undertaken under these programmes in the two SMTs.

7.2 *Activity Mapping: Unbundling the Functions Under Urban Domestic Water Supply*

In order to further scrutinise the similarities and dissimilarities in the implementation arrangements of the four different projects, an Activity Map was prepared. The Activity Mapping technique was developed to appraise the inter-governmental assignment of functions for rural local governments in India. However, such an

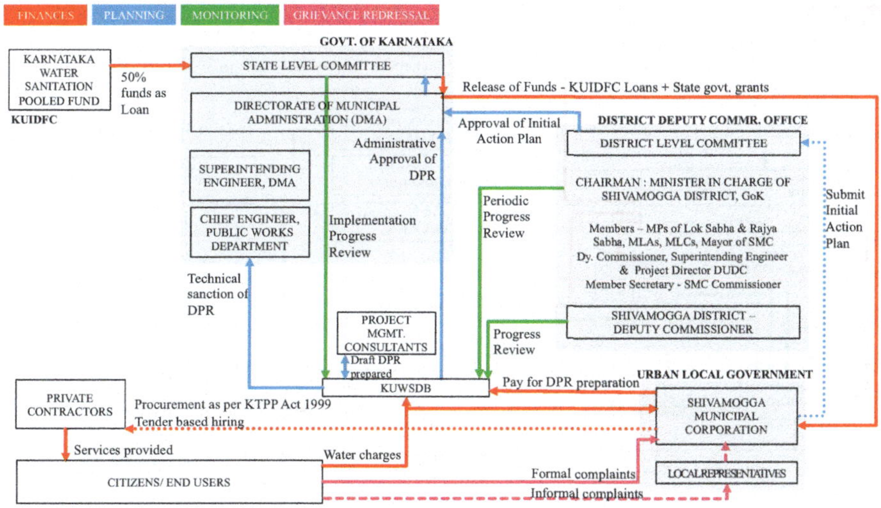

Fig. 3 Implementation framework for water supply projects under Mukhya Mantri Nagarothana Yojana, Special Grant of Rs. 100 crore Program in Shivamogga Municipal Corporation

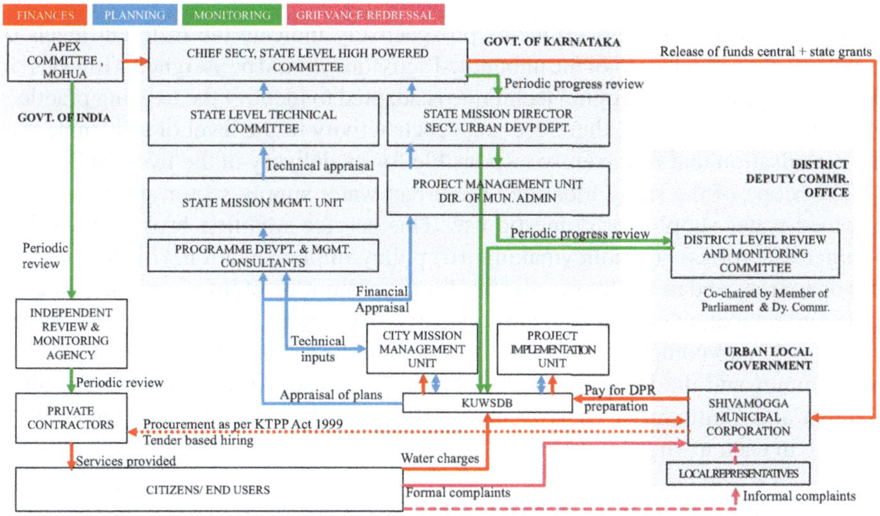

Fig. 4 Implementation framework for water supply projects under Atal Mission for Rejuvenation and Urban Transformation (AMRUT) in Shivamogga Municipal Corporation

activity map has never been prepared before for the function of domestic water supply in urban areas or any of the other functions carried out by ULGs.

The first step in the activity mapping process is the identification of an individual sector and a particular service under that sector. Next, the service is unbundled or broken down into smaller components, namely activities. Finally, the activity maps

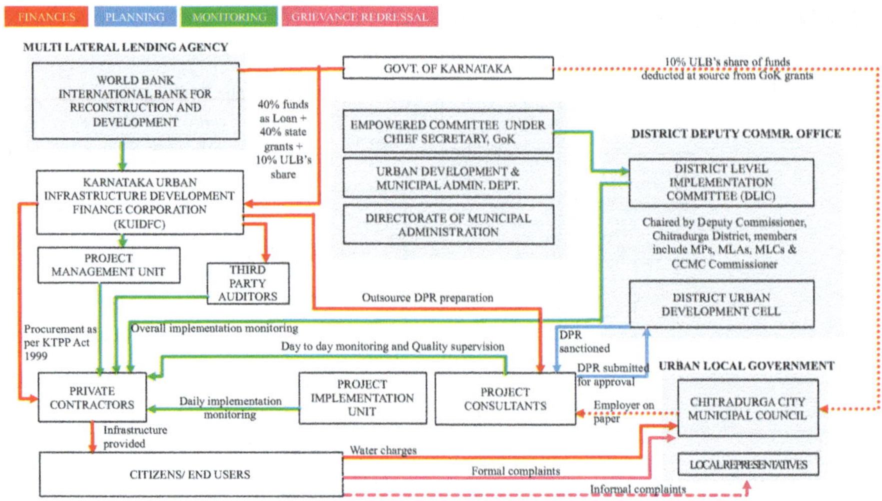

Fig. 5 Implementation framework for water supply projects under the Karnataka Municipal Reforms Project (KMRP)—Municipal Investment Component in Chitradurga City Municipal Council

prepared for rural local governments prospectively indicate the different levels of government to which each of the unbundled activities should be assigned. However, in this study, the activity mapping technique is adapted to identify the existing practices of urban governance and, therefore, map each activity to the level of government or the organisation that is currently responsible for its delivery in the two SMTs.

The scope of the study identified the urban water supply sector and the service of piped water supply for domestic use. This service was first broken down into 5 larger categories—(i) policymaking, (ii) policy implementation, (iii) monitoring, (iv) operations and maintenance and finally, (v) promotion of information, education and communication campaigns. Subsequently, these categories were further unbundled into 41 sub-components (Table 4). Based on the review of legislation, policies, programmes and the in-depth analysis of the four case study projects in the Shivamogga and Chitradurga, activity maps indicating the devolution of functions in practice in each town were prepared.

8 Challenges and Constraints

Based on the critical review of relevant policies, programmes and legislations from these periods certain key themes from Karnataka's history of urban reforms were identified. Coercive MFI-led reforms have resulted in the limited devolution of function, top-down policy shifts towards increased private sector participation, misdirected measures for institutional capacity development and an increase in bureaucratic decision-making structures parallel to elected local governments. Substantive

Table 4 Brief overview of activity mapping for urban domestic water supply. Source: Compiled by author from various sources

S. No.	Function	Level of government
Policy Making		
1	Articulate legislation on rights and obligations of entities involved in water supply service provision	State
2	Prepare policy on water resource allocation	State
3	Establish targets for service levels and benchmarks	State
4	Develop and approve appropriate technologies for water supply provision	Centre
5	Develop approaches and contractual principles for private sector participation	State
6	Take decisions on allocation of finances	Centre and State
7	Formulate and approve financial models and business plans	State
8	Prepare financial management and accounting guidelines	State
9	Prepare guidelines for tariff setting and collection	State
10	Arbitration and dispute resolution	State
11	Determine staffing requirements	State
12	Develop recruitment guidelines for staff	State
Project Planning and Asset Creation		
13	Project accounting	Local
14	Identify schemes and locations, estimate costs and formulate projects	State
15	Preparation of Detailed Project Reports	State
16	Technically appraise and approve schemes proposed	State
17	Financially appraise and approve schemes proposed	State
18	Procure additional loans/ grants for approved projects from the capital market and national/ bilateral/ multilateral banks	State

(continued)

Table 4 (continued)

S. No.	Function	Level of government
19	Form partnerships with private/ public sector agencies for the implementation of water supply schemes	State and Local
20	Grant/Renew licenses to private service providers	Local
21	Construct/Award contracts for the execution of bulk water supply/major schemes outside ULB limits	State in Shivamogga Local in Chitradurga
22	Construct/Award contracts for the execution of schemes within ULB limits	Local
Monitoring and Evaluation		
24	Periodically collect water samples for tests	State
25	Capture data on service levels	State
26	Collate and analyse performance indicators	State
27	Monitor and supervise the progress, quality of work and target achievement	State
28	Periodically report progress and status of schemes	State
29	Public grievance redressal	Local
30	Appraisal of progress reports, contract monitoring and corporate oversight	State
31	Asset Valuation	Local
Operations and Maintenance		
32	Award management and/or service contracts for operation and maintenance of infrastructure	State in Shivamogga Local in Chitradurga
33	Appoint operators wherever necessary	
34	Operate and maintain drinking water schemes	
35	Tariff setting and revision	Local
36	Billing and collection of water charges	State in Shivamogga Local in Chitradurga

(continued)

Table 4 (continued)

S. No.	Function	Level of government
37	Establish water testing laboratories for control of chemical and biogenic impurities	State
38	Periodically chlorinate open wells and treat water	State
39	Pay for electricity charges required for water supply	Local
Promote Information, Education and Communication campaigns		
40	Capacity building measures for technical staff	State
41	Public education on hygiene and water use	Local

public accountability through representative democracy at the local level has given way to nominal citizen participation. Further, state and national governments were not unwilling victims to MFI-led policy shifts but were active participants in facilitating policy transfer in exchange for external funding.

Based on the four project cases reviewed and the activity map delineating the present intergovernmental assignment of functions, the following challenges faced by SMC and CCMC in ensuring effective urban local governance were identified. It was found that despite promises of institutional development as a part of infrastructure investments, ULGs in SMTs remain critically challenged and constrained in effective governance of domestic water supply. While the process of planning for urban infrastructure remains centralised with minimal participation from ULGs, they are nevertheless required to share the financial burden and reduce risks to the MFIs and private investors. Regardless of the constitutional mandate for devolution of functions to the ULG and the important role of its elected council and standing committees, ad hoc institutional arrangements have been put in place to make critical decisions for each project. On the whole, urban development in SMTs is driven by fragmented efforts of short-term private consultants who collaborate with parastatal agencies to implement individual projects. As the ULGs are considered lacking in capacities to monitor these parastatal agencies and private consultants, several key monitoring and evaluation functions are delegated to the district administration, which essentially acts as the local agent of the state government. Finally, irrespective of size and previous achievements ULGs in SMTs are treated uniformly in policies, legislations and recommendations. Thus, the opportunity to provide targeted improvements to technical, managerial and financial capacities was missed time and again.

The major constraints to effective local governance are in the form of limited involvement of ULGs in shaping state-level legislation on urban development, restrictions on external borrowing without state and central government guarantees, absence of executive authority with the elected representatives of the ULG Council as well as restricted public access to information on urban development processes. While some of these constraints can be overcome through potential changes in legislation and institutional arrangements, others, such as the constitutional provisions of the Indian federal system, act as limits to the possible recommendations for a revised framework for effective local governance.

9 Towards a Revised Framework

In the 25 years since the 74th Constitutional Amendment Act 1992 has come into effect, both state and national government policies and reports have continued to pay lip service to the need for democratic decentralisation and effective devolution of functions, finances and functionaries to urban local governments. However, the examination of Karnataka's history of urban reforms and review of governance in two SMTs in the state showed not only the complete lack of functional and financial autonomy of the ULGs but also the increasing importance of the district administration and state-level parastatal agencies in urban development and service provision. Therefore, there is a dire need to rethink the existing arrangements of urban governance and identify an effective means to ensure successful democratic decentralisation and devolution even in SMTs.

Based on the findings from the assessment of urban governance in SMTs, certain strategic recommendations for effective urban domestic water supply governance in SMTs have been suggested.

i. The clarity in the inter-governmental assignment of functions to local governments vis-à-vis state governments and parastatals for urban planning and development is ensured through legislative amendments and periodic preparation of activity maps. Apart from keeping in mind the principles of economies of scope and scale, heterogeneity in the assignment of functions for ULGs of various sizes and environmental profiles must be considered seriously [17].
ii. On the basis of the devolution of functions, key institutional arrangements are put in place for the long-term through suitable legislation to plan, implement and manage assets created as a part of large-scale urban infrastructure development projects. In order to build municipal capacities over time, it is essential that these institutional arrangements are not modified for each project.
iii. While the ULG itself need not be involved in all aspects of the production and provision of key services such as water supply, it is essential to ensure that all service providers, from both private and public sector, operating within the ULG's jurisdiction remain accountable to the directly elected ULG Council.
iv. Suitable checks and balances to the autonomous functioning of local governments are put in place through multiple effective accountability relationships with citizen groups, local and state-level elected representatives, senior managers and frontline workers at parastatal water utilities as well as the media.

v. To transform the disaggregated information collected through each project, an open framework for sharing data with independent scholars and the general public can be created. This will assist in the generation and sharing of applicable knowledge on urban areas, their governance and development trajectories, thus dramatically increasing the demand for public accountability.
vi. Rethinking political and institutional hierarchies to provide formal and informal arrangements for inter-municipal co-operation and collaboration to negotiate with powerful state-level policymakers and monopolistic parastatal agencies.

With the number of SMTs in Karnataka showing an increasing trend, it is necessary to rethink urban governance frameworks so that citizens in these towns have equitable access to not just basic urban infrastructure, public services and opportunities but also platforms for democratic participation.

Acknowledgements The authors are sincerely thankful to CEPT University for providing financial as-sistance for this Directed Research Project.

References

1. Satterthwaite, D., & Hardoy, J. (1986). *Small and intermediate urban centres; Their role in national and regional development in the Third World.* Hodder and Stoughton.
2. Raman, B., Prasad-Aleyamma, M., Bercegol, R. D., Denis, E., & Zérah, M.-H. (2015). *Selected readings on small town dynamics in India.* USR 3330 "Savoirs et Mondes Indiens".
3. Satterthwaite, D., & Tacoli, C. (2003). *The urban part of rural development: The role of small and intermediate urban centres in rural and regional development and poverty reduction.* International Institute for Environment and Development.
4. Kundu, A. (2011). *Trends and processes of urbanisation in India.* London: Human Settlements Group International Institute for Environment and Development (IIED).
5. Denis, E., & Zérah, M.-H. (2017). *Subaltern urbanisation in India an introduction to the dynamics of ordinary towns.* New Delhi: Springer.
6. Sharma, K. (2012). Rejuvenating India's small towns. *Economic and Political Weekly*, 63–68.
7. Bercegol, R. D. (2017). *Small towns and decentralisation in India.* Springer India .
8. Ministry of Urban Development, Government of India. (2005). *Urban infrastructure development scheme for small and medium towns.* Government of India.
9. Office of the Registrar General & Census Commissioner, India. (2018). *District census hand book—Town amenities.* Retrieved April 2019, from https://censusindia.gov.in/2011census/dchb/DCHB.html.
10. Directorate of Municipal Administration, Karnataka. (2006). *Acts & rules.* Retrieved April 2019, from Directorate of Municipal Administration, Karnataka. https://www.municipaladmn.gov.in/sites/municipaladmn.gov.in/files/final_kmab_rules_2006_0.pdf.
11. Chaubey, P. (2003). *Urban local dodies in India: Quest for making them self-reliant. National seminar on municipal finances.* New Delhi: Indian Institute of Public Administration.
12. Baindur, V., & Kamath, L. (2009). *Reengineering urban infrastructure: How the World Bank and Asian Development Bank shape urban infrastructure finance and governance in India.* New Delhi: Bank Information Center, South Asia.
13. Walters, V. (2013). *Water, democracy and neoliberalism in India: The power to reform.* Routledge.

14. World Bank. (2018). *Projects and operations*. Retrieved April, from World Bank: https://projects.worldbank.org.
15. Asian Development Bank. (2019). *Projects and tenders*. Retrieved April 2019, from Asian Development Bank: https://www.adb.org/projects.
16. Baindur, V., & Budhya, G. (2011). *Unpacking the conditions and conditionalities through urban sector reforms in Karnataka*. Bangalore: Action Aid Regional Office.
17. Commission, S. A. R. (2007). *Sixth report: Local governance*. New Delhi: Government of India.
18. Kundu, A., & Sarangi., N. (2005). Employment guarantee: Issue of urban exclusion. *Economic and Political Weekly*, 36–42.
19. Lanjouw, P., & Murgai., R. . (2011). *Perspectives on poverty in India*. Washington: World Bank.
20. Ministry of Urban Affairs & Employment, Government of India. (1995). *Integrated development of small and medium towns*. Retrieved April 2019, from Ministry of Housing and Urban Affairs, Government of India. https://mohua.gov.in/upload/uploadfiles/files/90.pdf
21. Bhagat, R. (2011). Emerging pattern of urbanisation in India. *Economic and Political Weekly*, 10–12.
22. Dupont, V. (2002). Le monde des villes. In M.-C. S.-Y. (Ed.), *Population et développement en Inde* (pp. 55–84). Paris: Ellipses.
23. Shaw, A. (2013). Emerging perspectives of small cities and towns. In R. N. Sharma, & R. S. Sandhu (eds.) *Small cities and towns in global era: Emerging changes and perspectives* (pp. 36–53). Rawat Publications.
24. Véron, R. (2010). small cities, neoliberal governance and sustainable development in the global south: A conceptual framework and research agenda. *Sustainability*, 2833–2848.
25. Dhaliwal, S. S. (2004). *Urban infrastructure development in small and medium towns*. New Delhi: Deep & Deep Publications Pvt. Ltd.
26. United Nations Population Fund. (2007). *State of world population 2007*. UNFPA.
27. United Nations, Department of Economic and Social Affairs, Population Division. (2014). *World urbanization prospects: The 2014 revision*. New York: United Nations.
28. Directorate of Municipal Administration, Karnataka. (2009, June). *Government order scheme guidlines*. Retrieved April 2019, from Chief Minister's Small and Medium Towns Development Programme (CMSMTDP). https://www.municipaladmn.gov.in/sites/municipaladmn.gov.in/files/G.O%20scheme%20guidelines%20English.pdf.
29. Sarkar, A. (2018). *World Bank's reformed model of development in Karnataka*. Bengaluru: The Institute for Social and Economic Change.
30. Mitra, S. G. (2010). *World Bank and urban water supply reforms in India: A case study on Karnataka*. Retrieved April 2019, from The University of Manchester Library: https://www.escholar.manchester.ac.uk/jrul/item/?pid=uk-ac-man-scw:92889.
31. Connors, G. (2005). When utilities muddle through: Pro-poor governance in Bangalore's public water sector. *Environment & Urbanization*, 201–217.
32. Hoque, S. F. (2012). *Urban water sector reforms in india: financing infrastructure development through market—Based financing and private—Public partnerships*. Singapore: Lee Kuan Yew School of Public Policy.
33. Nadhamuni, S. (2012). *An approach to Integrated Urban Water Management (IUWM) The Mulbagal Experience*. Bengaluru: Arghyam.
34. Lele, S., Srinivasan, V., Jamwal, P., Thomas, B. K., Eswar, M., & Zuhail, T. M. (2013). *Water management in Arkavathy basin: A situation analysis*. Bengaluru: Ashoka Trust for Research in Ecology and the Environment.
35. Lele, S., Madhyastha, K., Sulagna, S., Dhavamani, R., & Srinivasan, V. (2018). Match, don't mix: implications of institutional and technical service modalities for water governance outcomes in south Indian small towns. *Water Policy*, 12–35.
36. Karnataka Municipal Data Society. (2017). *Home*. Retrieved April 2019, from Shimoga City Corporation: https://www.shimogacity.mrc.gov.in/.
37. Ministry of Urban Development. (2009). *Handbook of service level benchmarking*. New Delhi: Government of India.

Feasibility Analysis of PV-Battery and PV-TES for Cooling Application in Buildings

Brijesh Pandey and Rangan Banerjee

Abstract Battery storage is often used in PV-grid connected system to offset the peak electricity demand but is an economically costly option. Thermal energy storage (TES) could be an economically viable option due to its less installation and O&M costs. Literature shows techno-economic feasibility of the PV-TES and PV-battery and, also reveals that PV-TES leads to less self-consumption of the PV compared to PV-battery due to poor control, which causes misalignment of the PV generation and TES operation. In this study, an advanced control strategy, i.e., Model Predictive Control (MPC) is proposed to operate PV-TES for cooling of the built environment and compare its performance with PV-battery. For the techno-economic comparison of the PV-battery and PV-TES using MPC, a co-simulation framework between EnergyPlus (Building Energy Simulation tool) and MATLAB for MPC deployment is used. An air-conditioned building having a floor area of 463 m^2 and residential operation schedule integrated with PV-battery (PV installed capacity 32 kW and battery capacity 86 Ah) and a TES tank (capacity 0.6 GJ) is modelled with EnergyPlus. A MATLAB MPC toolbox is used to formulate and train the MPC model. Feasibility of the considered energy storage systems has been quantified by PV-ratio, Energy storage self-consumption (ESSC) and levelised cost of storage (LCOS). The present study addresses the key question of cost-effectiveness, attractiveness and ready to make essential contributions to flexibility (in demand response of power to cooling) of PV-TES over PV-battery.

Keywords PV · Thermal energy storage · Demand response · Battery · Model predictive control

B. Pandey (✉) · R. Banerjee
Department of Energy Science and Engineering, IIT Bombay, Mumbai, India
e-mail: brijeshpandeyby715@gmail.com

A. Jana and P. Banerji (eds.), *Urban Science and Engineering*, Lecture Notes in Civil Engineering 121, https://doi.org/10.1007/978-981-33-4114-2_25

1 Introduction

The use of air conditioners and electric fans to stay cool accounts for 20% of the total electricity consumption worldwide in the building sector [1–3]. This trend is likely to increase in the future due to the shifting of the world's economic growth to the south of the world, that is, tropical countries and an increase in living standards. Nearly 70% of all the air conditioners are globally in residential buildings and has a major share in the household's electricity bill. Solar photovoltaic (PV) installed on the rooftop of the residential buildings can contribute in reducing the grid electricity demand and consequently electricity bill. However, electricity generated by PV and electricity demand of the households does not match always and hence energy storage becomes essential to utilize the PV-generated electricity.

Luthander et al. [4] has investigated that for the off-grid households, energy storage is a necessity and he further mentioned that energy storage strategies are popularly being adopted in on-grid households to increase the PV self-consumption and provide services to the electricity grid. In literature, mainly batteries are mentioned as a popular energy storage option to store PV-generated electricity and supply it to the household's utilities when the demand arises [5, 6]. Very few studies are mentioned pertaining to the use of batteries for cooling application in buildings [7]. Thermal energy storages are popular options to store PV-generated electricity for the heating/cooling application in buildings [8]. However, the integration of energy storage options with the PV systems and cooling systems is decided not only on the performance of the energy storage solution but especially on its economic viability, self-consumption and efficiency in storing electricity supplied by solar PV [9].

Merei et al. [10] performed a sensitivity analysis for a supermarket deriving energy need from the PV integrated battery storage system in Germany. It has been observed that the battery storage option is viable only when the prices fall below 200 euro/kWh. Further, it has been concluded that for different commercial load profiles, battery storage might be feasible. Arteconi et al. [11] analysed a case study where PV system integrated with the chilled water tank has been used for demand-side management of an industrial building. Building requires cooling during the daytime on weekdays and PV system is used to charge the chilled water tank on weekends. The author has used a dynamic simulation model to predict the performance of the system and economic calculations. It has been observed that system's cooling electricity demand increases; however, the energy cost decreases because of the cheap PV energy used for charging the chilled water tank. While economic calculation, it has been observed that the initial cost of the chilled water tank was not considered. Adding the initial investment cost of the thermal storage system may increase the payback period of the PV integrated cooling systems.

Parra et al. [12] have conducted the techno-economic analysis for the PV-coupled battery and hot water tanks for a dwelling in UK. They have reported that among the batteries Li-ion battery has lower levelised cost than lead-acid (PbA) battery due to greater round-trip efficiency and cycling capability. They further observed that the best economic case was for the hot water tanks with a size ranging between 100 and

200 l, which were able to achieve the internal rate of return higher than the discount rate, especially when the hot water tank was already in the dwellings. But the author has reported that PV self-consumption is low for the hot water tank compared to the PV coupled battery system due to poor control of transferring the excess power generated by solar PV to the hot water tank.

It is clear from the literature that due to a mismatch of supply of power generated by PV system and demand profile of the building, the control stability will be challenged and the control states in the air conditioning system will experience a rather serious fluctuation before reaching a new control balance. Therefore, an advanced model predictive control (MPC), considering its robustness on the control, would be preferable to be used for the optimal control issues of the central air conditioning system during supply and demand mismatch of the PV coupled energy storage system for the cooling application in buildings. It is also evident from the literature that techno-economic analysis of the PV integrated thermal energy storage systems and battery storage system has been reported individually, but there is little research available comparing the performance and economics of the PV integrated thermal storage system vis-à-vis PV integrated battery storage system for cooling application in buildings. It is reported in the literature that chiller plant control is crucial for the integration of a thermal energy storage system with the PV.

The objective of this study is to apply the model predictive control strategy for the chiller plant loop control in the PV integrated thermal energy storage system and compare its performance and economics with the PV-battery system. A residential reference building is analysed with the PV-thermal energy storage system and PV-battery system with the proposed methodology for the warm and humid climatic conditions. A major portion of the building's energy demand accounts for the cooling energy demand. An energy simulation model has been developed with the help of EnergyPlus, a whole building simulation tool.

2 Scenario Description

In this paper, a grid-connected house in the warm and humid climate of Mumbai is analysed with regards to the integration of PV coupled energy storage. The house considered is one storey and has a gross floor area of roughly 463 m^2 with a typical residential operation schedule. The air conditioning need of the house is served through 20 kW water-cooled variable air volume chiller for all days with cooling setpoint ranging from 22 to 26 °C. The typical operation schedule of the house used in this study is shown in Fig. 1.

It can be observed from the figure that the maximum load occurs during night-time when solar energy is not available. Two energy storage options are considered in this study viz., battery (electrical energy storage) and ice tank (thermal energy storage) for cooling application in the house. For the PV coupled battery system battery is sized for eight hours of power backup for the cooling application. Excess power generated from the PV is stored in the battery through an inverter and is used during

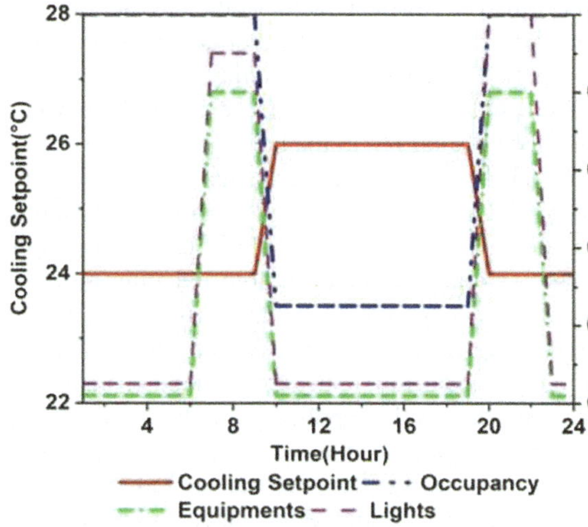

Fig. 1 Typical operation schedule of building

night-time when the solar energy will not be available. In the case of PV coupled TES system, an ice tank of capacity (0.6 GJ) is connected to the central chiller system. Excess energy generated by PV is used to charge the ice tank by turning the central chiller system on. During night-time, when the solar energy will not be available, electricity will be used from the grid and the central chiller will be off. Only the pump and fan will be in operation to circulate the heat transfer fluid from the ice tank to meet the cooling demand of the built environment. As chiller accounts for the major part of the cooling electricity consumption, integration of the thermal energy storage system with the chiller and use of PV electricity to charge the thermal energy storage to meet the cooling demand shifts the peak cooling electricity demand to off-peak hours.

Annual simulation has been performed for the Mumbai climatic conditions. The building load profile, building components data, weather data for the location of Mumbai, PV system size as well as battery and ice tank size serve as inputs for the simulation model.

3 Simulation Models

The simulation model of the PV coupled battery and thermal energy storage along with the building has been modeled using EnergyPlus 9.2. Figure 2 shows schematics of the different simulation models. Figure 2a shows that the model contains the PV array, battery and a grid-connected inverter that supplies electricity to the central chiller as well as to the grid in case of excess generation. In case of insufficient solar power availability, the deficit power between the supply from solar PV and demand from the house is fulfilled by the grid electricity. The battery stores excess

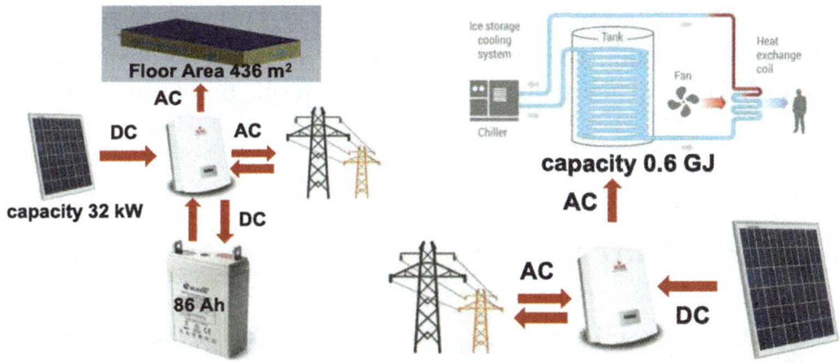

(a) (b)

Fig.2 Schematic of **a** PV-battery and **b** PV-TES system for grid-connected house

power generated by solar PV until it is charged at 90% state of charge instead of supplying all excess electricity to the grid and supplies the load to the house until it is discharged at 10% of state of charge. The grid supplies to the house when battery is fully discharged. This phenomenon increases the self-consumption of solar PV.

Figure 2b shows that the model contains PV arrays, an inverter and an ice tank connected to the central chiller plant. The inverter supplies excess electricity to the chiller plant during sunshine hours to charge the ice tank so that it can be used during night-time to meet the cooling demand. During night-time, electricity is provided by the grid to the house to run the utilities. During night-time chiller plant remains off, and only the pump and fan are switched on to circulate the heat transfer fluid from the ice tank to meet the cooling demand. The thermal load on the ice tank connected to the chiller is determined by the inlet flow rate, inlet fluid temperature and outlet fluid temperature. A positive load indicates a request for cooling and the tank discharges. A negative value indicates a request for charging the ice tank. A zero load indicates that no cooling is required, and flow bypasses the tank.

3.1 *Building Model Validation*

The building considered in this study has been taken from the EnergyPlus reference building template [13]. PV is modeled using the simple model where usable electrical power produced by PV can be given as

$$p = A_{surface} \times f_{active} \times G_T \times \eta_{cell} \times \eta_{invert}$$

where, $A_{surface}$ is the net area of PV surface, f_{active} is the fraction of surface area with an active solar cell, G_T is the total solar radiation incident on the PV surface, η_{cell} is the module conversion efficiency and η_{invert} is the DC (direct current) to AC (alternating current) conversion efficiency. The battery model of Manwell et al. [14]

and the detailed ice storage tank model of Strand et al. [15] have been incorporated into the EnergyPlus model. The detailed PV-battery and PV-TES integrated building models have been validated individually with the respective example template of EnergyPlus.

3.2 Model Predictive Control

In order to increase the self-consumption of PV system, an advanced control mechanism model predictive control [16] is implemented to control the chiller power. Dry bulb temperature, solar insolation on the house, the internal heat gain due to the equipment, occupants and lights affect the occupant's thermal comfort. In this study, these variables are used as disturbances to the MPC model and a predefined setpoint is provided as a reference value to the MPC model. MPC solves the objective function, that is, minimization of error between the observed variable (zone mean air temperature) and reference (setpoint) by manipulating the chiller power over the prediction horizon. The manipulated power demand of the chiller helps in utilising the PV-generated electricity efficiently and thus increases the PV self-consumption. MPC formulation can be written as

$$\begin{aligned} &Minimize(z) = \sum_{i=1}^{prediction\,horizon} \left(T_{predicted} - T_{reference}\right)^2 \\ &0 \leq Chiller\ power \leq 5.7\,\text{kwh} \\ &22\,^{\circ}\text{C} \leq Observed\ variable\,(air\ temperature) \leq 26\,^{\circ}\text{C} \end{aligned} \tag{1}$$

4 Performance and Economic Indicators

The impact of the energy storage system on the cooling application is quantified by two different performance indicators, viz. PV-ratio and energy storage self-consumption.

4.1 PV-Ratio

PV ratio, defined in Eq. (2), is the ratio between the amount of annual PV generation supplied to the ES system (i.e. ES charge) and the total annual PV generation.

$$PV_{Ratio} = \frac{Annual\ PV\ generation\ supplied\ to\ ES}{Total\ annual\ PV\ generation} \tag{2}$$

4.2 Energy Storage Self-consumption

Energy storage self-consumption defined in Eq. (3) is the ratio of requirement met by the ES system's discharge to the annual demand (electricity in case of battery and chilled water in case of ice tank).

$$ES_{Self\ Consumption} = \frac{Discharge\ of\ ES}{Cooling\ demand} \tag{3}$$

4.3 Levelised Cost of Energy Storage

Levelised cost of energy storage defined in Eq. (4) is calculated for the comparison of the economic benefits of the PV integrated TES and battery systems. LCOS is defined as the ratio of the life cycle cost of the energy system to the lifetime energy discharge of the energy storage system and can be written as

$$LCOS = \frac{Investment\ cost + \sum_i^t \frac{O\,\&\,M\ Cost}{(1+r)^i}}{\sum_i^t \frac{E_{dis}}{(1+r)^i}} \tag{4}$$

The numerator of Eq. (4) assumes that all investment costs are incurred in the first year, and it is summed with recurring cost of each year (i) up to the system's lifetime (t), discounted by the discount rate (r). In denominator, discharge power of the energy storage systems is discounted by the discount rate up to the systems' lifetime. For the LCOS calculation discount rate is assumed as 10%, lifetime of the system is assumed as 20 years. Investment and maintenance cost for battery is considered as 350 and 6.5 ($/kWh), respectively. The considered investment cost for ice storage tank is 2254 $ and O & M cost is 5% of the investment cost.

5 Results and Discussion

Figure 3 shows the dynamic simulation data of PV-battery and PV-TES system for the summer design week (19th May to 23rd May) of Mumbai. From the figure, it can be observed that in the case of PV-battery system battery is charged during the sunshine hours and it supplies electricity to chiller by discharging during night-time. The figure also shows that during early morning battery is not able to meet the electricity demand for cooling and deficit electricity to run the chiller is supplied through the grid. Figure 3 shows that in the case of PV-TES system thermal energy storage, that is, ice tank decouples the chiller operation from the cooling electricity demand, which means that the chiller operates only when the solar power is available. The

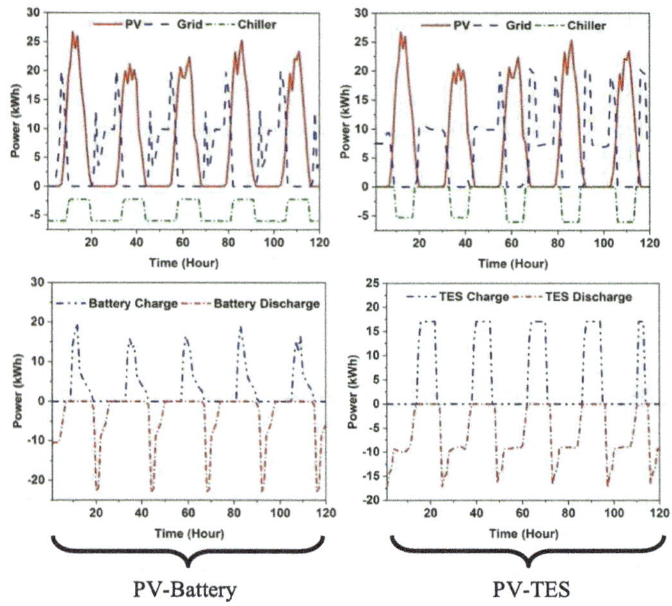

Fig. 3 Dynamic simulation data for the summer week of Mumbai for PV coupled battery (left) and PV coupled TES (right)

figure shows that during night-time, power needs to be imported from the grid. The ice tank is mainly charged once the chiller turns on and peaks in the afternoon when solar insolation remains at maximum. In the night-time, when demand for cooling arises, heat transfer fluid is flown through the ice tank and supplied to the occupant's zone with the help of a pump and fan to meet the cooling demand.

Figure 4a shows the self-consumption of energy storage for battery (electricity energy storage) and ice tank (thermal energy storage). It can be observed that for fixed cooling demand of house energy storage, self-consumption is lower for PV coupled TES compared to PV coupled battery systems. The reason behind the low

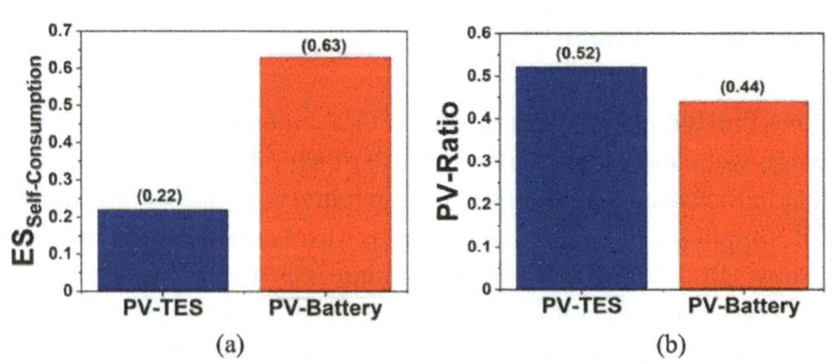

Fig. 4 **a** PV self-consumption and **b** PV ratio for the PV-TES and PV-battery system

self-consumption of energy storage for PV-TES system is the loss of energy from the tank surfaces. The self-consumption of the PV-TES system can be increased by better insulation of the TES system and through optimisation of the chiller operation and its power consumption.

Figure 4b shows the PV ratio of PV-battery and PV-TES system. It can be observed that for fixed PV-generated electricity, the amount of electricity used to charge the energy storage system to meet the cooling demand is higher for the PV-TES system compared to the PV-battery system. The reason behind this is the low round trip efficiency of the battery.

Figure 5a, c show the parametric study of self-consumption of the PV-battery and PV-TES system. Figures show that to meet the cooling demand completely by discharging of the energy storage system, i.e., full self-consumption of energy storage system battery capacity should be greater than 50 kWh and TES capacity should be

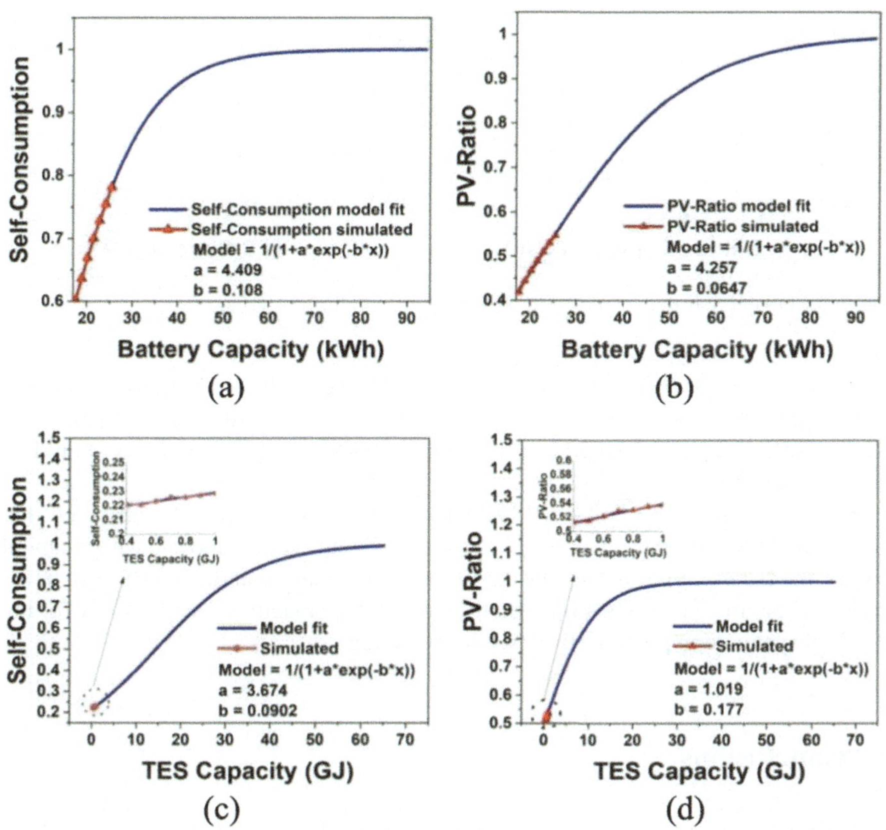

Fig. 5 Energy storage self-consumption for battery and TES (**a**, **c**) and PV ratio for the PV-battery and PV-TES system (**b**, **d**)

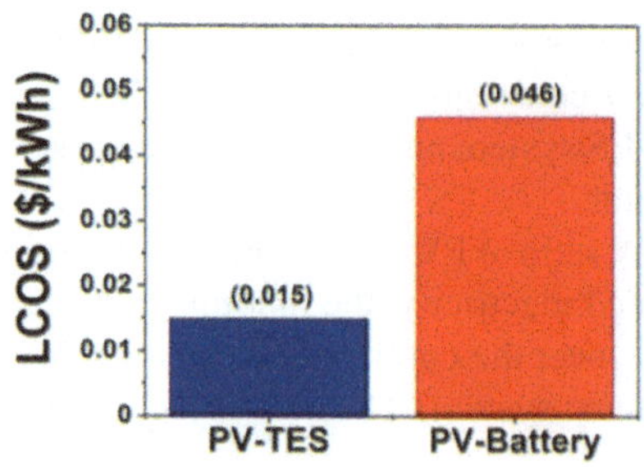

Fig. 6 LCOS of PV-battery and PV-TES system

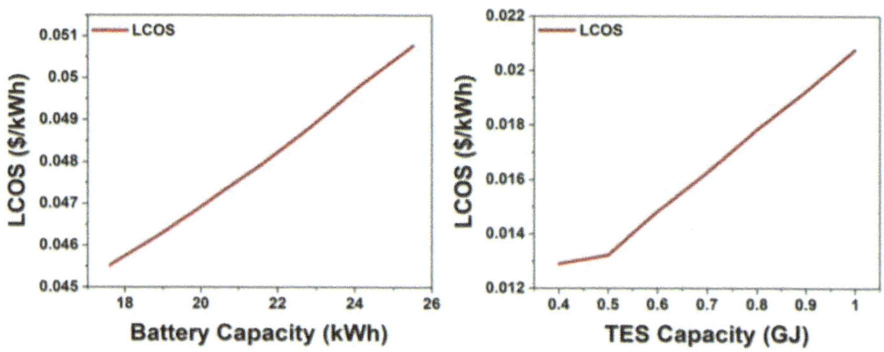

Fig. 7 LCOS of PV-battery and PV-TES system

greater than 50 GJ. Figure 5b, d show the parametric study of PV ratio for the PV-battery and PV-TES systems. Figures show that to fully utilize the PV generated power, battery capacity should be greater than 80 kWh and TES capacity should be greater than 25 GJ.

Figure 6 shows the LCOS for PV-battery and PV-TES. It can be observed that LCOS for TES system is lower than PV-battery system. This is due to the high capital cost of the battery. Figure 7 shows the LCOS of PV-battery and PV-TES system for various capacities. It can be observed that for all the capacities, LCOS is low for the PV-TES system compared to the PV-battery system.

6 Conclusions

In this paper, two storage options, i.e., battery and ice tank coupled with photovoltaic (PV), are compared for the cooling application in buildings. Both the energy storage options are applied to a residential building in a warm and humid climate using EnergyPlus, a whole building simulation tool. An advanced model predictive control strategy has been also proposed for the better operation of the chiller to increase the self-consumption of PV. Results show that energy storage self-consumption is higher for the PV-battery case compared to the PV-TES. This is because of the heat loss from the ice storage tank. It can be improved through better insulation and sizing of the thermal storage tank. Results also show that levelised cost of storage is higher for the PV-battery system compared to the PV-TES system due to the high capital cost of the battery. The space required for the PV coupled TES is higher compared to the PV coupled battery system to meet the cooling demand due to low energy density of the ice storage tank. Latent heat storage using high energy density material as thermal storage media would be an interesting alternative.

References

1. http://www.indiaenvironmentportal.org.in/files/file/The_Future_of_Cooling.pdf. Retrieved February 28, 2020.
2. Santamouris, M. (2016). Cooling the buildings—past, present and future. *Energy Building, 128*, 617–638. https://doi.org/10.1016/j.enbuild.2016.07.034.
3. Ürge-Vorsatz, D., Cabeza, L. F., Serrano, S., Barreneche, C., & Petrichenko, K. (2015). Heating and cooling energy trends and drivers in buildings. *Renewable and Sustainable Energy Reviews, 41*, 85–98. https://doi.org/10.1016/j.rser.2014.08.039.
4. Luthander, R., Widén, J., Nilsson, D., & Palm, J. (2015). Photovoltaic self-consumption in buildings: A review. *Applied Energy, 142,* 80–94.
5. Braun, M., Büdenbender, K., Magnor, D., & Jossen, A. (2009). *Photovoltaic self-consumption in Germany using lithium-ion storage to increase self-consumed photovoltaic energy*. Kassel: ISET.
6. Parra, D., Walker, G. S., & Gillott, M. (2014). Modeling of PV generation, battery and hydrogen storage to investigate the benefits of energy storage for single dwelling. *Sustainable Cities and Society, 10,* 1–10.
7. Wang, X., & Dennis, M. (2015). Influencing factors on the energy saving performance of battery storage and phase change cold storage in a PV cooling system. *Energy and Buildings, 107,* 84–92.
8. Partnership, E. R. (2013). *The future role of energy storage in the UK*. Main Report. (Tech. Rep.). Energy Research Partnership.
9. Christoph, L., Oktoviano, G., Thomas, R., & Chandan, S. (2018). *Levelised cost of thermal energy storage and battery storage to store solar PV Energy for cooling purpose*. https://doi.org/10.18086/eurosun2018.04.09.
10. Merei, G., Moshövel, J., Magnor, D., & Sauer, D. U. (2016). Optimization of self-consumption and techno-economic analysis of PV-battery systems in commercial applications. *Applied Energy, 168,* 171–178.
11. Arteconi, A., Ciarrocchi, E., Pan, Q., Carducci, F., Comodi, G., Polonara, F., et al. (2017). Thermal energy storage coupled with PV panels for demand side management of industrial building cooling loads. *Applied Energy, 185,* 1984–1993.
12. David, P., Gavin, S., & Walker, M. G. (2016). Are batteries the optimum PV-coupled energy storage for dwellings? Techno-economic comparison with hot water tanks in the UK. *Energy and Buildings, 116*, 614–621. https://doi.org/10.1016/j.enbuild.2016.01.039.
13. https://www.energy.gov/eere/buildings/commercial-reference-buildings. Retrieved February 2, 2020.
14. Manwell, J. F., & McGowan, J. G. A lead acid battery storage model for hybrid energy systems. *Solar Energy, 50*(5), 399–405.

15. Strand, R. K. (1992). *Indirect ice storage system simulation.* M.S. thesis, Department of mechanical and industrial engineering, University of Illinois.
16. Eduardo, F. C., & Carlos, B. A. (2007). *Model predictive control* (2nd ed.) London: Springer.

Inducing Dynamism in the Process of Planning Through Crowdsourced Geo Tagged Photograph

Vipul Parmar

Abstract Millennials think of urban areas to be both physically and digitally accessible with a wholesome view of the same available in their palm. However, the traditional urban planning process is incompatible with addressing the emerging requirement of this technologically advanced era. Technological interventions have changed how individuals interact with their surroundings. Cities are always planned for fulfilling the requirements of people socially, economically, and culturally. Urban planners must have adequate and comprehensive knowledge about the current situation of the city to make an effective plan. Though this paper, a systematic method is described to know about the neighbourhood, the point of interest of people in the neighbourhood, and planning for it as per need for the people. This paper seeks to explore sentiment, which is linked with different places of the city. In the study, initially, data were extracted for Indore city from open pages of Instagram, and then especially the survey was done for selected areas. Crowdsourced geotagged photographs can be used to know about the public's points of interest in a neighbourhood that may signify some potential or problem which needs to be addressed. Such an approach facilitates urban planning at a micro-level. It is a more people-centric approach that increases the probability of acceptance of subsequent planning interventions by the public. This collection of the crowdsourced database from various social media platforms periodically can provide dynamism to the conventionally static nature of planning.

Keywords Crowdsource · Geotagged photograph · Point of interest · Sentiment · Neighbourhood

V. Parmar (✉)
SPA Bhopal, Bhopal, India
e-mail: vipsparmar31@gmail.com

A. Jana and P. Banerji (eds.), *Urban Science and Engineering*, Lecture Notes in Civil Engineering 121, https://doi.org/10.1007/978-981-33-4114-2_26

1 Introduction

Millennials think of urban areas to be both physically and digitally accessible with a wholesome view of the same available in their palm. Urban planning is the study related to an urban area, concerned with effective uses of land and provides a healthy environment for people living in the urban area. Cities are growing very rapidly and these changing conditions are challenging the urban planning processes to be more efficient and effective [1]. To face challenges, such as urbanisation, globalisation, and environmental changes, there is a need for more inclusive and informed planning [2].

Cities are always planned for fulfilling the needs of the people—socially, economically, and culturally. Urban planners must know the current situation of the city to make an effective plan. This can be possible by making the planning process more participatory. Participatory urban planning is a user-centric approach that gathers information such as how people experience and use place, what are the issue in their environment and why it is so [1]. It involves the participation of all kinds of users irrespective of ages, gender, caste, and economic status. Participation plays an important role in exchanging knowledge and raises voices together for solving the problem of localities [3, 4, 5].

Technological advancement has completely changed the scenario worldwide. The development of information and communication technology (ICT) has changed the way of interaction and communication between the people and forming a virtual city in which information flows frequently [6]. Data and information are generated by the people, which can be utilised for knowing the characteristics of the city [7]. Big data, data analytics, or data mining has enhanced the feedback system, which helps to fulfil the demand as per the need of the user [8].

The increasing use of social media and smartphones leading people to share their routine habits online and leaving their digital footprints in the urban environment [9, 10]. These footprints can be used to make urban planning more participatory as well as dynamic. For example, Google uses real-time location data to analyze vehicular movement and provides user live traffic details. Location-based social media datasets can be used to analyze the travel pattern and understand the spatial interaction of people in cities [11, 12].

Technological interventions are one of the important factors in the development of any country. Developing countries such as India are enhancing their capabilities with the use of the technological intervention. In India, the era of the fourth industrial revolution (2010–present), technology becomes fully integrated into the daily lives of the people, such as the Internet of Things, Artificial intelligence, cloud technology, smartphones, social media, etc. These intervention has completely changed how people interact with each other. Indian monthly data usage is expected to rise fivefold from 3.9 GB in 2017 to 18 Tb by 2023 [13]. India's social network subscribers were 326.1 million in 2018, and it will reach 447.9 million in 2023 [14].

Social media are the platform where people can share their thoughts, ideas, issues, emotion without any hesitation. This datasets generated by people is collecting by businessmen, politicians, and other stack holders to improvise the facility and attract the customers to purchase the product. This leads to the continuous evolution of markets, and it became a challenge for urban planners to plan for it. There is no shortage of smart city projects globally, but participation by the millions who work, live in cities is missing. The datasets generated by people can be collected and converted into useful information, which can give more meaningful insight about different places of the city. Participation of the people helps planners to plan effectively and data generated by social media platforms can be one of the new approaches to enhance the participation of the public directly or indirectly. Improvement in communication technology and the availability of internet service at a cheaper rate has resulted in generating an enormous amount of data every second. Continuous up-gradation of this real-time information can help to make the planning process more dynamic and participatory.

2 Related Work

2.1 Spatial Mapping of Social Media Dataset

Social Media Dataset as a source of the user-generated database that has been interpreted for many purposes such as event detection, urban socio-spatial inequality, or understanding the behavior of people. Different scholars have used social media datasets to understand the occurrence of any emergency event, to know about the characteristic of the place. In the field of urban planning, geospatial mapping plays an important role. Locational data which are linked with the post on social media can be used for spatial mapping of the activity. Such a kind of mapping was done to know the popular area of sport for the Alicante city of Spain [15]. Similarly, land use detection was done through the classification of messages, phone calls using K-mean algorithms for a complete day for Beijing city of China [9]. This study better makes in predicting the dynamic land use pattern of the city. Frias-Martinez and Frias-Martinez [15], proposed a technique to determine the urban land use pattern by clustering the areas with similar kinds of activity patterns. The clustering model can cluster the similar kind of activities and helps in understanding the dynamics of the city [15, 16]. Spatial analysis of the social media dataset enables urban planners to learn about the various locations, the different events in the city.

2.2 Photostory with Hashtags

People use social media platforms to share their thought, view, and ideas in the form of images, videos, or text. Usually, each post over social media platform consists of some textual expression. The text associated with the image or video provides a reference to the posted media. Kress and Leeuwen [17] has introduced the concept of photostory with hashtags[1] which have been used worldwide to take participate in certain kinds of events or to bring similar kinds of people to discuss the special issue. There are very less researches available on how hashtags are used by different social bodies for collaborating with people all over the world [19]. PON (photostory of neighbourhood), it is a competition which provides a perspective of the people about their surroundings. It consists of different sets of hashtags that are to be considered during posting any kinds of images related to the urban area [20]. This also helps in finding out a common point of interest of the people in their surroundings. The text-based story enables the planner to gain the perception of the people for their surroundings.

2.3 Sentiment Mapping

In the field of urban planning, space and place are often considered to be synonym but they are different from each other. Space can be an area, but places are spaces with emotions and feelings. Understanding of such sentiments related to different places assist planner to plans for space into a place. Sentiment mapping allows identifying sense associated with the places, which added qualitative aspect in the field of urban planning. It can be done after performing a sentiment analysis. Sentiment analysis relies on analysing the human language to figure out the polarity of the sentence. It can be possible by certain algorithms that are used for finding opinions and sentiment [21]. People share their feeling on the social media platform in the form of statements as well as hashtags. Such kind of information can be utilised to understand the behavior of the people. It can also be used for influencing the people [6]. Different places in the city are associated with different kinds of feelings or emotions. Fathullah and Willis [22], the study has shown that how people feel in different places when they travel from one point to another in the neighbourhood which further helps in improvising the location where people feel depressed due to surrounding. Such kinds of intervention are very important when planning for any area.

3 Research Methodology

3.1 Data Collection

Social media platforms are a digital platform where people share different kinds of information related to them. It may be a personal message/image or a post related to different places in the city. Before directly proceeding towards the extraction of the dataset from the social media platform, a survey was performed through google

[1] Hashtags are semiotic codes started with #, for example #food [18].

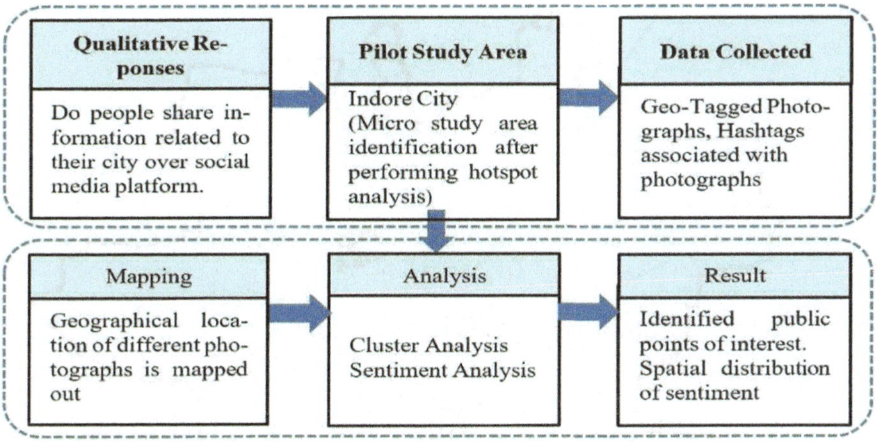

Fig. 1 Research methodology

forms, and results were analysed for carrying out further study. Approximately 40,000 social media posts were extracted from Instagram open pages related to the Indore city. Spatial mapping was done and a micro-study area was identified for study. A Survey of 450 people[2] was conducted using EPI collection after the identification of a micro-study area where the further analysis was performed. Figure 1 showing a framework through which research was conducted.

3.2 *Identification of Micro Study Area*

To understand the Dynamic of the City in detail, it is required to have much more dataset from different social media platforms, that is, based on only 40,000 posts, it is not possible to tell about the dynamic of the city. Due to privacy concerns, it is difficult to extract more datasets from these platforms. Therefore, hotspot analysis was performed on the extracted data point from the geotagged crowdsourced photograph.

From Fig. 2, it can be seen that there are three zones where the density of points is high, and if a particular area has high dot density, mix-up of all kinds of activities and has been changed its character in the past few years, is considered for further study. Therefore, from these three zones, ward no 74—Vishnupuri was selected because the center and northeast zone were already commercialised. Vishnupuri is another developing area of Indore city. Due to the high concentration of coaching

[2]Sample size is calculated by considering a 95% confidence level and 5% marginal error.

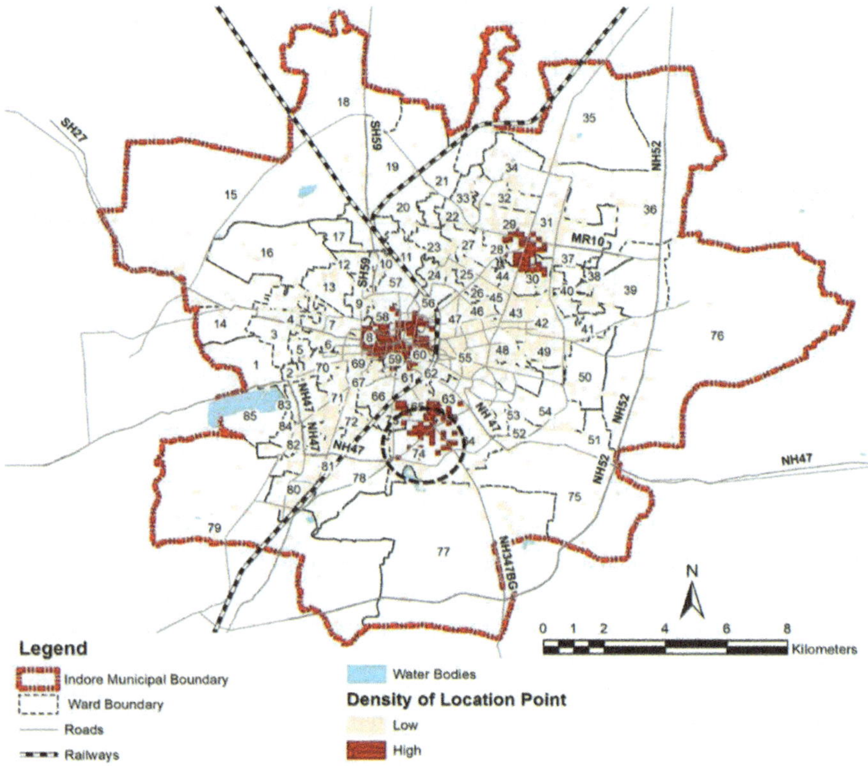

Fig. 2 Hotspot analysis of the geotagged location of the photograph for selection of micro-study in Indore City

institutions in this area, it is also known as a student hub.[3] A further detailed analysis was performed on this selected micro study area to find out the point of interest of the people in the neighbourhood.

4 Research Findings and Proposed Framework

4.1 Qualitative Response

From Fig. 3, it clear that 76% of respondents from the primary survey shares information related to the city in their post while 24% of people do not share anything related to the city. Post related to city consists of recreational areas where people spent their quality time. This post can be extracted from social media platforms and

[3]It was known during the qualitative survey.

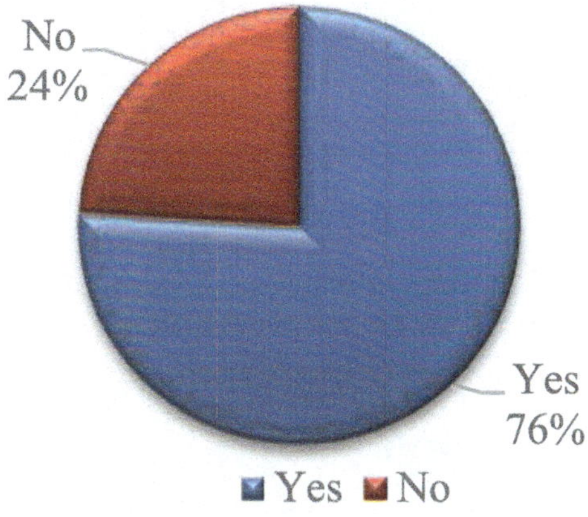

Fig. 3 The percentage share of people share posts related to the city on their social media account

text encoded with the post can be used for knowing the importance of a place or the feeling associated with that place. This helps to urban local body to identifies the issue and monitor the city's need from the people's feedback, which are indirectly posted over the internet.

The text associated with the post will be helpful when performing text-based classification and analysis. People write a caption with the post, but nowadays, people used hashtags, which are a single word or combination of few words which are utilised directly to express the feeling and emotion attached to that post. From Fig. 4, it can be seen that 65% of people use hashtags, that is it can be concluded that different types of hashtags classification be used for analysis.

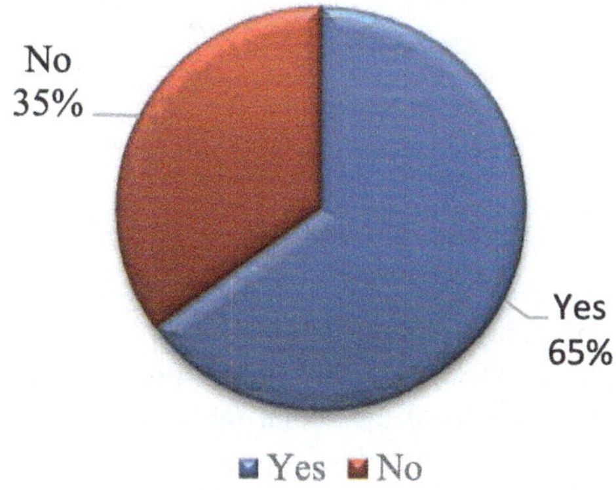

Fig. 4 Percentage of a share of people use hashtags in their post

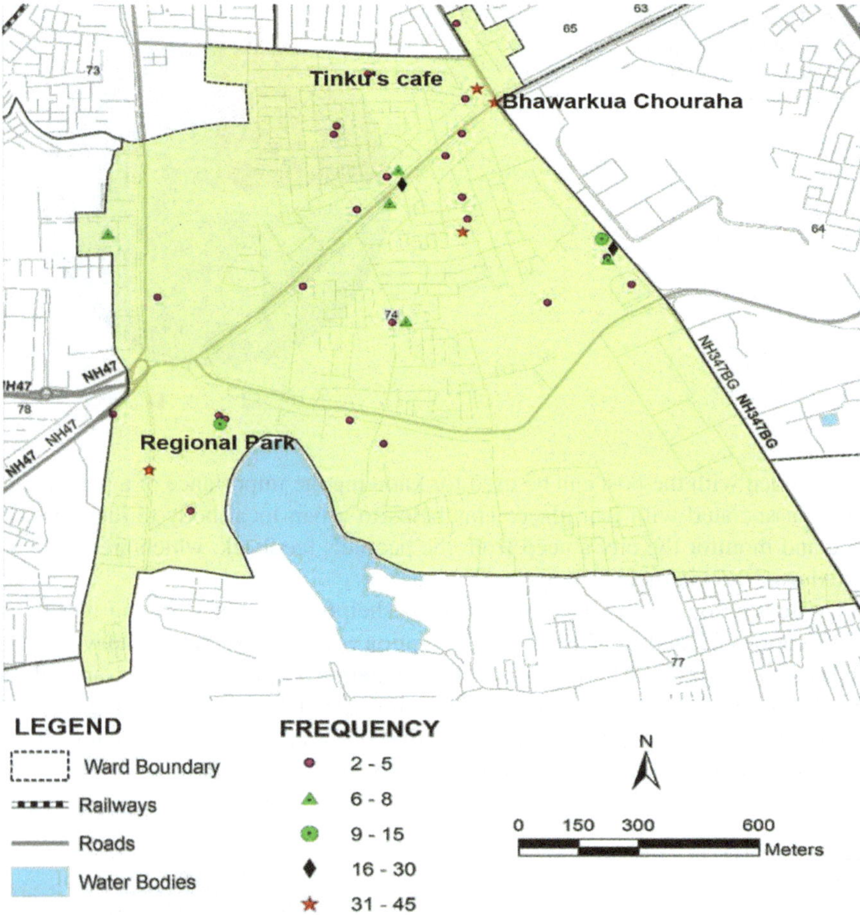

Fig. 5 Point of interest of the people based on collected dataset

4.2 Point of Interest of People

The point of interest[4] of the people in the location, which is identified as the most prominent area of any ward. For the identification of the Point of Interest, data extracted from the social media platform is not sufficient, so for a more accurate and effective result, a primary survey was conducted. Based on the occurrence of a particular location from social media as well as from the primary survey, response frequency is calculated for different places or location points of the city.

From Fig. 5 it can be seen that Bhawarkua Chouraha, Tinku's Café, and Regional Park in the neighbourhood are having the highest number of frequency. These points are the main point of interest of the people. But this interest varies from person to

[4]Point of interest is the places or location which has been shared by people repetitively.

Fig. 6 Traffic congestion

person, that is, further analysis of text associated with them leads to knowing about issues related to these places.

Bhawarkua Chouraha is a point of interest of the people because most of the coaching institutes are situated around this. The students find out this attractive because all the facilities are available there, but few tags also tell about the problem of the traffic congestion and accident. Figure 6 consists of hashtags which tell about traffic congestion at the Bhawarkua Chouraha. Similarly, Fig. 7 consists of tags related to the parking problem near the coaching institution which is situated on the Bhawarkua Chouraha.

Tinku's cafe is another point of interest for the people. This is famous for its food. From the primary survey, it is known that student spent their quality time over there. This café acts as an attraction point for other café's. During the visit, it was identified that more then 10–15 new café has been set up in the last few years after Tinku's café.

Regional Park is one of the recreational areas developed by Indore Municipal Corporation. It has many fun activities, such as boating. People usually go over there to enjoy themselves in the evening and during holidays. This place consists of tags related to boating, hangout, greenery, Happy, fun, etc.

Fig. 7 Parking problem

4.3 Sentiment Analysis

Sensitivity analysis is performed based on text associated with the post of a particular location. Due to the availability of lots of hashtags and captions, it is classified into three categories, that is, positive, neutral, and negative, for the long statement analysis, polarity check was done with the help of VADER and TextBlob using python. Similarly, hashtags were also classified and weightage was given to the statement.

It can be seen that from Fig. 8, an activity such as food/café has more positive responses or sentiments concerning other areas. Regional park has more positive sentiments as compared to other areas of the surroundings. Most of the areas have neutral responses since, due to the lack of availability of data points, fine-tuned map generation is not possible. It can be concluding that availability of green spaces as well as places for hanging out such as food shops/café enhance the positivity of the surrounding area.

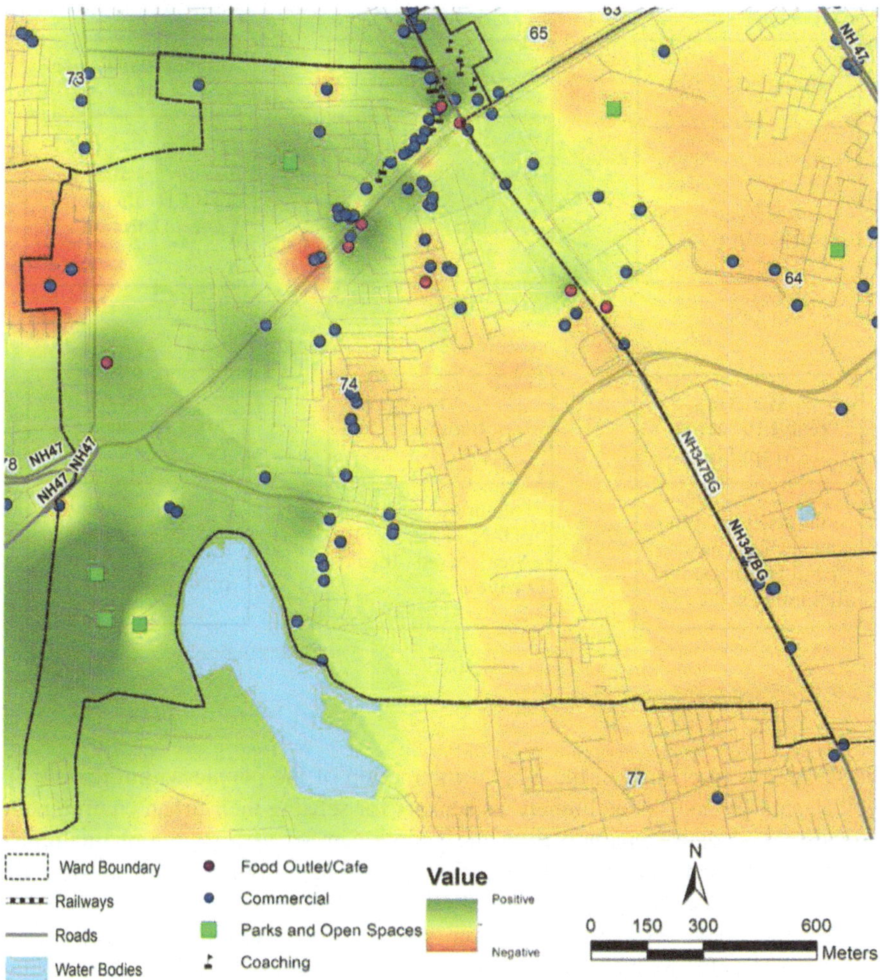

Fig. 8 The sentiment are extracted from the caption as well as positive hashtags

4.4 *Proposed Framework*

Data Collection and Information generation strategy

Social media is a platform where people share their thoughts, feeling and personal stuff frequently. There are lots of posts which contain details related to the city. There are many places which are frequently visited by people or places where people spent their quality time with friends and families. Planning, according to the needs of the people, is one of the major tasks in the field of urban planning. Different places have different characteristics, which leads to the requirement of a special type of area-based intervention in that particular locality. As day by day, improvement in

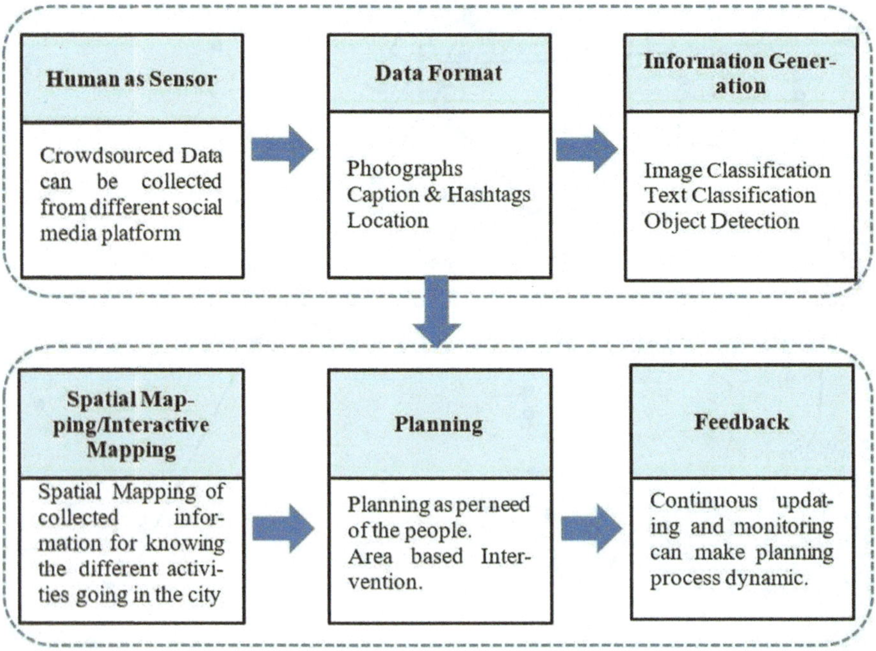

Fig. 9 Proposed framework

the ICT leads to change in the way of interaction of the people within the city as well as smartphones made humans as sensors for sensing their urban environment. Crowdsourced data generated from the people can act as a source of qualitative information which can be used for planning the city. Figure 9 showing that the initial stage is to collect the crowdsourced data from different social media platforms such as Facebook, Instagram, and other, which people use more frequently. Data collection mainly require information such as a photograph, its location, textual caption attached to the images. Since privacy is one of the major concern by most of the researchers but only allowing the content without revealing the actual identity of the person, this dataset can be used.

During the study, it has been noted and analysed that there are divides in using the caption in the form of hashtags by the elder people. So only the text classification method is not sufficient to extract the information. There is a need for image classification techniques. Deep learning and artificial intelligence is the key to extracting more information from the image generated by the people. There are certain negative sides to social media. Bots are encoded computer program which generates text and information which can reduce the quality of data and hurdle in using the social media dataset as a tool. But with the advancement in technology, bots can be identified with the use of a certain algorithm. This helps to improve the quality of the data set.

Interactive Mapping

Interactive spatial mapping of the data set is another step for the proposed framework. Spatial mapping of the collected points allows the identification of the activities in different parts of the city. Interactive mapping of this data point encourages people to share more valuable comments related to the city. It was noted that people are willing to share more posts and information as they see their information is utilised for meaningful purposes. Spatial mapping of these points with their caption or activity can be compiled and an interactive map for the city can be prepared, which helps in area-based planning intervention as per the need of the people. Through this kind of mapping, the urban local body can also provide suggestions to the different stakeholders and businessmen in selecting the appropriate sites. It will help in the economic growth of the city as well as deciding the growth direction of the city. Sentiment Mapping used for finding out the sense of different place which can be used for improving the location which spreads negativity. This will also improve the health of people and increase their productivity.

Database Management

Data is the foundation for any kind of researched work, or any kind of intervention is required in the field of planning. Technological advancement provides real-time data set that helps in detailing any activity at a very micro level very effectively. Maintaining privacy is one of the important concerns when dealing with the crowd-sourced dataset. From the primary survey, it was concluded that people are willing to share the data if it is used for the development of the city. Continuous extraction and maintains of the data set is one of the important aspects two understand the evolution of any particular locality, to identifies the changes in the activity, to know about the point of attraction. Spatial mapping of such data points will help in knowing the dynamics of the city. As can be seen from the above analysis, which was performed for a very small area, has given in-depth knowledge about the city at a very cheaper cost as compared to extensive primary survey exercise. Periodically maintains and updating such a dataset is a very important constraint for understanding the dynamics of the city. It also enhances the feedback mechanism and increases the transparency between the Urban local body and the people. While dealing with such kind of data, any kind of personal information must not be used. Only spatial references and posts which provide details about different activities in the city should be extracted or collected from people.

Data Validation Authority

Crowdsourced data generated over the internet have lots of information, but it is in an unstructured format and requires technical skills and resources to convert the data into meaningful information. The Command and Control Centre of a smart city has the capabilities and resources to handle such an amount of data. They can extract the dataset and can convert that data into useful information with the help of text-based classification through machine learning algorithms that can further be spatially mapped out. This mapped layer can be used by the urban local body or Town and

Country Planning Organization for making plans for any particular area. This body will also act as a data validation authority.

5 Conclusion

Social media platforms are one of the important parts of peoples' day to day life. Peoples are sharing their thought, invitation, feeling over this platform in the form of images, videos, or text. These datasets are utilised by different stakeholders for their business growth. Such kind of dataset can also be used in the field of urban planning. As it can be seen that crowdsourced geotagged photographs with caption or hashtags can be used to know about the public's points of interest in a neighbourhood, signify some potential or problem which needs to be addressed. Such an approach facilitates urban planning at a micro-level. Being a more people-centric approach increases the probability of acceptance of subsequent planning interventions by the public. It also brings down the gap between the urban local body and people.

A crowdsourced dataset is also used for understanding the behavior of the crowd towards any kind of event, product or place. It can be noted that spatial mapping of sentiments which are related to different places helps the planner to understand the relation between different land use and feeling associated with them. It will help in making and designing the places which spread positivity in the city, which increases the productivity of the people. The collection of the crowdsourced database from various social media platforms periodically can provide dynamism to the conventionally static nature of planning.

References

1. Lopez-Ornelas, E., Abascal-Mena, E., & Zepeda-Hernández, S. (2017). Social media participation in urban planning: A new way to interact and make decisions. In: *The international archives of the photogrammetry, remote sensing and spatial information science* (Vol. XLII/W3, pp. 59–64).
2. Evans-Cowley, J. (2011). Micro-participation: The role of microblogging in planning. *Social Science and Research Network.*
3. Wilson, A., Tewdwr-Jones, M., & Comber, R. (2019). Urban planning, public participation, and digital technology: App development as a method of generating citizen involvement in local planning processes. *Urban Analytics and City Science.*
4. Cilliers, E. J., & Timmermans, W. (2014). The importance of creative participatory planning in the public placemaking process. *Environment and Planning B: Planning and Design 2014, 41*, 129–413.
5. Nair, P. (2017). *Cities for citizens, by citizens: Public participation in urban planning.*
6. Mora, H., Pérez-delHoyo, R., Paredes-Pérez, J. F., & Mollá-Sirvent, R. A. (2018). Analysis of social networking service data for smart urban planning. *Sustainability.*
7. Celiketen, M., LeFalhar, G., & Mathioudakis, M. (2016). Modeling urban behavior by mining geotagged social data. *Transactions on Big Data, Advance Online Publication*, 1–14.

8. Rathore, M. M., Paul, A., Ahmad, A., & Rho, S. (2015). Urban planning and building smart cities based on using inter of things using big data analytics. *Computer Networks.*
9. Wang, Y., Wang, T., Tsou, M. H., Li, H., Jiang, W., & Guo, F. (2016). Mapping dynamic urban land use pattern with crowdsourced geo tagged social media (Sina–Weibo) and commercial points of interest collections in Beijing, China. *Sustainability.*
10. Bollier, D., & Rapporteur. (2016). *The city as platform-how digital networks are changing urban life and governance.* The ASPEN Institute.
11. Liu, B., Yuan, Q., Cong, G., & Xu, D. (2014). Where your photo is taken: Geolocation prediction for social images. *Journal of the Association for Information Science and Technology*, 1232–1243.
12. Marsal-Llacuna, M.-L., & Gesa, R. F. (2015). Modeling citizen's time-use using adaptive hypermedia surveys to obtain an urban planning, citizen-centric, methodological reinvention. *Time & Society*, 1–23.
13. *Ericsson mobility report.* (2017).
14. *Statista.* (2019). Retrieved November 2019, from https://www.statista.com/statistics/278407/number-of-social-network-users-in-india/.
15. Frias-Martinez, V., & Frias-Martinez, E. (2014). Spectral clustering for sensing urban land use using twitter activity. *Engineering Application of artificial intelligence, 35*, 237–245.
16. Cranshaw, J., Schwartz, R., Hong, J. I., & Sadeh, N. (2012). The livelihoods project: Utilizing social media to understand the dynamics of a city. In: *Sixth International AAAI Conference on Weblogs and Social Media.*
17. Kress, G., & Leeuwen, T. van. (2006). *Reading images: The grammar of visual design.* Routledge.
18. *Techopedia.* (2019, December 1). Retrieved from https://www.techopedia.com/definition/15075/hashtag.
19. Roux, J.-H. L., & Cilliers, E. J. (2013). The participatory planning paradigm shift: Comparing disciplines and methods. In: *49th ISOCARP congress.*
20. Niksic, M., Tominc, B., & Gorsic, N. (2018). Revealing resident's shared value through crowd-sourced photography: Experimental approach in participatory urban regeneration. *Urbani izziv, 29* (supplement).
21. Bo Pang, L. L. (2008). Opinion mining and sentiment analysis. *Foundations and Trends in Information Retrieval, 2,* 1–135.
22. Fathullah, A., & Willis, K. (2018). Engaging the senses: The potential of emotional data for participation in urban planning. *Urban Science, 2*(4), 98. https://doi.org/10.3390/urbansci2040098.

A Review on Methods for Estimating Quantities of Fecal Sludge Produced in Urban India

N. Chandana and Bakul Rao

Abstract Urban slums and unplanned settlements in India depend on highly on-site sanitation systems (OSS). Fecal Sludge (FS) produced in OSS is one of the most dangerous pollutants that have the potential to spread many diseases. The Government of India is emphasizing proper collection, treatment, and disposal of FS by launching various programs and schemes (such as NUSP, AMRUT, smart city mission, UIDSSMT, etc.). The private sector plays a major role in the collection of FS, while very little if any, treatment exists for the FS collected. The design and sustainability of any FS treatment depend on the accurate knowledge of the quality and quantity of FS at the city/town level. This paper systematically reviews the existing methods of FS quantification and its drawbacks in Indian conditions. In the literature on FS, two theoretical approaches (i) FS production method (FSP), (ii) FS collection methods (FSC) are being used for quantification. Most of the Indian practices use the FSP method for FS quantification, which is similar to conventional wastewater quantification. However, in India, the data on the amount of excreta produced and sludge accumulation rate (SAR) in different OSS are seldomly found. This results in the usage of data available in developed countries leading to over/underestimation of FS on which treatment facilities are designed. This study concludes that the design of FSM for Indian conditions requires basic research on excreta produced and its degradation properties in different OSS.

Keywords Fecal sludge · Fecal sludge management · Fecal sludge quantification · Fecal sludge collection method · Fecal sludge production method

N. Chandana · B. Rao (✉)
Centre for Technology Alternatives for Rural Areas, Indian Institute of Technology Bombay, Mumbai 400076, India
e-mail: bakulrao@iitb.ac.in

N. Chandana
e-mail: chandana2n@gmail.com

A. Jana and P. Banerji (eds.), *Urban Science and Engineering*, Lecture Notes in Civil Engineering 121, https://doi.org/10.1007/978-981-33-4114-2_27

1 Introduction

On-site sanitation systems (OSS) are partial treatment units designed for degradation of human excreta and the sludge produced, thereby termed as FS is spreading microbes that cause diseases [1–6]. In urban India, around 30 million households (HHs) depend on OSS and the FS thus produced is either not emptied or legally dumped. Having understood the problem, India is focusing on building the FS collection, transportation, and treatment plants. In India, desludging of OSS is usually done by private vendors; they mostly dump the FS in water bodies or on barren land and in a storm/sewer lines where manholes are accessible. Further, it should be noted that one of the principal reasons for the breakdown of sewage treatment plants (STPs) is the unauthorized dumping of FS in sewers. Usually, the design of STPs caters to treat wastewater containing 250–600 mg TSS/l, while FS has concentrations of 12,000–52,500 mg TSS/l, and mixing these high loads often leads to failure of the STPs [7]. This shows that the sanitation sector is lagging due to a lack of knowledge of FSM. The status of FS treatment in India is in its developing stage and requires precise knowledge on load and quantities reaching the treatment. Unlike wastewater, FS characteristics and load, differ in a town or city, so it is important to note that treatment technologies need to be designed on accurate load and quantities for the sanitation system to be sustainable.

In this paper, we begin by explaining the importance of FS quantification, followed by a critical review of FS quantifying methods existing in the literature. The factors affecting each method in Indian conditions and a framework to select an appropriate method based on existing knowledge are discussed. The article concludes by summarizing the key challenges and limitations of present FS quantification methodologies in Indian conditions.

2 Quantification of FS from OSS

In the absence of FSM practices, any efforts of improved sanitation facilities would still degrade the surface and groundwater resources [5, 7–9]. The viability of any FSM design for a city/town level depends on the appropriate calculation of FS getting generated, collected, and reaching the treatment facility. The major challenge in FSM is the scattered data available and experiences being concentrated in a few parts of the world. Also, as it is a relatively new field, much of the existing knowledge remains with the practitioners without much documentation. The volume of FS produced depends on (i) socio-economic profile of the inhabitants, that is, income, number of inhabitants, water usage and food habits (ii) type of OSS, that is, the retention time of sludge, construction quality (lined/partially/unlined/improved/collapsed/abandoned/watertight) of containments, and soil characteristics (bearing strength, slippage, water retention capacity), and (iii) climatic conditions. All the above parameters result in significant variation in the volume and characteristics of FS and making its estimation difficult. Arriving at accurate or near to accurate volumes is important; it would determine the collection, transportation, treatment, and end-use of the FS. In literature, two theoretical approaches (i) Fecal Sludge Production method and (ii) Fecal Sludge Collection method, have been used [10, 11].

3 Fecal Sludge Production (FSP) Method

FSP method estimates the FS on the HH level and is further divided into two sub-methods, one based on contribution directly from the population, and the other is based on the number of OSS that exist and likely to be constructed in the service period of the treatment facility. According to the author's opinion, the FSP method is suitable for a new/planned city.

3.1 FSP—Population Method

In the first method, the load on the FS treatment is calculated by considering the population and average excreta produced by them, as shown in Eq. (1). This method depends on the average value of excreta produced, which varies daily because of dietary habits (i.e., high fiber will lead in high quantities of excreta) and on factors such as liquid consumption, the physical activity carried out by the user and climate.

$$V_{fs} = \frac{AMe * P_{oss}}{N_{oss} * E_{oss}} \tag{1}$$

V_{fs} Volume of FS produced
AM_e average of excreta produced
P_{OSS} population using OSS
N_{OSS} inhabitants per OSS
E_{OSS} Number of OSS emptied per day

This method does not account for the FS dependency on water availability, water used for cleaning, and greywater mixing or not. Also, the sludge accumulation rate depends on degradation characteristics and frequency of desludging (if the desludging rate is high, time for degradation is less, the sludge accumulation would be more or less equal to the FS produced).

3.2 FSP—OSS Method

In this method, the number of OSS that is existing and likely to be constructed through the enforcement of pollution control law/rule is taken into account. As shown in Eq. (2), FS is quantified by multiplying the volume of FS generated by the emptied OSS numbers. This method is better than the FSP-population method, as it takes into account degradation and the other materials entering the OSS. However, this method does not account for FS loss in collection and transportation due to the inaccessibility of OSS.

$$V_{fs} = V_{Fsoss} * E_{oss} \tag{2}$$

$V_{fs\ oss}$ Volume of FS produced in OSS
E_{OSS} Number of OSS emptied per day.

3.3 FSP—Combined Method

This method uses an average of excreta sludge accumulation rate per person in FS quantification for the population, as shown in Eq. (3).

$$V_{fs} = SAR_{OSS} * P \tag{3}$$

SAR_{OSS} Sludge accumulation rate of OSS in l/person/day
P Total population

The FSP method does not account for the FS not collected fully, loss during the collection, and directly disposal into the environment. In addition to the FS produced, accessibility of the OSS also plays an important role in quantification of FS to treatment, as not only the location of OSS but width of the road (if using a truck, roads need to be wide enough to accommodate the truck or sludge emptying equipment), also affect the quantification. This could lead to overestimating of FS volumes reaching the treatment facility and leading to over design of the treatment facilities.

4 Fecal Sludge Collection (FSC) Method

The FSC method estimates the FS quantity at the collection system and can be applied to only places having an FS collection system. As shown in Eq. (4), FS is quantified based on the volume of FS reaching the disposal point. In this system, the volume of FS produced depends on FSM infrastructure, regulations (i.e., legal discharge location available or not, discharge fees structure), and enforcement of rules to stop illegal dumping available.

$$V_{fs} = V_{FsEc} * E_{oss} \tag{4}$$

$V_{fs\,Ec}$ Volume of emptying vehicle container
E_{OSS} Number of OSS emptied per day

This method depends on the quality of data available and is suitable for saturated towns or cities, which have the infrastructure for FS collection. Besides, previous researchers mentioned that the FS quantified based on existing data will be an underestimation, as it is expected to increase due to the market for FS [7, 10].

5 FS Treatment Technologies and Method Used for FS Quantification in India

In 2009–10, the ministry of urban development of India conducted a rating of cities on sanitation parameters and found that none of the 423 cities surveyed met the benchmark of sanitation [12]. After this, the Indian government is developing a different sanitation service benchmark based on the requirement of places. In this context, few cities where the FS collection system is already available are selected to build FS treatment technologies, as shown in Table 1. The advisory note on septage management in Urban India [12] has mentioned both FSC and FSP method for quantification of the FS and CPHEEO manual has suggested to be 40–53 l/person/year (0.1–0.15 l/person/day) for dry pits and 95–114 l/person/year (0.26–0.31 l/person/day) if

Table 1 FS treatment technologies in India and its method of FS quantification

FS technology, location	The method used for FS quantification
DEWATS, Devanahalli, Karnataka	6 KLD—FSC excreta based
Sludge Drying Beds with Co-composting, Bansberia, West Bengal	12 KLD—FSC excreta
DEWATS—Bhubaneswar, Odisha	FSP, used SAR as 0.41–0.55 l/person/day
DEWATS—Phulera-Sambhar, Khandela and Lalsot Rajasthan	FSP method, SAR as 0.21 l/person/day
Unnao City, Uttar Pradesh	FSP method mix of excreta and accumulation method, SAR as 0.21 l/person/day
Co-treatment in STP, Puri, Odisha	FSP-method, used SAR as 0.41–0.55 l/person/day
Planted Drying Beds, Leh, Jammu & Kashmir	12 KLD—Trial runs
USAB, Brahmapuram, Cochin	100 KLD—Trial runs
Pyrolysis, Narasapur, Andhra Pradesh	15 KLD—trial runs
Geobags, Warangal, Telangana	10 KLD—trial runs
Pyrolysis, Warangal, Telangana	15 KLD—trial runs

Note SAR-Sludge accumulation rate; *Source* [14–17]

desludging is done in 2 years, whereas for 3 years desludging 67–80 l/person/year (0.18–0.22 l/person/day) [13]. As shown in Table 1, the gap in basic research on SAR has led to the usage of different values by the existing treatment technologies, which may lead to the non-sustaining of technologies due to variation in load.

6 Factors Affecting the Selection of Appropriate Quantification of FS

The methods explained above are the majorly used quantifying methods of FS. Many technical factors need to be considered while selecting the quantifying method for the FS, some of which are described below.

6.1 Accurate Quantity of Excreta Calculation

The weight of excreta produced governs the (i) volume of excreta received at OSS, (ii) volume of FS accumulation, and (iii) final volume of FS for the treatment. A literature survey revealed that only very limited information exists on the generation rate of human excreta in India. The data from other parts of the world [1, 2, 18–24] is represented in Fig. 1 and the studies are mostly concentrated during the past 30–40 years. For developed countries, the mass of excreta was reported as 140–500 grams/cap/day, whereas for developing countries, it was 80–250 grams/cap/day [2, 18, 19]. India [20] has reported 155 grams/cap/day in 1972, whereas [22] has reported 19–1510 grams/cap/day in 1975, which are largely varying and there are no recent studies found in this regard. As the diet habits and living conditions of Indians have been changed a lot from 1975, detailed research on current excreta produced by the population due to lifestyle needs to be conducted.

6.2 OSS Characteristics

Usually, OSSs are designed such that the excreta gets biologically stabilized depending on a range of geophysical factors and biological processes [25]. The size of OSS varies with factors such as the size of the HH, income level, type of soil, groundwater level, plot size, weather conditions, and also the local regulations. Previous researchers [11, 18, 24, 26] have reported a sludge accumulation rate, as shown in Table 2, which varies all over the world. It should be noted that Indian standards have shown sludge accumulation rates around 40–53 L/cap/year for pit latrines and 120 L/cap/year for septic tanks, which is very less compared to ground reality [5, 27, 28], as the living standards have been changed from then.

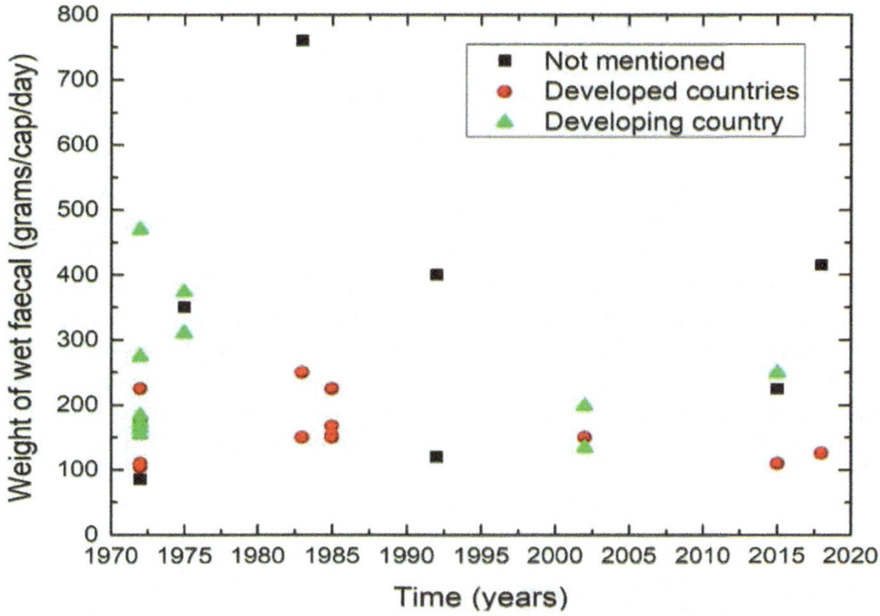

Fig. 1 Mass of excreta reported by the previous researcher

Table 2 Sludge accumulation rates reported by previous researchers and standard manuals

SAR in L/person/year	Type of OSS	Country	References
270	Pit latrines	Uganda	[10]
280	Septic tank	Kampala	[10]
297	Cesspool	Thailand	[26]
133	Septic tank	Thailand	[26]
178	2 Cesspools	Thailand	[26]
179	Commercial septic tank	Thailand	[26]
40–90	All	All	[24]
40–53	Pit under dry conditions	India	[11, 13]
80–114	Pit under wet conditions	India	[11]
120	Septic tank	BIS standards	[11]
230	Septic tank	USEPA	[11]
85.3	Septic tank		[29]
25.5	VIP	South Africa	[30]
29.2	Septic tank	South Africa	[30]

6.3 Emptying Frequency

India stands 4th in the world for not emptying the OSS [5]. The sludge produced in OSS depends on the emptying frequency, as per the [29] the sludge accumulation rate falls from 0.254 l/ca/day over 6 months to 0.178 l/ca/day after 60 months. The longer the sludge ages, it results in thick and stabilized sludge, which cannot be pumped out by normal vacuum trucks, thereby leading to manual scavenging and the sludge getting dumped in open areas results in not reaching the treatment plant. The main stakeholders in FSM are residents and their awareness of OSS emptying plays a critical role in eliminating FS mismanagement. The financial burden of emptying the OSS has led to not emptying the OSS or unscientific way of disposal of the FS. Previous studies [28, 31, 32] have claimed that unauthorized desludging operators in India often charge high user fees (₹3051–₹5340). Hence, while quantifying the FS, emptying frequency need to be considered as an essential parameter for a sustainable sanitation service chain.

6.4 Policy and Regulations

The 74th Constitutional Amendment Act of 1992 bought reforms to the sanitation sector by transferring its management from state to local bodies. However, the availability of data about the types and sizes of OSS lacks in the literature due to the lack of importance given to the sanitation sector until recent times. In 2008, though NUSP has highlighted the safe disposal of FS, very little attention has been given to its management. Further, in 2013 'The Prohibition of Employment as Manual Scavengers and their Rehabilitation' act emphasized the usage of safe truck-based emptying of FS. The National Policy on Fecal Sludge and Septage Management was published in 2017, emphasized the safe practices and rules to be followed for truck-based services. However, this document does not cover the existing practices and improvising on them. Further, the OSS constructed depends on the locally available material and training of the builder masons [16]; hence a primary survey has become a must in designing FSM.

7 Framework for Selection of Appropriate Method for Quantification of FS

Based on the reviews of the existing method of the quantification of FS and relevant factors affecting them, a decision tool as shown in Fig. 2 has been developed to assist the practitioners in finding the possible option for quantification of FS at a city/town level, based on local geophysical and societal factors. If a new city/town is planned, then the FSP-FS accumulation method is best suited as the population in the city can

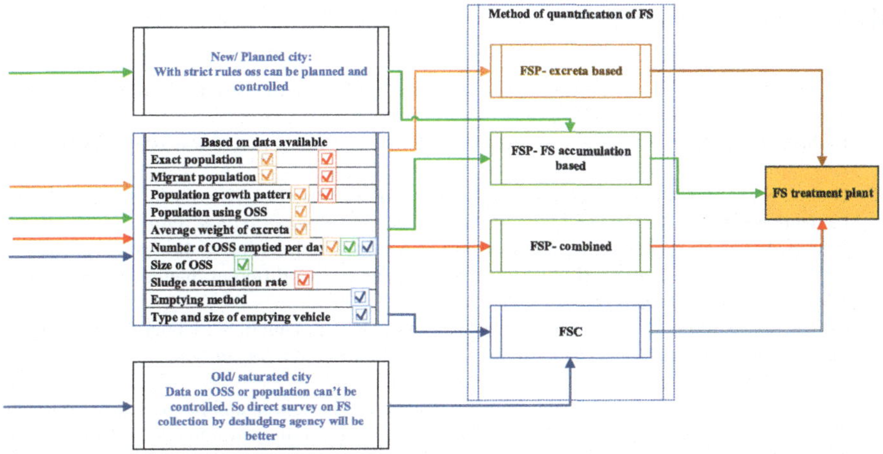

Fig. 2 Decision tool in selecting the FS quantification method

be monitored. Whereas in the case of the old and saturated city, FS quantification can be done by the FSC method, as the quantification of loads based on the population will be difficult. The application of this framework will be discussed in the next paper.

8 Conclusion

FS quantification is a challenging task in India because of the large gap in data regarding the sanitation sector. An accurate FS quantification is difficult due to the following challenges:

- Excreta produced varies daily due to dietary habits, physical activity, and climate conditions.
- The FS quantity varies widely depending upon OSS.
- Traditional emptying systems have less efficiency in empty dense and thick sludge
- Accessibility of the OSS as not only the location of OSS, but the width of the road also affects the collection.
- The method of quantification must not only satisfy the current requirements but also must cater to the future demand.

It is crucial to select the proper method of quantification of FS with regards to accurate data available with locals as the sustainability of the FS treatment depends on it. It is hoped that this critical review and the accompanying decision-making framework may help to disseminate knowledge towards better FS quantification.

References

1. Burkitt, D. P. (1973). Some diseases characteristic of modern western civilization. *British Medical Journal, 1*(5854), 274–278.
2. Richard, F. (1983). *Sanitation and disease health aspects of excreta and wastewater management, sanitation*. Wiley.
3. Diener, S., et al. (2014). A value proposition: Resource recovery from faecal sludge—Can it be the driver for improved sanitation? *Resources, Conservation and Recycling, 88,* 32–38.
4. Knopp, S., et al. (2008). Diagnosis of soil-transmitted helminths in the era of preventive chemotherapy: Effect of multiple stool sampling and use of different diagnostic techniques. *PLOS Neglected Tropical Diseases, 2*(11), 331.
5. Coffey, D., Spears, D., & Vyas, S. (2017). Switching to sanitation: Understanding latrine adoption in a representative panel of rural Indian households. *Social Science and Medicine, 188,* 41–50.
6. Crockera, J., et al. (2016). Building capacity for water, sanitation, and hygiene programming: Training evaluation theory applied to CLTS management training in Kenya. *Social Science and Medicine, 166,* 66–76.
7. CSE. (2011). *Policy paper on septage in India*. C.f.S.a. Environment. Delhi: Centre for Science and Environment.
8. Biswas, P. K., & Mandal, K. (2010). Drinking water in rural India: A study of deficiency, quality and some social implications. *Water Policy, 12*(6), 885–897.
9. Panigrahi, M. K., et al. (2013). Defecation frequency and stool form in a coastal eastern Indian population. *Journal of Neurogastroenterology and Motility, 19*(3), 374–380.
10. Strande, L., et al. (2018). Methods to reliably estimate faecal sludge quantities and qualities for the design of treatment technologies and management solutions. *Journal of Environmental Management, 223,* 898–907.
11. Bhitush, L., et al. (2017). Septage management. In A. A. Parrey (Ed.), *A practitioner's guide*. Centre for Science and Environment.
12. NUSP. (2013). *Advisory note on septage management in urban India*. M.o.U.d.g.o. India.
13. CPHHEO. (2012). *Manual on sewerage and sewage treatment. Part A Engineering*. C.P.H.A.E.E. Organization.
14. Society, C. (2018). *Detailed project report implementation of faecal sludge and septage management solutions for Bagru, Rajasthan.*
15. Society, C. (2017). *Detailed project report faecal sludge management solutions for Unnao City, Uttar Pradesh.*
16. SCBP. (2018). *Detailed project report of implementation of FSM in Bagru, Rajasthan*. National Institute of Urban Affairs.
17. Rohilla, S. K., et al. (2017). *SFD promotion initiative Bansberia India*. C.f.S.a. Environment.
18. Roni, P., et al. (2018). Review of synthetic human faeces and faecal sludge for sanitation and wastewater research. *Water Research, 132,* 18.
19. Rose, C., et al. (2015). The characterization of feces and urine: A review of the literature to inform advanced treatment technology. *Critical Reviews in Environmental Science and Technology, 45*(17), 1827–1879.
20. Burkitt, D. P., Walker, A. R. P., & Painter, N. S. (1972). Effect of dietary fibre on stools and transit-times, and its role in the causation of disease. *The Lancet*.
21. Davies, G. J., et al. (1985). Bowel function measurements of individuals with different eating patterns. *Gut, 1*(1), 10.
22. Rakesh, K., & Tandon, B. N. (1975). Stool weights in North India. *The Lancet*.
23. Schouw, N. L., et al. (2002). Composition of human excreta—A case study from Southern Thailand. *The Science of the Total Environment, 286,* 155–166.
24. WHO. (1992). *A guide to the development of on-site sanitation*. W.H. Organization.
25. Still, D., & Foxon, K. (2012). *Tackling the challenges of full pit latrines* (Vol. 2).

26. Koottatep, T., et al. (2012). Assessment of faecal sludge rheological properties. In *Environmental engineering*. School of Environment, Resources and Development Asian Institute of Technology.
27. Barani, V., et al. (2018). Characterization of fecal sludge as biomass feedstock in the southern Indian state of Tamil Nadu. *Gates Open Research, 2*.
28. Mehta, M., Mehta, D., & Yadav, U. (2019). Citywide inclusive sanitation through scheduled desludging services: Emerging experience from India. *Frontiers in Environmental Science, 7*.
29. Gray, N. F. (1995). The influence of sludge accumulation rate on septic tank design. *Environmental Technology, 16*(8), 795–800.
30. Norris, G. A. (2000). *Sludge build-up in septic tanks, biological digester and pit latrines in South Africa*. Water Research Commission.
31. Centre, U. M. (2015). *Standard operating procedure (SOP) for fecal sludge management for municipalities in Gujarat*.
32. Srinivas, C., Maliki, R., & Safaris, A. (2017). *Towards a model sanitation city: Operationalizing FSM regulations in Warangal*. Bill & Melinda Gates Foundation.

The Trade-Off Between the Economic and Environmental Impact of Conventional and Green Building Materials

Sushibala Nambram, K. Narayanan, and Arnab Jana

Abstract The building stock in India is estimated to grow by five-fold from 2005 to 2030. Building causes significant environmental impact, and it is imperative to reduce the environmental impacts of buildings to achieve Sustainable Development Goal 11. This study integrates Building Information Modeling (BIM) with Life Cycle Assessment (LCA). This integration supports novel decision-making paradigms of building design as it considers sustainability parameters along with traditional engineering performance parameters. The BIM-LCA integration method was applied to a multi-storied residential building in Mumbai, India. The BIM model was created in Autodesk Revit, and then outputs data were converted to an appropriate input format by recomputing the materials of construction. Further, the converted material of construction in terms of physical mass units was used for analysis in OpenLCA. The environmental impacts of the building with different materials of non-load-bearing wall assemblies are calculated to determine the best building assembly options from the economic and environmental perspectives. A comparative analysis of building assemblies through its life cycle stages was conducted. The results provide a robust decision space to make an informed decision by building designers to select optimal designs, considering the trade-off between economic and environmental impact. Green building assemblies are found to be costlier than conventional assemblies in the product stage and construction stage of the building.

S. Nambram (✉) · A. Jana
Centre for Urban Science and Engineering, Indian Institute of Technology Bombay, Powai 400076, India
e-mail: sushibala@iitb.ac.in

A. Jana
e-mail: arnab.jana@iitb.ac.in

K. Narayanan
Department of Humanities and Social Sciences, Indian Institute of Technology Bombay, Powai 400076, India
e-mail: knn@iitb.ac.in

A. Jana and P. Banerji (eds.), *Urban Science and Engineering*, Lecture Notes in Civil Engineering 121, https://doi.org/10.1007/978-981-33-4114-2_28

Keywords Building information modeling · Life cycle assessment · Sustainable building

1 Introduction

The construction industry is at the center of concerns regarding environmental impact as construction is a very resource-intensive and non-environmental friendly process. The building sector is recognized as the most critical consumer of natural resources. The significant impacts are on the depletion of natural resources due to the extraction of resources and the emissions of greenhouse gases as a result of fossil fuel combustion. It is reported by earlier researchers [1–3] that globally, "it depletes 32–40% of natural materials, consumes 40% of the total primary energy, 12–15% of the world's freshwater resources, generates 25% of all wastes and emits 40–50% of greenhouse gas (GHG)."

The building stock in India is estimated to grow by five-fold from 2005 to 2030 [4]. Building causes significant environmental impact, and it is imperative to reduce the environmental impacts of buildings to achieve Sustainable Development Goal 11. The use of sustainable building materials has become the main focus of research and development in achieving the goal of sustainable construction, to minimize the environmental impacts of the construction industry. Life Cycle Assessment (LCA) provides a comprehensive approach for evaluating the environmental impacts of building materials, buildings, and products through all of its life cycle stages. The use of raw materials during building construction, maintenance, and renovation emits harmful substances throughout the building's life cycle, exerting a direct impact on the environment [5].

This study contributes to existing research on integrating Building Information Modeling (BIM) and LCA, which is in a nascent stage. The integration of BIM and LCA supports novel decision-making paradigms of building design as it considers sustainability parameters (energy use, resource depletion, emissions, and waste) along with traditional engineering performance parameters (cost, safety, and buildability) of BIM-enabled building performance analysis [6]. The combination of BIM and sustainable design strategies has the potential to produce high performance, sustainable design alternatives.

The Indian construction industry is growing at a swift pace accompanied by a large volume of investment, consumes a large volume of natural resources, manufactures a large volume of materials and products, generates employment, and impacts the environment. Du Plessis [7] opined that the construction industry could make a responsible contribution toward protecting the environment by using sustainable building materials. The goal of sustainable construction can be realized by giving more attention to the design and selection of sustainable building materials that complement the environment and improve quality of life, user health, and comfort.

Traditional building materials, including steel, concrete, aluminum, and glass, are high in embodied energy content. The selection of sustainable building materials is a difficult and challenging task in the construction industry. The LCA method can be used as an analytical decision support tool for the selection of sustainable material. The main objective is to determine, among the mentioned alternatives of building assemblies, the option with the lowest environmental impact and the lowest associated economic costs. In order to contribute to sustainability, it is intended to understand which option is more suitable, conventional, or Energy Conservation Building Code (ECBC) compliant building envelope.

2 Literature Review

ISO 14040 defines LCA as "A technique for assessing the environmental aspects and potential impacts associated with a product by compiling an inventory of relevant inputs and outputs of a product system, evaluating the potential environmental impacts, and interpreting the results of the inventory analysis and impact assessment phases." BIM is defined as "a set of interacting policies, processes, and technologies generating a methodology to manage the essential building design and project data in digital format throughout the building's life cycle" [8]. The literature recognizes the advantages of BIM-LCA integration [9–12]. The LCA can be used as an early-stage decision-making tool to assess design choices that impact the environment [13, 14]. Due to the unique nature of the operational and embodied energy use scenarios as well as building codes and regulations applicable for various geographic and administrative regions, it is evident a comprehensive study needs to be conducted to quantify the actual benefits of material substitution decisions.

3 Data

The combination of BIM and sustainable design strategies has the potential to produce high performance, sustainable design alternatives. The BIM-LCA integration method was applied to a multi-storied residential building in Mumbai, India. The subject building is (stilt + 32 floors) with 224 residential units of 27.87 m^2 carpet area each. We created a BIM model (Fig. 1) in Autodesk Revit and converted the outputs data to an appropriate input format by recomputing the materials of construction and converting it to physical mass units for LCA in OpenLCA.

We compiled a functional database of conventional and green building materials and building assemblies commonly used in residential buildings in Mumbai. The state schedule of rates and ecoinvent databases are used as the source of costs and life cycle inventory data for materials, respectively. In case an exact match is unavailable, we selected the material from the ecoinvent database based on the material properties which are closer to the actual material used in the construction of the building. The functional unit considered is the whole building with a life cycle of 50 years. This study considers three early life cycle stages out of five life cycle stages of EN

Fig. 1 Building Information Model (BIM) and Building Energy Model (BEM) in autodesk revit

15978:2011 standard: product stage, construction stage, and use stage. We recomputed the materials of construction and converted them to physical mass units to suit our analysis. Secondary data are collected and compiled from available literature on the Sustainable Building Materials Research and LCA.

4 Method

An LCA of a building involves typically evaluating its whole life cycle, including all the stages in the assessment: raw material supply, manufacture of construction products, the construction process stage, use stage, demolition, and when the materials are disposed of or recycled. For this study, we considered the life cycle stages up to (1) product stage (2) construction process stage and (3) use stage only. The functional unit is defined as the "a multi-storied residential building with 224 residential units of 27.87 m^2 carpet area each, for a life cycle 50 years in India," aiming to provide shelter and protection from weather events. Earlier LCA research by Asif et al. [15] and Reddy et al. [16] have divided the whole building based on the different types of construction materials, and impact assessment was conducted for each type of material. Cavalliere et al. [17] recently proposed a methodology to integrate useful data in BIM models, which focuses on the identification of the life cycle impacts of external walls. Sharif and Hammad [18] used the LCA and LCC methodologies to decrease the environmental and economic impacts of a building renovation, building on the material take-off generated by a BIM model. openLCA 1.7 software is used as the means to conduct the LCA in this study. "openLCA is the world's leading, high performance, open-source LCA software" [19]. openLCA offers the most extensive collection of data sets and databases worldwide for LCA software, some for purchase, some for free; altogether, almost 100,000 different data sets are available. Ecoinvent 3.5 database is used for inventory analysis in the study.

4.1 Product Stage and Construction Process Stage (A1 to A5)

Three life cycle stages: production stage, construction process stage, and use stage (A1-B6) are considered as the system boundary for the study. The embodied energy is the energy required to produce the building materials, and it is related to the production stage (A1–A3) of the life cycle stages. Construction process stage energy (A4–A5) is consumed in the transport of materials and construction installation process. Furthermore, the in-use energy is energy consumption related to the use stage (B6) is considered in this study. We compiled a functional database of conventional and green building materials and building assemblies commonly used in residential buildings in Mumbai. The state schedule of rates and ecoinvent databases are used as the source of costs and life cycle inventory data for materials, respectively. In case an exact match is unavailable, we selected the material from the ecoinvent database based on the material properties which are closer to the actual material used in the construction of the building.

Table 1 shows the actual material used in the construction of the residence and corresponding material selected from the database for this study. Actual transportation distances, 65 km from Vasai to Bhendi Bazaar and 11 km from Wadala to

Table 1 Technical specification comparison

Material	Technical specification	Database from Ecoinvent 3.5
Plum concrete	Mix 1:3:6 M10	Poor concrete, at plant
Plain cement concrete	Mix 1:2:4 M15	Normal concrete, at plant
Concrete	Mix: M35 (IS:456)	Concrete, exacting, at plant
Reinforcement	TMT HYSD bars-Fe-500 grade (IS 1786)	Reinforcing steel, at plant
Masonry	Bricks	Brick, at plant
Cement mortar	1:6 (1 cement: 6 sand)	Cement mortar, at plant
Cement mortar plaster	1:4 (1 cement: 4 sand)	Light mortar, at plant
Cement mortar plaster	1:3 (1 cement: 3 sand)	Lime mortar, at plant
Gypsum plaster	Internal gypsum plaster	Cover coat, mineral, at plant
Flooring	Indian patent stone	Poor concrete, at plant
Insulation	EPS and XPS	Polystyrene foam slab, at plant

[a]*Note* Bricks (minimum crushing strength 85 Kg/cm^2 and water absorption maximum 20%)

Table 2 Material and quantities

Material	Quantity in m^3	Quantity in kg	Transport distance (place of production to construction)	Means of transport
Poor concrete, at plant	77.00	168,630.0	11.0	Lorry >38t
Normal concrete, at plant	483.00	1,149,540.0	11.0	Lorry >38t
Concrete, exacting, at plant	3,850.00	939,400.0	11.0	Lorry >38t
Reinforcing steel, at plant	–	490,000.0	65.0	Lorry 20–28t
Brick, at plant [a]	1,979.04	3,067,512.0	65.0	Lorry 20–28t
Cement mortar, at plant	538.16	1,119,372.8	11.0	Lorry >38t
Light mortar, at plant	210.42	391,381.2	11.0	Lorry >38t
Lime mortar, at plant	91.00	160,160.0	11.0	Lorry >38t
Cover coat, mineral, at plant	151.20	105,840.0	11.0	Lorry >38t
Polystyrene foam slab, at plant	62.8 (XPS) 219.2 (EPS)	2135 (XPS) 3288 (EPS)	11.0	Lorry 20–28t

[a] *Note* The brick dimension considered is 230 mm × 110 mm × 70 mm

Bhendi Bazaar, are considered. The quantity of construction material transported and assumed means of transport is as given in Table 2.

4.2 *Use Stage (B1 to B5)*

We have considered use, maintenance, repair, refurbishment, and replacement of building material as negligible and hence is not part of the computation and analysis.

4.3 *Operational Energy Use (B6)*

Mumbai is in a warm humid climate zone, and the thermal discomfort level is extremely high; therefore, it is assumed that the occupant of the subject building will use HVAC for maintaining the thermal comfort level in mixed mode. Accordingly, the Building Energy Model was generated in Autodesk Revit (Fig. 1), and simulation was conducted using Insight 360 to quantify the heating and cooling load and energy performance of the building. The detail of the simulation studies is given in an earlier study by the author [20]. The energy efficiency of a building is expressed in terms of the Energy Performance Index (EPI). EPI is defined as annual energy consumption in kWh divided by m^2 of the area of the building, excluding storage area and parking in the basement. The EPI of the subject building with the conventional building envelope was 343 kWh/m^2/year. A reduction of 42 kWh/m^2/year was achieved after adopting the ECBC compliant building envelope. The potential energy saving from adopting prescriptive energy-efficient measures under ECBC resulted in a 12.24% reduction in EPI of the residential building.

5 Results and Discussion

The Life Cycle Impact Assessment was conducted using the Recipe (H) method with weights and normalization. Figure 2 shows the impact of building materials on the 17 midpoint impact categories. Reinforcing Steel and Concrete have the highest impact on eight midpoint impact categories each, and brick has the highest impact on one impact category. Figure 3 shows the impact of building materials on the three endpoint impact categories.

Concrete has the highest impact on ecosystem quality and human health. Reinforcement steel has the highest impact on resources. Substituting the conventional

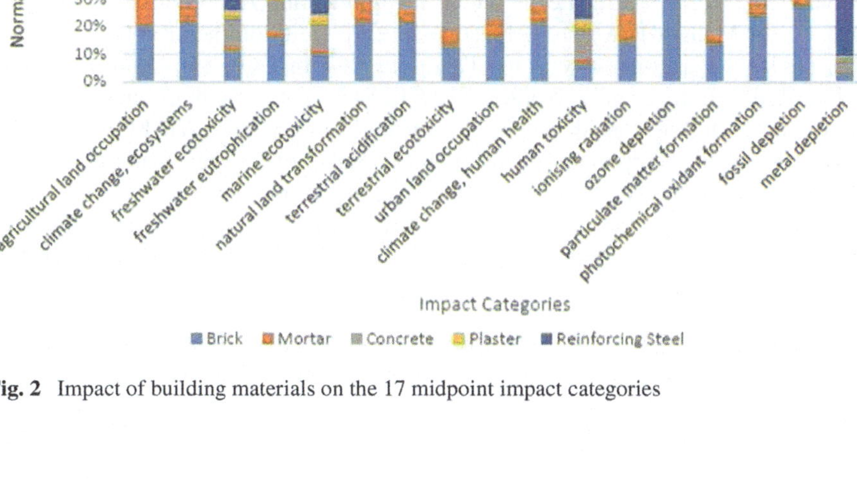

Fig. 2 Impact of building materials on the 17 midpoint impact categories

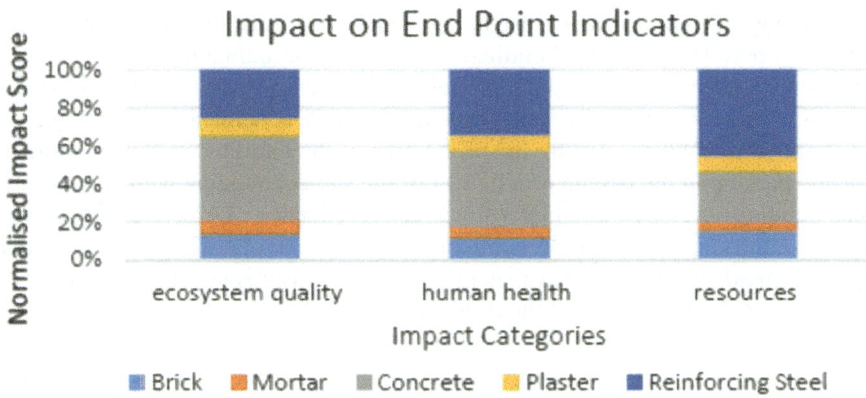

Fig. 3 Impact of building materials on the three endpoint impact categories

envelope with ECBC compliant envelope resulted in an increase of overall impact score from 274,053 to 275,732, i.e., a 0.6% increase in impact score. Due to the non-availability of data in the ecoinvent 3.5 database, we have excluded construction materials for waterproofing work and earthwork in the analysis. We have also excluded material for formwork as it is used temporarily during the construction phase only and removed after that.

6 Conclusion

With ₹5,237,561 increase in the cost of construction and a negligible 0.6% increase in environmental impact score, EPI reduction 42 kWh/m^2/per annum can be achieved by substituting conventional envelope with ECBC compliant one. This substitution will result in 262,164 kWh avoided electricity capacity generation per annum. Accounting for reported 23.12% Aggregate Technical & Commercial (AT&C) losses of the year 2015–16 for Maharashtra state [21], state-specific emission factor is 0.001009 ton CO_{2e} per kWh of end-use electricity. Therefore, 264 ton CO_{2e} emissions per annum for the whole building can be avoided. ₹1,654,254 saving per annum can be achieved for the whole building considering electricity price of ₹6.31 per kWh. LCA is a useful analytical tool for comparison of different building construction material over the different lifespan for eco-efficiency. However, we have considered the only three early life cycle stages in our study. This study needs to be expanded to Cradle to Grave and needs to consider other properties of building construction material like longevity and durability of materials.

Acknowledgements This study is conducted under the Ministry of Human Resource Development, Government of India funded project (14MHRD005) at the Centre of Excellence: Frontier Areas of

Science and Technology, Centre for Urban Science and Engineering, Indian Institute of Technology Bombay.

References

1. Mokhlesian, S., & Holmén, M. (2012). Business model changes and green construction processes. *Construction Management and Economics, 30*(9), 761–775. https://doi.org/10.1080/01446193.2012.694457.
2. Ramesh, T., Prakash, R., Kumar Shukla, K., & Shukla, K. K. (2013). Life cycle energy analysis of a multifamily residential house: a case study in Indian context. *Open Journal of Energy Efficiency, 2*(01), 34–41. https://doi.org/10.4236/ojee.2013.21006.
3. Yeheyis, M., Hewage, K., Alam, M. S., Eskicioglu, C., & Sadiq, R. (2013). An overview of construction and demolition waste management in Canada: A lifecycle analysis approach to sustainability. *Clean Technologies and Environmental Policy, 15*(1), 81–91. https://doi.org/10.1007/s10098-012-0481-6.
4. USAID. (2014) BEE: HVAC market assessment and transformation approach for India. Technical report, PACE-D technical assistance program (2014). https://www.climatelinks.org/resources/hvac-market-assessment-and-transformation-approach-india
5. Balaras, C. A., Droutsa, K., Dascalaki, E., & Kontoyiannidis, S. (2005). Heating energy consumption and resulting environmental impact of European apartment buildings. *Energy and Buildings, 37*(5), 429–442. https://doi.org/10.1016/j.enbuild.2004.08.003.
6. Eleftheriadis, S., Mumovic, D., & Greening, P. (2017). Life cycle energy efficiency in building structures: A review of current developments and future outlooks based on BIM capabilities. *Renewable and Sustainable Energy Reviews, 67*, 811–825. https://doi.org/10.1016/j.rser.2016.09.028.
7. Du Plessis, C. (2007). A strategic framework for sustainable construction in developing countries. *Construction Management and Economics, 25*(1), 67–76. https://doi.org/10.1080/01446190600601313.
8. Succar, B. (2009). Building information modelling framework: A research and delivery foundation for industry stakeholders. *Automation in Construction, 18*(3), 357–375. https://doi.org/10.1016/j.autcon.2008.10.003.
9. Ajayi, S. O., Oyedele, L. O., Ceranic, B., Gallanagh, M., & Kadiri, K. O. (2015). Life cycle environmental performance of material specification: a BIM-enhanced comparative assessment. *International Journal of Sustainable Building Technology and Urban Development, 6*(1), 14–24. https://doi.org/10.1080/2093761X.2015.1006708. http://www.tandfonline.com/action/journalInformation?journalCode=tsub20.
10. Basbagill, J., Flager, F., Lepech, M., & Fischer, M. (2013). Application of life-cycle assessment to early stage building design for reduced embodied environmental impacts. *Building and Environment, 60*, 81–92. https://doi.org/10.1016/j.buildenv.2012.11.009.
11. Kreiner, H., Passer, A., & Wallbaum, H. (2015). A new systemic approach to improve the sustainability performance of office buildings in the early design stage. *Energy and Buildings, 109*, 385–396. https://doi.org/10.1016/j.enbuild.2015.09.040.
12. Soust-Verdaguer, B., Llatas, C., & García-Martínez, A. (2016). Simplification in life cycle assessment of single-family houses: A review of recent developments. *Building and Environment, 103*, 215–227. https://doi.org/10.1016/j.buildenv.2016.04.014.
13. Meex, E., Hollberg, A., Knapen, E., Hildebrand, L., & Verbeeck, G. (2018). Requirements for applying LCA-based environmental impact assessment tools in the early stages of building design. *Building and Environment, 133*, 228–236. https://doi.org/10.1016/j.buildenv.2018.02.016.
14. Soust-Verdaguer, B., Llatas, C., García-Martínez, A., & Gómez De Cózar, J. C. (2018). BIM-based LCA method to analyze envelope alternatives of single-family houses: Case

study in Uruguay. *Journal of Architectural Engineering*, *24*(3), 1–15. https://doi.org/10.1061/(ASCE)AE.1943-5568.0000303.
15. Asif, M., Muneer, T., & Kelley, R. (2007). Life cycle assessment: A case study of a dwelling home in Scotland. *Building and Environment*, *42*(3), 1391–1394. https://doi.org/10.1016/j.buildenv.2005.11.023.
16. Reddy, S. M., Palaniappan, S., Pinky, L., & Reddy Sivakumar Palaniappan, S. M. (2012). Application of life cycle asessment for a residential building construction. *International Symposium on Life Cycle Assessment and Construction*, *1*, 213–222. http://demo.webdefy.com/rilem-new/wp-content/uploads/2016/10/db498e08acafa711205c3485aa7749f0.pdf.
17. Cavalliere, C., Dell'Osso, G. R., Pierucci, A., & Iannone, F. (2018). Life cycle assessment data structure for building information modelling. *Journal of Cleaner Production*, *199*, 193–204. https://doi.org/10.1016/j.jclepro.2018.07.149.
18. Sharif, S. A., & Hammad, A. (2019). Simulation-based multi-objective optimization of institutional building renovation considering energy consumption, life-cycle Ccst and life-cycle assessment. *Journal of Building Engineering*, *21*, 429–445. https://doi.org/10.1016/j.jobe.2018.11.006.
19. openLCA.org | openLCA is a free, professional Life Cycle Assessment (LCA) and footprint software with a broad range of features and many available databases, created by GreenDelta since 2006. http://www.openlca.org/.
20. Nambram, S., Jana, A., & Narayanan, K. (2020). Analysis of energy saving and emission reduction potential through the energy efficient building design of a residential building in a warm humid climate. In *Advances in Energy Research*, (vol. 1, chap. 58, pp. 609–620). Springer, Singapore (2020). https://doi.org/10.1007/978-981-15-2666-4_58.
21. Aggregate Technical & Commercial (AT&C) Losses in power sector | data.gov.in (2017). https://data.gov.in/catalog/aggregate-technical-commercial-atc-losses-power-sector?filters%5Bfield_catalog_reference%5D=3089361&format=json&offset=0&limit=6&sort%5Bcreated%5D=desc.

Transport Carbon Footprints—Examining the Neighbourhoods through Residents' Travel Behaviour

Case Study of Ahmedabad

Aashlesha Gupte and Rahul Shukla

Abstract The Transportation Sector is a key contributor to the global challenge of Climate Change. Increase in demand for transportation, increase in the use of the private mode of transportation, increase in trip lengths, low occupancy rate and lower levels of fuel efficiency result in higher Greenhouse Gas (GHG) Emissions. This research attempts to examine the Neighbourhoods with regard to the Carbon Footprints generated by capturing the Residents' Travel Behaviour in the context of Ahmedabad, India. The Neighbourhoods were selected based on spatial location, density and socio-economic groups along the public transport corridors. A sample of 270 was recorded by a stratified random sampling method to analyse Travel Behaviour. An integrated analysis of Transportation Energy Demand, CO_2 Emissions and Neighbourhood Planning was carried out. The study revealed that the modal share and the average trip lengths varied across distinct socio-economic groups and purpose of trips. Vehicle Kilometres Travelled (VKT) increase with an increase in age and ownership of motorized vehicles. Personal household income and the ownership of motorized vehicle also have an assertive relation to the mode of transport; so with higher income, the use of motorized vehicles for daily trips is also high.

Keywords Travel behaviour · Neighbourhoods · Carbon footprint · Low-carbon development strategies

1 Introduction

Today, 55% of the world's total population resides in urban areas and is estimated to escalate to 68% by the year 2050 [1]. Along with the upsurge in urban population, there has also been an accelerated growth in private motorized vehicles, particularly in the metropolitan cities. The increase in the demand for the private mode of

A. Gupte · R. Shukla (✉)
Institute of Architecture and Planning, Nirma University, Ahmedabad, India
e-mail: rahul.shukla@nirmauni.ac.in

A. Jana and P. Banerji (eds.), *Urban Science and Engineering*, Lecture Notes in Civil Engineering 121, https://doi.org/10.1007/978-981-33-4114-2_29

transport is a factor of urbanization and economic growth, which ideally should be accompanied by supporting factors like scientific traffic management system and efficient public transportation [2]. But this isn't the scenario in India, resulting in several transportation issues namely traffic congestion, accidents and safety, air and noise pollution and many other adverse environmental impacts. Among this, the predominant issue is of Transport Carbon Emissions.

As a consequence of burning fossil fuels over the past several decades, Transport Carbon Emissions have risen faster and are projected to rise even more rapidly in the future, if the current trend continues. According to the statistics of International Energy Agency (IEA)—CO_2 Emissions Database (2019 edition), road transport significantly contributes to the growth of Carbon Emissions and is responsible for 23% of Global Carbon Emissions [3].

The relation between urban form, travel behaviour and carbon emission is established at both aggregate studies and disaggregate levels, through the literature. Urban form and transportation system are interconnected with each other. The term Urban form has been defined in the literature by several authors as the spatial pattern of urban components such as road network, built forms and land use [4, 5]. It is the spatial imprint of an urban transport system as well as the adjacent physical infrastructures. The transportation system is this context is defined as the travel behaviour of residents and is computed as distances travelled by individuals in a day and the mode used for travel for work and non-work purposes. The basic parameters used to determine the travel behaviour are distance travelled, mode of travel, purpose of travel, time taken and occupancy. Transport Carbon Footprints are defined to be a collective outcome of the modal split, distance travelled by different modes of transport for various purposes, time taken to travel the particular distance and mode-specific emission factors [6].

The relationship between Urban Form and Travel Behaviour can be studied at different levels of planning such as region, district, metropolis, city, ward, neighbourhood and along transport corridors [7]. Every neighbourhood type will have a definite density, built urban components, land use zoning, road network patterns and public transport system, which would support the design of a Transit Oriented Neighbourhood. Thus, the urban form parameters that influence the travel modes and vehicle kilometres travelled are expected to contribute to a great extent in the variation of transport carbon emissions.

Mixed land use development provides stress on high density and a higher degree of mix land use [8, 9]. It is a critical factor of many developments, including traditional neighbourhood planning, transit-oriented development (TOD), new urbanism, inclusive communities [8, 10], etc. These developments are a resilient way to foster a sustainable environment [11, 12]. According to American Public Transportation Association, the presence of public transport amenities within walking distance facilitates an individual to use public transport facility and hence reduces the Transport Carbon Footprints [13].

As socio-demographic characteristics, travel and activity patterns, and travel behaviour of individuals are interrelated, travel behaviour is said to be a derived function of income, socio-demographics and perception and attitude towards urban life. Income is a predominant determinant of travel behaviour in both developed and developing countries. The travel behaviour of residents differs with regard to different socio-economic groups influencing their energy consumption pattern, expenditure pattern and Transport Carbon Footprints [14]. Income has a major influence on the affordability of a house and to own motorized vehicles [14]. Thus, it is important to identifying the difference and understanding the dynamics of travel behaviour within different socio-economic groups within a city, which results in better planned neighbourhoods for all the segments of the society [14].

The study aims to analyse the Neighbourhood's Transport Carbon Footprints by capturing the Residents' Travel Behaviour and correlating it with other Urban Planning and Transportation Planning tools.

2 Study Area—Ahmedabad

Ahmedabad is the 7th largest metropolitan city and one of the key emerging urban centres in India. Ahmedabad city's transport network has grown in a circular ring pattern from the city centre extending towards the periphery of the city limits. The road network has evolved as a ring radial form, comprising 5 rings and 17 well-defined radials. According to Ahmedabad Municipal Corporation (AMC), the total road length of the city is 2,580 km, and the present mean trip length is 5.5 km. There exist two public transport modes in the city, Ahmedabad Municipal Transport Service (AMTS) and Bus Rapid Transit System (BRTS).

Through baseline analysis, six neighbourhoods were selected along the Transport Oriented Zone (as a controlling factor), varying in distance from the city centre, built form density and socio-economic groups (Table 1 and Fig. 1).

Table 1 Details of selected neighbourhood sites

Particulars	Paldi	Ambawadi	Prahlad nagar	Chandkheda	Bopal	Science City
Distance from the city centre (km)	3	5	9	11	13	16
Area (km^2)	2.2	2.3	2.4	2.5	2.5	2.7
Population density (persons/hectare)[a]	192	217	181	60	120	98

[a]Ward densities

Source Compiled by Author

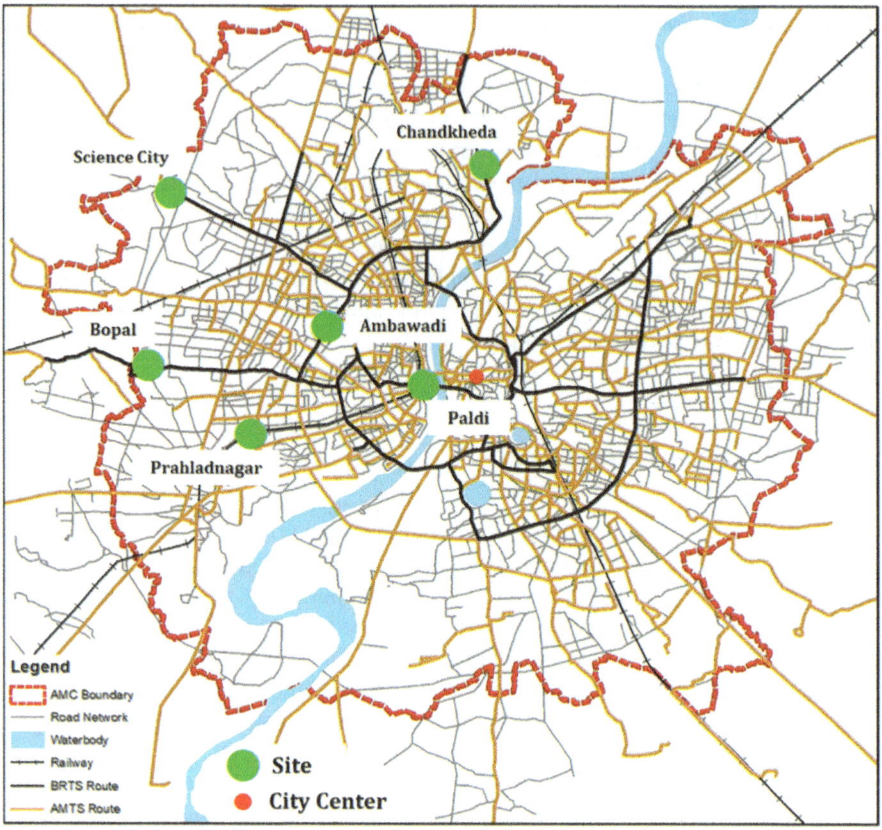

Fig. 1 Location of selected Neighbourhoods in Ahmedabad. *Source* Compiled by Author

3 Methods

3.1 Data Collection

The data for this research was collected at three levels: City, Neighbourhood and Household through various sources. At the City and Neighbourhood levels, census demographics, road network, macro-scale land use, public transport (i.e., AMTS and BRTS) nodes, routes and ridership information was collected. Household surveys were carried out at each of the six Neighbourhoods to understand the neighbourhood and travel characteristics.

A sample of 270 was recorded by a stratified random sampling method to analyse Travel Behaviour and Carbon Emissions of the six selected Neighbourhoods of Ahmedabad. The recorded information included Household Details, Travel Characteristics and Neighbourhood Perception. The secondary information was integrated with land use, built form and travel behaviour collected through the household

Table 2 Emission factors (mode-wise)

Particulars	Emission factors
Two-wheeler—Scooter	0.0387
Two-wheeler—Motorcycle	0.0597
Four-wheeler—Hatchback	0.140
Four-wheeler—Sedan	0.290

Source Emission Factors provided by WRI in India Specific Transport Emission Factors [16]

surveys of the selected neighbourhoods. The resulting database was further used in multivariate statistical analysis.

3.2 Carbon Emission Calculation Method

The variables included in the study to calculate carbon emissions were Passenger Transportation Activity (in vehicle-kilometre travelled—VKT), Modal Structure and Emission Factors (in grams of CO_2 released per kilometre vehicle travelled). Maximum Emission Factors were considered for computing the Carbon Footprints generated for the selected neighbourhoods (Table 2).

The total CO_2 equivalent emissions for trips made in a day from a neighbourhood (i) by total vehicle (j) are computed as follows:

$$C_i = \sum j\{[VKT(2W) * EF(2W)] + [VKT(Car) * EF(Car)]\} \quad (1)$$

where C_i = Total CO_2 emitted by households "j" from the neighbourhood "i", VKT = Vehicle Kilometres Travelled by a household in a day and EF = Emission Factor for India specific road transport emissions mode-wise.

4 Results

The unit of analysis for this research is considered as individuals, for the reason that specific attributes can be incorporated into the analysis, such as daily travel trips, mode preference and vehicle kilometre travelled.

From the analysis mentioned in Table 3, it can be inferred that private vehicles comprise more than 80% modal share within the city. And approximately 10% are from public transport and a mere 5% from Intermediate public transport. The average trip length of two-wheelers and four-wheelers from the surveyed households are more than the mean trip length of Ahmedabad city. Also, it can be interpreted that public transport also contributes considerably to a good distance of approximately 8.5 km as a mean trip length of residents.

Table 3 Average modal share, trip length and travel time

Mode	Mode share (%)	Trip length (km)
Bus	11	8.7
Auto rickshaw	05	5.2
Two-wheelers	45	6.8
Four-wheelers	39	12.6

Source Compiled by Author

In Table 4, income groups are defined as follows: Low (up to 50,000), Medium (50,000–150,000) and High more than 1.5 lakh. The high- and middle-income groups showed private vehicle-oriented travel behaviour in the surveyed neighbourhoods.

Table 4 Household and neighbourhood characteristics

Location of study areas from city centre	Near centre		Between centre and outskirt		Near outskirt	
	Paldi	Ambawadi	Prahlad nagar	Chandkheda	Bopal	Science City
Area (Ha)	2.2	2.3	2.4	2.5	2.5	2.7
Population density	High	High	High	Medium	Medium	Low
Income group	Low medium	Medium high	Medium high	Low medium	Medium high	Medium high
Average household size	3.9	4.2	3.6	4.5	4	5.3
Mixed land use (%)	22	25	18	15	12	8
Residential (%)	47	40	45	60	60	57
Commercial (%)	8	12	15	4	7	5
Transport (%)	15	16	13	11	14	15
Institutional (%)	4	6	3	7	–	–
Open space (%)	4	1	6	3	7	15
Public transport connectivity	Yes	Yes	Yes	Yes	Yes	Yes
BRT	Yes	Yes	Yes	Yes	No	Yes
Number of bus routes	14	10	8	4	2	1

Source Compiled by Author

Along with the ratio of the number of vehicle owned versus the number of people having access to driving licence was more. Thus, it can be established that higher the income, higher the ownership of private vehicles, more the number of travel trips.

The analysis shows that people belonging to low- and middle-income groups have access to opportunities such as work and education at shorter distances. In the study, we have analysed the variation in trip length, and modal share for work and education trips has a negative relation to income groups. Trip length and mode choice vary according to different income groups [14]. Since the low-income groups cannot afford private modes of transport, their travel behaviour is more inclined towards non-motorized or public transportation (Table 5).

Through the integrated analysis of transport, energy demand and CO_2 emissions, the results were as follows (Table 6).

Table 5 Travel characteristics

Location of study areas from city centre	Near centre		Between centre and outskirt		Near outskirt	
	Paldi	Ambawadi	Prahlad nagar	Chandkheda	Bopal	Science City
Average vehicle ownership (per household)	20	48	47	50	32	65
Average trip length	5.3	6.5	8.2	8.5	13.5	17.8
Two-wheeler	5.5	6.8	3.5	3.5	3.2	2.5
Four-wheeler	7.0	9.5	10.5	12.3	15.8	18.0
Public transport	5.5	6.3	6.5	6.0	8.5	7.5

Source Compiled by Author

Table 6 Neighbourhood carbon emission summary

Location of study areas from city centre	Near centre		Between centre and outskirt		Near outskirt	
	Paldi	Ambawadi	Prahlad nagar	Chandkheda	Bopal	Science City
Population density	High	High	High	Medium	Medium	Low
% of mixed land use	High	High	Medium	Low	Low	Low
Carbon footprints (kg CO_2/km) (per individual)	59.316	34.626	239.994	182.009	295.176	463.318
Two-wheeler	22.356	11.874	19.194	6.9696	12.936	7.452
Four-wheeler	36.96	22.752	220.8	175.04	282.24	455.84

Source Compiled by Author

The neighbourhoods such as Paldi and Ambawadi, located within the proximity of city centre, CBDs and workplace produced comparatively lesser carbon emissions compared to the peripheral neighbourhoods. The neighbourhoods Science City, Bopal, Prahlad nagar and Chandkheda were recorded to have high-income group residents with higher dependency on private vehicles and generated higher carbon emissions.

Science City, Bopal, Prahlad nagar and Chandkheda, peripheral neighbourhoods with higher residential population density (with an exception of Science City), high-income groups and lesser rate of public transport services recorded to have the maximum generation of Transport Carbon Footprints.

5 Conclusion

Ahmedabad is developing rapidly with the rapid increase in motorized vehicles. Hence, it is extremely crucial to understand the current urbanization trends to create a framework for the implementation of Sustainable Development Goals. Neighbourhood Planning being the backbone of urban planning and development is an appropriate scale to assess the relation between urban form and transportation system.

Thus, it is necessary to look at the role of urban planning, transport and policy, and their consequences on reducing travel demand, dependency on motorized vehicles and carbon emissions [15]. Urban and Transportation Planning Department needs to ensure that the undesired mode shift is detained and that the future urban expansion does not result in an increase in the travel distances for the basic purpose of work and education.

The study revealed that the modal share and the average trip lengths varied across distinct socio-economic groups and purpose of trips. Vehicle Kilometres Travelled (VKT) increase with an increase in age and ownership of motorized vehicles. Personal household income and the ownership of motorized vehicle also have an assertive relation to the mode of transport; so with higher income, the use of motorized vehicles for daily trips is also high.

In the above case study of six neighbourhoods, the Transport Carbon Footprints reflected to have a negative relation to the pattern and distribution of city centres. Residential population density of the neighbourhoods and the availability of public transport infrastructure influenced in generating Transport Carbon Footprints.

There exists a need to attain positive travel behaviour, especially in developing countries as a matter of concern from the perspective of Low-Carbon Sustainable Transport. A positive travel behaviour can be achieved by the reduction in the usage of the private mode of transport, reduced vehicle kilometres travelled, traffic congestion and enhancement in transit ridership [7].

The demand for transport can substantially be reduced by adopting appropriate land use policies and sustainably planned neighbourhoods. Reducing the need for motorized transport can be achieved by internalizing the trip within the neighbourhood and providing mixed land use activities. For promoting Low-Carbon Travel

Behaviour, medium- and high-density neighbourhoods should be encouraged [17]. Along with that, mixed land use zoning is another necessary policy tool to be implemented so as to internalize their trips within the neighbourhoods.

References

1. Rong, P., Zhang, L., Yang, Q., Qin, X., & Yaochen Qin, H. L. (2016). *Spatial differentiation patterns of carbon emissions from residential energy consumption in small and medium-sized cities: A case study of Kaifeng*. Retrieved from http://www.dlyj.ac.cn/EN/abstract/abstract38520.shtml.
2. Vijayalakshmi, S., & Raj, K. (2019). *Income and vehicular growth in India: A time series econometric analysis* (No. 439). Bangalore.
3. IEA. (2019). CO_2 emissions database (2019 ed.). In *International Energy Agency*. https://doi.org/10.1670/96-03N.
4. Bourne, L. S. (1982). *Internal structure of the city : Readings on urban form, growth, and policy*. Oxford University Press.
5. Munshi, T. G. (2013). Built form, travel behaviour and low carbon development in Ahmedabad, India. In *Faculty of ITC*. https://doi.org/10.13140/2.1.1217.9845.
6. Wu, X., Tao, T., Cao, J., Fan, Y., & Ramaswami, A. (2019). Examining threshold effects of built environment elements on travel-related carbon-dioxide emissions. *Transportation Research Part D, 75,* 1–12. https://doi.org/10.1016/j.trd.2019.08.018.
7. Kumar, P. P., Ravi, C., & Parida, M. (2020). Identification of neighborhood typology for potential transit-oriented development. *Transportation Research Part D, 78,* 102186. https://doi.org/10.1016/j.trd.2019.11.015.
8. Bahadure, S., & Kotharkar, R. (2015). Assessing sustainability of mixed use neighbourhoods through residents' travel behaviour and perception: The case of Nagpur, India. *Sustainability, 7*(September), 64–89. https://doi.org/10.3390/su70912164.
9. Grant, J. (2002). Mixed use in theory and practice: Canadian experience with implementing a planning principle. *Journal of the American Planning Association, 68*(1), 71–84. https://doi.org/10.1080/01944360208977192.
10. Planning, Development and Management of Sustainable Cities. (2014).
11. Evans, G., & Foord, J. (2007). *The generation of diversity: Mixed-use and urban sustainability*. Retrieved from www.vivacity2020.org.
12. Jacobs, J. (1961). *The death and life of Great American Cities*.
13. American Public Transportation Association. (2008). *Public transportation reduces greenhouse gases and conserves energy—The benefits of public transportation*.
14. Jain, D., & Tiwari, G. (2019). Explaining travel behaviour with limited socio-economic data: Case study of. *Travel Behaviour and Society, 15*(December 2018), 44–53. https://doi.org/10.1016/j.tbs.2018.12.001.
15. Dulal, H. B., Brodnig, G., & Onoriose, C. G. (2011). Climate change mitigation in the transport sector through urban planning: A review. *Habitat International, 35*(3), 494–500. https://doi.org/10.1016/j.habitatint.2011.02.001.
16. WRI. (2015). *India specific road transport emission factors. 1*(1.0), 36.
17. Ewing, R., & Cervero, R. (2001). Travel and the built environment: A synthesis. *Transportation Research Record: Journal of the Transportation Research Board, 1780*(1), 87–114. https://doi.org/10.3141/1780-10.

A Descriptive Analysis of Rehabilitation and Resettlement Policies in India Through Policy Feedback Approach

K. Sen Sharma and V. Jothiprakash

Abstract Policy Feedback Approach (PFA) considers existing policies as inputs which reshape the political environment and eventually shape the forthcoming policies and their outcomes. PFA is a globally established policy analysis tool which hasn't so far been used for India's involuntary resettlement scenario. Through our study, we found some positive and negative feedback effects on this policy and identified some points which have a long-term diminishing effect on it. Our descriptive analysis of Land Acquisition Rehabilitation and Resettlement (LARR) timeline of India through PFA's six lenses recognizes positive feedbacks like resistance, awareness, participation, consent clauses, etc. and a few negative ones like stakeholders' imbalance, clash of interests, vagueness, etc. The paper concludes with prospects to develop a policy analysis tool aspiring better efficiency in resettlement policies.

Keywords Policy feedback · Land acquisition · Resettlement · Rehabilitation

1 Introduction

Rehabilitation and resettlement (R&R) plans have evolved over decades since independence in India, originating from the archaic Land Acquisition Act 1894 to Right to Fair Compensation and Transparency in Land Acquisition, Rehabilitation and Resettlement Act (RFCTLARR) in 2015. Over the years, these plans have proven to be the perfect examples of how policies have the power and ability to confer properties and resources towards some interest groups over others. The timeline of LARR in India unfolds complex layers with each passing year resulting in more elaborate policies. But this increasing complexity doesn't ensure an increase in efficiency. The Indian scenario of rehabilitation and resettlement (R&R) fits the definition of 'policyscape' [1], referring to the landscape heavily weighted by past policies which

K. Sen Sharma (✉)
Centre for Urban Science & Engineering, Indian Institute of Technology Bombay, Bombay, India
e-mail: kanchansensharma@gmail.com

V. Jothiprakash
Department of Civil Engineering, Indian Institute of Technology Bombay, Bombay, India

A. Jana and P. Banerji (eds.), *Urban Science and Engineering*, Lecture Notes in Civil Engineering 121, https://doi.org/10.1007/978-981-33-4114-2_30

have established as institutions themselves. LARR caters to the urban needs of the mass population but directly and intensely affects a smaller population (mostly poor). Research and critique of public policies are usually done through a very idealistic lens, through which the policy is supposed to be perfect and must cater to everyone's needs. Despite being unrealistic, this approach catalyses the improvement in policy-making. The term 'policy' has a range of multifaceted definitions, like theory/model, programme, output, outcome, and process.

By using very simple terminology, Colebatch states that policy is what the government chooses to do, or chooses not to do and both remain equally important and that the things that are not done create the base for upcoming policies [2]. The future of a policy depends on the currently running version of it and the feedback and lessons learnt through it. If the policy manages to gain the attention of the government towards its journey and feedback, it's the first and most important step towards its evolution [3]. The policies built on the foundations of inequality often widen the planning and implementation gap. These gaps are very evident in the case of Indian Policies and have often emerged as major hindrances on the path of development. Policy Feedback Approach considers existing policies as catalysts to reshape the political environment which eventually results in new policy outcomes [4]. To understand and pursue the policy feedback approach, it's important to understand the basic terminologies it's based on. Studies by Elinior Ostrom state three different levels of specificity when it comes to ideas of researchers in policy studies namely Frameworks, Theories and Models [5]. With each step getting further refined and precise, Ostrom defines framework as a 'metaphorical language' used for comparison of theories, Theories as specific elements of frameworks and finally Models as a set of assumptions concerning a limited set of variables and parameters [5]. Policy scientists have used another interesting term named 'metaphor'. The policy analysis in this paper deals with involuntary resettlement policies due to land acquisition. The main concern here is whether and how resettlement policies can be studied through this approach and the major take-outs from this study. The interesting factor about the whole 'policy package' around Land Acquisition and rehabilitation and resettlement is the dual nature of it being both universal and catering to the urban needs of the mass population as well as directly and intensely affecting the lives of a smaller population (mostly poor). Political systems often distribute resources to individuals and groups through policy incentives creating the basics of political science [6]. With LARR being such a policy which decides the incentives for the affected people by the authorizing state, PFA seems an apt lens for its analysis. We have also established our own definitions of 'positive' and 'negative' feedbacks of policies. Understanding how some agencies benefit from these would yield clarity on the general public's situation as public policies encompass all citizens and affect them in some or the other way [7]. The definitions of PFA are quite general and broad with little to no specifics, which is both good and bad for the study. On the one hand, it doesn't clearly dictate the rules of analysis but on the other hand, it also gives us the freedom to apply it for varying political contexts. Each policy choice when made has its own political consequences, however, it's hard to make specific claims about how, when and where these consequences occur [6]. PFA to some extent sheds light on a few of these.

Daniel Beland presents the whole concept of PFA in a relatively organized manner by discussing six research streams which are interlinked yet distinct. The whole concept can be perceived with 6 different faces of policy feedback as State Building, Interest Groups, Lock-In Effects, Private Institutions, Political Participation, and Ideational and Symbolic Legacies [8]. We shall carry out our discussion on the skeletal base of these 6 streams of PFA. Partha Chatterjee's acclaimed work establishes the idea that the political society of India revolves around the welfare policies of the government, including the R&R policies which try to provide relief to the people displaced by corporate capital projects which consider them unwanted [9]. Whether the PAP have so far been treated as obstacles, liabilities or opportunities that tell us the story of LARR in India. Hence, given the solidarity of several researchers regarding the impact of LARR on politics and vice versa, it's a bit surprising how Policy Feedback Approach being used to analyse this policy is an unexplored path. The reason we're inclined towards PFA is that studying a policy like LARR strictly as contemporary polity tends to distort its understanding [10], while studying it as a multi-streamed dynamic process can recognize the problems alongside their counteracts better.

2 Streams of Policy Feedback

2.1 State Building

Whenever a new policy is to be launched, the state makes use of its existing capacities and administrative arrangements to implement it efficiently, and these new policies have the capacity to inflate the power and capacity of the state in question [8, 11]. This segment of policy feedback deals with the way in which authorities use policies as agencies to expand and develop their administrative structures, but this expansion may or may not proceed as per their intentions. At times, policies have even led to the formation of strong allies which act in favour or opposing the government's ideals. Major social policies entrench the formation of fragments of state or political bodies which tend to shelter the rights of stakeholders either due to political reasons or simply for the sake of their own ideologies [8].

The Land Acquisition Act (LAA) 1894 was used as a tool to maintain its autonomy over land and the population. Post independence, this act provided the government with the ideal instrument to implement development plans, infrastructure projects and other pro-progressive agendas [12, 13]. The basic philosophy behind LAA 1894 was to systematize the state's power to confiscate land from unwilling private owners as the Eminent Domain and to regulate its exercise for 'public good' and to regulate compensation. The initial policies of LAA lacked a rigid definition of what 'public purpose' meant in terms of the acquisition. These policies definitely helped the state strengthen its entitlement to private property. But the 'escape routes' or bendable definitions of public purpose exempted them from transparency and clarity of state's alignment towards land dispossession movements [14]. Over the years as India underwent democratic awakening, the Act became more perceptive to the affected population and their interests. From stringency to sensitivity, the state building made the power dichotomous and every state started developing its own laws for acquisition. But the central government's authority still prevailed over conflicts. The 2007 Bill stated central government as the 'appropriate authority' for projects of 'public purpose' and multi-state ones. For internal acquisitions within, the state governments had the scrutinizing power. But as violent protests increased, the need for a uniform policy was felt and the 2011 Bill further empowered the central Land Acquisition Act exempting 16 other national acts, later reducing to 13. State Building also occurred through the formation of dispute settlement authorities and project development authorities. These departments branch out the state government's administrative capacities to organize the acquisition of land and project proceedings. The increased weight of PAP's consent and the Government's authority over them creates a feedback timeline of intersection and overlaps. In the 2007 Bill, the Land Acquisition Disputes Settlement Authority was directed to be set up. The government strengthened its own position even while handling grievances declaring that no civil court shall have jurisdiction to entertain any dispute relating to land acquisition. In this respect, the Collector or the Authority is empowered by or under this Act. It was not until the Act of 2013 which finally made a strong statement about the consent of the PAP. The public-private partnership (PPP) projects would require 70% consent from the PAP. For private projects, 80% of consent was mandatory. But as soon as this Act was declared, it was taken for amendment consideration before it could complete a year. Later, the acquisitions for the purposes of defence, rural infrastructure, affordable housing, industrial corridors and social infrastructures including public–private partnership (PPP) on a government-owned land were exempted from both the consent clause and the social impact assessment (SIA).

2.2 Interest Groups

The most significant arena of policy feedback research emphasizes how policies, through their resources and incentives, affect social groups and their mobilization [6]. The impacts of these policies bring some specific groups together with similar political goals and capabilities. As per researchers, the policy programs designed and targeted for the poor mostly don't form strong constituencies against the policy and form vulnerable interest groups [15]. The quasi-universal social schemes usually emerge with the most robust and politically impactful constituencies [8, 15]. However, this relationship between the targeted groups and political impacts must be kept in the background and shouldn't prominently affect the policy design. Policies apart from providing incentives to the favouring groups also create countermobilizations [11] but the frequency and the circumstances remain illustrative [6]. The policy feedback study of interest groups has 2 different approaches. One can be the study of policies and correlating them to the behaviour of a specific group. The other is the identification of incentives used by such groups to overcome their collective issues and then determine if those were availed through the policies in question [6]. LARR and resistance from PAP have always gone hand in hand. This policy creates a few distinct groups, i.e. the acquirers/investors/benefitting group and the displaced/affected/losing group. The clashes of interests between these two groups fuel the policy transformations. It has been pointed out by researchers that in India, land losers (especially farmers and cultivators) have been resisting more aggressively to dispossession for private organizations, than to public sector projects. This resistance is seen as a bottleneck for the liberalized economic growth model of the country [14]. LARR and its altercations with the tribal and agricultural communities is a well-documented subject. The protests and outcries of specific groups has also pushed LARR into the centre of Indian politics and caused amendments [14]. These protests have led to amendments regarding fair compensation, livelihood restoration, acquired land being left unused, etc. It's also been observed that active involvement and a good organized society of the affected people always lead to better R&R implementation [16]. Apart from visible and substantive interest groups, there are several other internal groups which are often neglected in R&R. As the social networks, community bonds and fabrics break, the marginalized population of the scheduled castes/tribes, the elderly and women are the ones who suffer the most. They are often unable to express deal in their own favour [17] as baseline surveys treat families as units. The male members are the most active participants of such surveys. Hence, their representation and consultation at each step of R&R implementation are extremely important.

2.3 Lock-In Effects

These are the negative impacts of old and existing policies on the new ones, which restrict the policy's progressive changes. The main studies and discoveries regarding these negative lock-in effects were conducted by Paul Pierson over the years [6, 10, 18]. The established policies, as mentioned before, lead to the formation of interest groups and an intricate economic and social network. The government officials or agencies which try to modify or replace these policies tend to face a lot of resistance, backlash and massive transition expenses [6]. The older the policy, the lesser politically vulnerable it becomes [18]. Change is often considered preferable in PFA. The lingering lock-in effect in LARR is evident through the unperturbed upholding of the archaic law LAA 1894 for so many decades before finally being challenged with a minor amendment in 1984 and a proper one finally in 2007. The dominance of an act over such a long period reflects a strong network, intent and inertia which were difficult to break free from. The act of 1894 was initially practised to abolish the system of '*Zamindari*' in the country which was basically the stocking up of land by the rich. The purpose was to fairly allocate the land to the poor in a just manner. Post independence, the Act tuned into the government 's primary tool to implement pro-development resolutions [12, 13]. But eventually again, the rich and the powerful found their way to exploit the act. The Act allowed the requiring bodies to divert and manipulate the acquisition even before the actual process commenced. The reason why the policy is unable to make progress and evolve for the better lies majorly in the fact that whoever is in opposition, readily resists it and whoever is in power tries to push it towards its older archaic versions, which are both dead ends. The best example is the unorganized involvement of NGOs in R&R. It has been observed by researchers how most of the NGOs are extremely vocal while opposing a development project, but hardly accept the serious opportunity the same project offers, to help and uplift the affected people [17]. Similarly, during the political reign of United Progressive Alliance (UPA), the opposing party of BJP was very vocal in opposing the Bill and actively participated in pushing forth the amendments regarding consent and SIA in the 2013 Act, although the same party when in power nullified several of those amendments through an ordinance which took place the very next year [19]. This cycle of protest, criticism and amendment also seems to lock-in the policy in a way.

As mentioned before, another form of lock-in effect can be seen in the state's refrain from using definite terms of compulsion. When it comes to the provision of alternative livelihood, land for land compensation or any other such incentives other than money, there are hardly any strict regulations in the act. A lock-in effect makes it very difficult to radically revamp the policy. The authorities often find it challenging to make solid claims regarding the deliverables of the R&R package and rather opt to provide soft assurance to avoid future litigations. This also exists to make the plan flexible to various fragments of PAP and provide options for exercising the resettlement smoothly.

2.4 Private Institutions

Since the 1990s, there have been extensive studies about the correlation between private institutions and public policies [8, 20–23]. The focus of these studies has been on how the private benefits which are widely desired and relied on by the citizens impact their expectations and political mobilization in comparison to public benefits [24]. Such private benefits become so obligatory to the people that they form sturdy political alignment for maintaining the existing private institutions which avail these benefits [24]. This may also be considered as a hindrance while enacting a National public policy. Hacker has used Health Care as an example to explain this phenomenon. In the U.S., large-scale private benefits of health had already emerged which resulted in divided interests and constituencies and caused hindrances in the path of National Health insurance and Medicare [24].

India's liberalized model of growth and economy hugely relies on the state's capability to provide large chunks of land to the private and foreign investors [14]. The Indian government's inclination towards massive private industrialists and using Land Acquisition as a tool to aid plot collection for their private industries, factories, IT parks, malls, etc. lately led to vast stretches of land acquired for Special Economic

Zones (SEZs). This deprives the farmers and cultivators of their agricultural land. The government's active involvement here is a result of codependency. Infrastructure, which was historically provided by the State for the public good, has gradually been privatized and turned into profit-making enterprises which now stand somewhere between public good and private commodity [25]. The state utilizes private agencies to enable the development of quality housing, infrastructure and residential colonies. The private companies too need the support of the state to be able to acquire large fragments of land without facing land 'hold outs' [25]. The conflict of interest between the peasantry and private companies initiates policy change. Apart from 'public purpose' and 'public good', another prominent term causing bewilderment in the policy's implementation has been the term 'company' to such extent that at some point (1984) there was barely any difference between acquisition for state or private enterprises [19]. Time and again organizations like Confederation of Indian Industries (CII) and Confederation of Real Estate Developers have criticized the LARR policies for making the acquisition process too long and complex. Although their claims do hold strong reasons, the criticisms by Private institutions have been felt to be a bit exaggerated by researches conducted by two renowned private financial institutions (ICICI and Kotak) [19]. Nonetheless, the demands of private institutions do have major impacts on the policy's discourse.

2.5 *Political Effects and Participation*

The focus here is how the policies affect the electoral participation of individuals and the political environment. The voting tendencies of elderly citizens, youth or even working-class population are affected by the policies that have been imposed upon the public. In the book *How Policies Make Citizens,* Campbell provides the empirical evidence of how the Social Security Policies have made the elderly people more active electoral participants. They are more actively aware of their political rights and more expressive and communicative to politics [26]. When Social policies are drafted targeting particular sections of the population, each section forms a different type of relationship with the authorities, resulting in different levels of political participation. In Marxist theories, the peasantry has often been considered as politically inferior, and their exploitation through the accumulation of their properties has even been termed as tragic but inevitable [27]. But it was hardly anticipated that the resisting movements by the very same class of people might turn out to be the biggest hurdle in the path of capitalism [25]. Primitive accumulation here didn't directly refer to involuntary acquisition, but could refer to anything which led to agricultural disruption and alienation of the affected people [25, 28]. The poor implementation of R&R and unfair treatment at times reflects people's electoral choices. The PAP incline towards the political parties which support their cause. But the infrastructure and other development projects when completed often cater to millions of people while drastically uprooting a few hundreds. This major difference in this number shows a significant effect on votes. The voters, who didn't lose anything to the project, often overlook the woes of those who did and perceive such projects as progress.

2.6 Ideational and Symbolic Legacies

Symbolic pollution occurs when the ideas and the symbolic effects of previous policies are still fresh in the minds of the public and the authorities. This section deals with the way these ideational and symbolic legacies influence political outcomes. For example, if the public has had some bad experience with the policies of a particular government, they might hesitate to support a member of the same political party. Also, there are some language and symbolic factors which create a sense of security or insecurity among the stakeholders [8]. The two major symbolic legacies which dominate the LARR timeline in India are Development and Criticism. Development or progress of the country is an ideational symbol every ruling body tries to put up as a facade for the people. Since LAA 1894, the approach of the Indian government towards development has been through the lens of collective industrialization and that it would eventually benefit all [29]. This hasn't much changed in recent times as well. The mass acquisition helps the authorities to conduct gigantic infrastructure projects (flyovers, ring roads, or metro rails), and these projects (irrespective of their completion status) do help to create a symbol that development is happening. The clear visibility of such projects is what makes a ruling party eager to make acquisition more hassle free. As mentioned before, this explains the NDA's desperate seek for amendment through Ordinance in 2014 [19]. Meanwhile, the other prominent symbol, i.e. Criticism and resistance, leaves a mark on the people and the civil society organizations. It's a common notion that whenever a project displaces people, injustice is bound to happen. This makes both the PAP and the NGOs more inclined to fight, resist and criticize LA rather than work towards making R&R more efficient through active participation, constructive suggestions and helping the authorities in things like maintenance and surveillance. There are several researchers who have concluded how active participation, leadership and organized representation from the PAP could make R&R and maintenance more efficient and how R&R can actually be used as a tool for the uplifting of the underprivileged [15, 16, 30, 31]. Hence, it's important to clear the symbolic pollution to change the outlook of people.

3 Discussion

The LARR policies, like most others, are subject to social structures, social diversities and political ideologies of the area they're applied to. Just as several other researchers before us, we too are trying to find the most generic aspects and factors which act as feedback driving forces for these policies to keep evolving. It's also noteworthy that the 6 streams of PFA enable a detailed understanding of the policy but they aren't mutually exclusive. There are several aspects which fall under multiple streams. For example, the PAP's identity-based war for their land [32] is both a result of a symbolic identity and an interest group formation on the basis of perceived injustice. Hence, instead of classifying into the 6 streams, a more efficient way of classification would be into positive and negative feedbacks. There are different approaches to classify a policy's endogenous forces into positive and negative feedback categories. The commonly known theory from which we would mostly derive our analytical apparatus for this classification is the Historical Institutional (HI) approach. This approach is based on the particular mechanisms which reshape the state and social actors' capacities over time in ways that alter the reversal or expansion of the policy [6, 11, 33]. Aligning our study with HI, positive feedbacks would be the ones which drive the policy towards change, but have a self-reinforcing positive impact on the policy and its effectiveness. And the feedbacks which keep the policy stable, but have a long-term depreciative effect, would be considered as negative. A few examples of positive feedbacks could be the justification of purpose and consent clauses, the formation of grievance redressing bodies and the clauses against leaving an acquired land unused over 5 years. The following are some negative effects observed. The usage of vague terms and regulations in policy documents increases the chance of its misuse and makes it harder to transform into a better policy. The lack of acknowledgement and representation of gender and other differences among PAP is another negative feedback. Another would be a great difference in the number of users of giant infrastructure projects and the PAP which weakens their position politically and restricts positive changes. Apart from this, the rampant criticism but lack of constructive involvement of public organizations and NGOs keeps the R&R from getting better. But as per our observations, even this taxonomy of positive and negative feedbacks does have blurred edges, which we hope to clarify with our further analysis in upcoming studies. Several feedbacks also have a neutral impact on the policy where the positive and negative balance out. It's important to establish these distinctions more clearly with supporting data, as there are several small outcomes/feedbacks which may turn out to be highly impactful, while some big ones may become irrelevant in the long run [10].

4 Conclusion

The aim of the study was to generate and enlist plausible explanations of policy progress over time. But apart from that, we came to the realization of how PFA provides a suitable foundation and guidance to engender a policy analysis tool. If the negative feedbacks are balanced with counter-positive feedbacks, the result might be a neutral or even positive one. The violent protests and resistance by the PAP are positive and change initiating feedbacks but for the policy to move forward, resistance must be accompanied by active participation and constructive organization of the PAP. Michael Cernea developed a world-renowned IRR model which has inspired this study, and it stands on the balancing out of impoverishments and risks by reconstructions [34]. It's this balance of the policy feedbacks which needs to be maintained. Policy framing literature has a shortage of analytic models. A more defined version of this equalization model should hopefully yield more distinctly comprehensible policy balancing recommendations. This balance would ensure the policy's sustainability, i.e. checking whether the policy has self-reinforcing dynamics of social adaptation and supports new interests.

References

1. Mettler, S. (2002). Bringing the state back in to civic engagement: policy feedback effects of the G.I. bill for world war II veterans. *American Political Science Review, 96*(2), 351–365. https://doi.org/10.1017/S0003055402000217.
2. Colebatch, H. K. (1998). Theorizing public organization: An Australian perspective. *International Journal of Organization Theory & Behavior*. https://doi.org/10.1108/IJOTB-01-03-1998-B003.
3. Kingdon, J. W. (1984). *Agendas, alternatives, and public policies*. Boston: Little, Brown.
4. Campbell, A. L. (2008). *Policy feedbacks and the political mobilization of mass publics*. Cambridge: Unpublished manuscript, Department of Political Science, Massachusetts Institute of Technology.
5. Ostrom, E. (1999). *Theories of the policy process boulder*. Colo Oxford: Westview Press.
6. Pierson, P. (1993). When effect becomes cause: Policy feedback and political change. *World Politics, 45*, 595–628.
7. Akdogan, A. A. (2011). *Historical traces of public policy discipline in turkey, public administration and public policies in Turkey, Editör: Filiz Kartal*. Ankara: Public Administration Institute for Turkey and Middle East Publishing.
8. Béland, D. (2010). Reconsidering policy feedback: How policies affect politics. *Administration & Society, 42*(5), 568–590. https://doi.org/10.1177/0095399710377444.
9. Chatterjee, S. (2007). Conceptions of space in India's look east policy: Order, cooperation or community? *South Asian Survey, 14*(1), 65–81. 10.1177.
10. Pierson, P. (2005). The study of policy development. *Journal of Policy History, 17*(1), 34–51. https://doi.org/10.1353/jph.2005.0006.
11. Skocpol, T. (1992). *Protecting soldiers and mothers: The political origins of social policy in the United States*. Cambridge, MA: Belknap Press.
12. Merrilat, H. C. L. (1970). *Land and the constitution in India*. New York and London: Columbia University Press.
13. Ramanathan, S., Allison, K. R., Faulkner, G., Dwyer, J. J. M. (2008). Challenges in assessing the implementation and effectiveness of physical activity and nutrition policy interventions as natural experiments. *Health Promotion International, 23*, 10.1093.
14. Levien, M. (2013). Regimes of dispossession: From steel towns to special economic zones, special issue: Governing the global land grab: The role of the state in the rush for land (Vol. 44), 10.1111.
15. Skocpol, T. (1990). Sustainable social policy: Fighting poverty without poverty programs. *The American Prospect, 1*, 58–70.
16. Patel, S., d'Cruz, C., & Burra, S. (2002), Beyond evictions in a global city: people-managed resettlement in Mumbai. *Environment and Urbanization, 14*(1), 159–172. https://doi.org/10.1177/095624780201400113.
17. Sharma, R. (2003). Involuntary displacement: A few encounters. *Economic and Political Weekly, 38*(9), 907–912.
18. Pierson, P. (1995). Fragmented welfare states: Federal institutions and the development of social policy. *Governance: An International Journal of Policy*. https://doi.org/10.1111/j.1468-0491.1995.tb00223.

19. Raghuram, G., & Sunny, S. (2015). *Right to fair compensation and transparency in land acquisition, rehabilitation and resettlement ordinance 2014: A process perspective*. Ahmedabad: Research and Publications Indian Institute of Management.
20. Berkowitz, E. D., & McQuaid, K. (1980). *Creating the welfare state: The political economy of twentieth-century reform*. CT: PWestport, Praeger.
21. Esping-Andersen, G. (1990). *The three worlds of welfare capitalism*. Cambridge, UK: Polity.
22. Quadagno, J. (1988). *The transformation of old age security: Class and politics in the American welfare state*. Chicago, IL: University of Chicago Press.
23. Titmus, C., & Judge et al. (1995). The university and the teachers. *Internationales Jahrbuch der Erwachsenenbildung, 23*(1), 297–298. https://doi.org/10.7788/ijbe.1995.23.1.297.
24. Hacker, J. S. (2002). *The divided welfare state: The battle over public and private social benefits in the United States, Cambridge*. Cambridge, UK: Cambridge University Press.
25. Levien, M. (2011). Special economic zones and accumulation by dispossession in India. *Journal of Agrarian Change, 11*(4). https://doi.org/10.1111/j.14710366.2011.00329.
26. Campbell, A. L. (2003). Participatory reactions to policy threats: Senior Citizens and the defense of social security and medicare. *Political Behaviour, 25*, 29–49. https://doi.org/10.1023/A:1022900327448.
27. Shanin, T. (1983). *Late marx and the Russian road*. New York: Monthly Review Press.
28. Hall, C. M. (2012). Second Homes, planning, policy and governance. *Journal of Policy Research in Tourism, Leisure & Events, 2015, 7*(1), 1–14. https://doi.org/10.1080/19407963.2014.964251.
29. Desai, M. (2011). Land acquisition law and the proposed changes. *Economic and Political Weekly, 46*(26/27), 95–100.
30. Burra, S. (1999). *Resettlement and rehabilitation of the Urban poor: The story of Kanjur Marg*. Mumbai: Society for the Promotion of Area Resource Centers.
31. Ploeg, L., & Vanclay, F. (2017). A human rights based approach to project induced displacement and resettlement. *Impact Assessment and Project Appraisal, 35*(1), 34–52. https://doi.org/10.1080/14615517.2016.1271538.
32. Baviskar, A. (1995) *In the belly of the river*: Tribal Conflicts Over Development in Narmada Valley. Oxford University Press Delhi.
33. Jacobs, A. M., & Weaver, R. K. (2015). When policies undo themselves: self-undermining feedback as a source of policy change governance. *An International Journal of Policy, Administration and Institutions*. https://doi.org/10.1111/gove.12101.
34. Cernea, M. (2000). Risks, safeguards and reconstruction: A model for population displacement and resettlement. *Economic and Political Weekly, 35*(41).

System Dynamics Approach for Urban Transportation System to Reduce Fuel Consumption and Fuel Emissions

Sandeep Singh and Challa Prathyusha

Abstract The world today is facing a severe environmental crisis. High fuel consumption by the various categories of vehicles in urban areas causes fuel emissions, worsening environmental conditions. This study aims to investigate and analyze the scenario-based System Dynamics (SD) models to reduce fuel consumption and fuel emissions. In this study, we built an SD model, including the factors influencing the transportation system, energy system, emissions system, and environment system. A broad range of policy scenarios was constructed for Chennai city, India, which considered criteria such as model split, fuel consumption, and fuel emissions. The SD simulation-based forecasting models are built considering three scenarios, such as do-nothing scenario, do-minimum scenario, and desirable scenario to project the SD parameters for the horizon year 2030. The scenarios of augmenting the proportion of public transportation and simultaneously restricting the proportion of private transportation by a model split of 10%:90% in the do minimum scenario and 20%:80% in the desirable scenario led to substantial reduction in the number of vehicles plying on the city roads. This has eventually resulted in pursuing a significant reduction in fuel consumption and fuel emissions. The results from SD simulation findings have led to the development of policies to regulate increasing fuel consumption and fuel emissions, based on the estimated figures.

Keywords Urban transportation system · System dynamics (SD) · Scenario analysis · Vehicular fuel consumption · Vehicular fuel emission

---

S. Singh (✉)
Department of Civil Engineering, National Institute of Technology Tiruchirappalli, Tiruchirappalli, India
e-mail: sandeepsingh.nitt@gmail.com

C. Prathyusha
Department of Civil Engineering, REVA University, Bangalore, India
e-mail: prathyushareddy1990@gmail.com

A. Jana and P. Banerji (eds.), *Urban Science and Engineering*, Lecture Notes in Civil Engineering 121, https://doi.org/10.1007/978-981-33-4114-2_31

1 Introduction

As the urbanization cycle accelerates, developing countries are under increased pressure to make savings in energy and reduce pollution, particularly for urban regions. Furthermore, energy consumption and CO_2 emissions from the public and private transport modes plays an important role in energy conservation and emission reduction. Travel demand models that can be used to forecast travel by mode, time of day, length of travel, or travel location must be built [1] immediately to examine the impact of one factor over the other. Enhanced capacity policies, which include mobility management strategies for conservation and emission reduction in both developed nations and developing nations, are required [2]. Collective actions and coordinated government initiatives are thus the immediate requirements for addressing the problems and reducing damages to improve productivity through the usage of renewable transport systems [3]. So, it is vital to comprehend the structures and the relationships between different transportation systems, energy systems, emissions systems, and environment systems of the urban region. As a result, the quantitative analysis of behavioral characteristics and the process of complex transportation of urban travelers have become important areas of study.

This work, therefore, aims primarily to study the energy and emission factors affecting the transport sector in road transport. In addition, we have developed three scenarios based on possible policies that could be adopted by the city administration to predict their potential for both reducing vehicle fuel consumption and reducing fuel emissions. The development of different alternative scenarios based on the System Dynamics (SD) simulation models to predict potential demands for the transport, energy, and emission sectors are also conducted. Later, the SD model scenarios are critically reviewed and evaluated in such a way that could tackle and minimize losses in transport, energy, and emissions. Finally, effective transport policies to ensure the implementation of a sustainable transport system are proposed.

2 Background Literature

In the past few years, a significant number of studies have been conducted into energy consumption and emissions.

Wang et al. [4] measured CO_2 and passenger car pollutants through 2000–2005, in order to find a reduction in the potential policies. The authors predicted the future trend in pollutant emissions from passenger cars under three different scenarios. Han and Hayashi [5] used the SD model to evaluate the influence of the supply chain and examined the effects of policy scenarios on traffic volume, modal share, energy conservation, and CO_2 emissions.

Rentziou et al. [6] anticipated urban passenger transport volume, energy consumption, and CO_2 emissions based on simultaneous equations. Song et al. [7] formulated a methodology for the development of mesoscopic fuel consumption and fuel emission models to assess the environmental impacts of Intelligent Transportation Systems (ITS) strategies.

Cheng et al. [8] explored three SD scenarios, which include fuel taxation, motorcycle parking management, and free bus service to potentially reduce vehicle fuel consumption and mitigate CO_2 emissions over a 30-year timeframe (from 1995 to 2025). The author suggested that the most efficient way of curbing growth in the number of private cars, the amount of fuel usage, and CO_2 emissions was both fuel taxing and motorcycle parking management.

Malik et al. [9] have carried out a comprehensive disaggregate-level analysis to quantify freight fleet emissions in Delhi, India. The effect of freight emissions on the environment of the city was quantified. This research work revealed that freight vehicles contribute 42.3% and 45.5% of the total PM_{10} and NO_x emissions, respectively. Interestingly, the light-duty freight vehicles were found to be a significant contributor to the overall pollution with 28.71% and 33.07% PM_{10} and NO_x emissions, respectively.

Akbari et al. [10] developed six scenarios using the SD technique and quantitatively analyzed them to figure out the best-performed scenario. They found that the comprehensive policy SD scenario-based model performed the best compared to all the other individual policies.

As can be seen from the abovementioned literature, not many studies have been made in the past that focused on the interconnection between transport, energy, emissions, and environmental systems. Furthermore, the implications of various transport policies have still not been studied through a systemic approach covering more facets of the urban transportation system. This has instigated the need for this research work considering the importance of co-existence among the 4E systems—'Engineering', 'Energy,' 'Emission,' and 'Environment.'

3 Study Methodology

A research approach to evaluate and examine the interrelationships among the transport, the energy, the emissions, and the environment is proposed in accordance with the principles of the SD methodology. In this study, using the SD technique, the effect of different scenarios on future fuel consumption and fuel emissions is predicted for 2018–2030. The three possible scenarios are the do-nothing scenario, do-minimum scenario, and desirable scenario. The collected data from the different sectors are used for the SD model conceptualization to establish the relationship between the variables of the urban road transportation system (public and private), fuel consumption, fuel emissions, and the environment sector. The analyses of the three different scenarios were carried out by assuming the modal split ratio between public and private modes of transport as 10:90 and 20:80 for the do-minimum scenario and desirable scenario, respectively. The results were compared under different scenarios to determine the benefits of the policy measures. In addition, based on the results

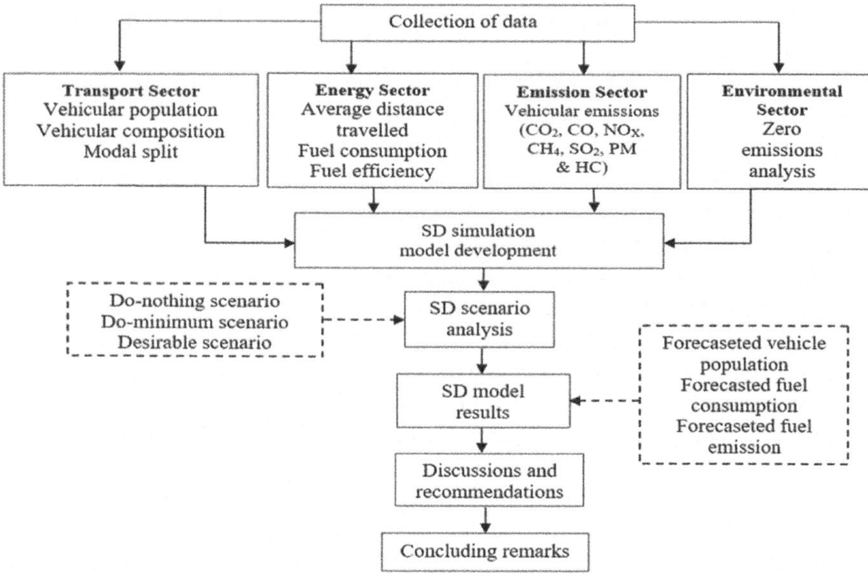

Fig. 1 Flowchart showing the study methodology

obtained and comprehensive discussions, important and relevant recommendations have been made for policy adoption. Figure 1 illustrates the study methodology.

4 Study Area

The city of Chennai, India, has been selected as the case study city to illustrate the application of proposed scenarios. The road traffic in the city is majorly occupied by private vehicles like Two-Wheelers (TW) and Cars (CR) and marginally by public transport vehicles like Metropolitan Transport Corporation Buses (BUS). However, the city roads are permanently hit by heavy traffic due to the density of the population and the frequency of TWs and CRs, which intensifies the concentration of emissions and energy consumption. Additionally, the rapid increase in the growth rate of the vehicular population, especially private vehicles (TW and CR), has increased fuel demand and emission levels. This has resulted in increased fuel prices due to fuel imports from other nations. Hence, it is considered alarming and imperative to study the impact of the increase in vehicles on fuel consumption and fuel emissions through a System Dynamics (SD) approach.

Table 1 Base year vehicle population

Class of vehicle									
Particulars	BUS	AR	TX	LCV	HCV	PB	MB	TW	CR
Count	7517	81,147	51,941	49,237	42,814	1677	3095	3,999,354	808,291
Percentage composition (%)	0.15	1.61	1.03	0.98	0.85	0.03	0.06	79.27	16.02

Source District statistical handbook Chennai district, 2016–2017 [11]

5 Data Collection

5.1 Transport Sector

The classified vehicle population data for the base year 2016–17 was acquired from the State Transport Department, Chennai, to forecast the horizon year (2030) vehicle population. The different classes of vehicles consist of public transport buses (BUS), auto-rickshaws (AR), taxis (TX), private buses (PB), mini-buses (MB), light commercial vehicles (LCV), heavy commercial vehicles (HCV), two-wheelers (TW), and cars (CR). Table 1 shows the various classes of vehicles along with their population of vehicles and the percentage composition in the city of Chennai.

Table 1 shows that the model split between public transport and private transport is 0.15%:95.29%. The present share of public transport is less than 1%, which is alarming. This classified vehicle population data is used to develop three different scenarios, such as a do-nothing scenario, a do-minimum scenario, and a desirable scenario. Furthermore, the SD models are developed on the basis of these data, and forecasts of fuel consumption and fuel emissions for the respective vehicle population have been made till the horizon year 2030.

5.2 Energy Sector

The energy sector is an important part of the transportation system's driving mechanism. However, there are concerns about this now and in the near future, given the limited availability of energy. The fuel consumption is measured according to the average journey distance and fuel efficiency. In the SD-model building and simulation development process presented in Table 2, the average distance traveled by the various classes of vehicles in km/day, the efficiency of fuel in km/liters, and the liter-year consumption of fuel are taken into account.

Table 2 Data regarding fuel efficiency and fuel consumption by different vehicle types

Parameter	BUS	AR	TX	LCV	HCV	PB	MB	TW	CR
Average distance traveled (km/day)	151	96	21	51	55	111	22	18	24
Fuel efficiency (km/liter)	4.1	21	13	14	4.33	5.0	8.7	53	12.9 (Petrol) 15.6 (Diesel)
Fuel consumption (liters/year)	13,415	1669	534	1330	4637	11,863	897	124	684 (Petrol) 652 (Diesel)

Source Ministry of Petroleum & Natural Gas, Government of India, Report of The Expert Group on a Viable and Sustainable System of Pricing of Petroleum Products [12]

Table 3 Pollutants emitted from the different vehicle types in grams/km

Class of vehicle									
Type of pollutant	BUS	AR	TX	LCV	HCV	PB	MB	TW	CR
CO_2	515.2	60.3	208.3	423.84	423.84	515.2	515.2	26.6	223.6
CO	3.60	5.10	0.90	1.61	1.61	3.60	3.60	2.20	1.98
NO_X	12.00	1.28	0.50	10.96	10.96	12.00	12.00	0.19	0.20
CH_4	0.09	0.18	0.01	0.05	0.05	0.09	0.09	0.18	0.17
SO_2	1.42	0.02	10.3	1.39	1.39	1.42	1.42	0.01	0.05
PM	0.56	0.20	0.07	0.33	0.33	0.56	0.56	0.05	0.03
HC	0.87	0.14	0.13	0.50	0.50	0.87	0.87	1.42	0.25

Source Ramachandra and Shwetmala [13]

5.3 Emissions Sector

India's transportation sector is the third-largest emitting sector for greenhouse gas (GHG), with the largest contribution from the highway transport sector. Chennai city, with a population of around 10 million, and more than 75% ownership of cars, is one of the largest emitters of various forms of pollutants. Table 3 provides data on the types of pollutants emitted from different vehicle types.

5.4 Environmental Sector

A clean vehicle strategy and a mobility management strategy can be adopted to reduce energy consumption and emissions of vehicles to zero levels. However, the implementation of both such strategies requires proper and comprehensive analysis and involvement of policymakers, government, and other stakeholders. Other emission mitigation measures can be the use of public bicycle sharing system, especially during winters, and improvement in the walking environment by setting up special driveways and pedestrian lanes, to meet the requirement of short-distance travel.

6 System Dynamics (SD) Model Development

System dynamics (SD) was first proposed by Forrester [14] to analyze dynamic complex system feedbacks. Based on computer simulation technology, this tool can analyze the relationships between various factors, simulate quantitative data, and provide information about the feedback structure and system behavior. This simplifies the understanding of the overall system and various relationships associated with policies related to the dynamic performance control of the system [15]. The SD model is not only capable of analyzing a system with several interrelated variables but is also capable of defining its complex patterns based on a limited collection of details [5]. The SD incorporates qualitative analysis with quantitative analysis and uses system synthesis logic to explain unknown behavioral features, making SD a better choice when dealing with non-linear, high-level dynamic time-varying processes. The SD models embodied the complex cycle of energy consumption and emissions and expressed the unpredictable nature of the key issues associated with urban transport systems. For these reasons, we have chosen an SD model to analyze the energy and emissions of the urban transportation system in Chennai city, India.

The vehicle population, fuel consumption, and fuel emission models have been established based on the proposed SD approach, which is presented in this section. The developed SD simulation models for scenario I (do-nothing scenario), scenario II (do-minimum scenario), and scenario III (desirable scenario) are shown in Figs. 2 and 3, respectively.

7 Scenario Analysis and Results

7.1 Forecasted Vehicle Population

The population is one of the key factors that affects passenger transport. The population growth has been one of the main contributors to average distance traveled, fuel consumption, and fuel emission. The existing growth rates in various vehicle groups have been assumed to continue up to the year 2030, according to the do-nothing scenario. On the basis of this, the vehicle population, fuel consumption, and fuel emissions values have been simulated and predicted.

The development of these scenarios focuses on reducing the use of personal TWs and CRs while increasing the use of BUS. The growth rate values of BUS (public): TW and CR (private) have been modified to 10:90 modal split for the horizon year 2030, in the do-minimum scenario. To achieve the 10:90 modal division between

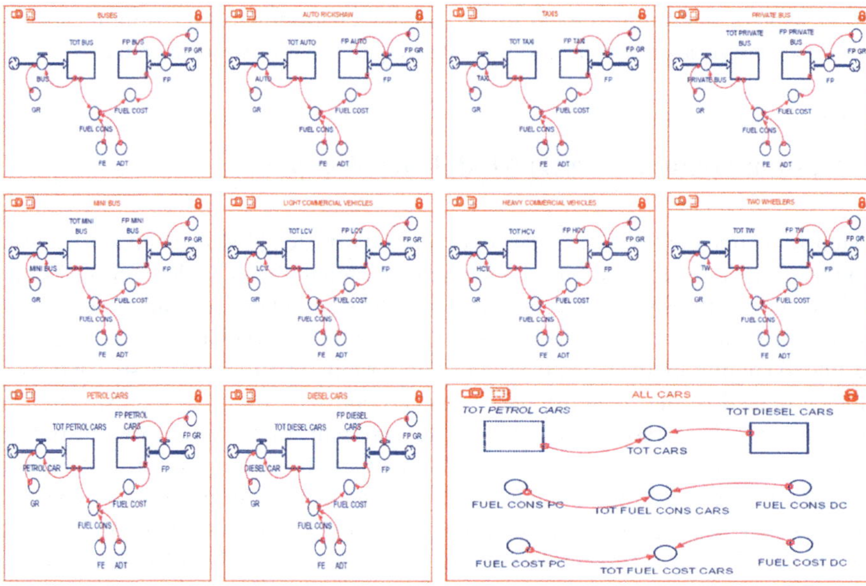

Fig. 2 SD simulation models for scenario I (Do-nothing scenario)

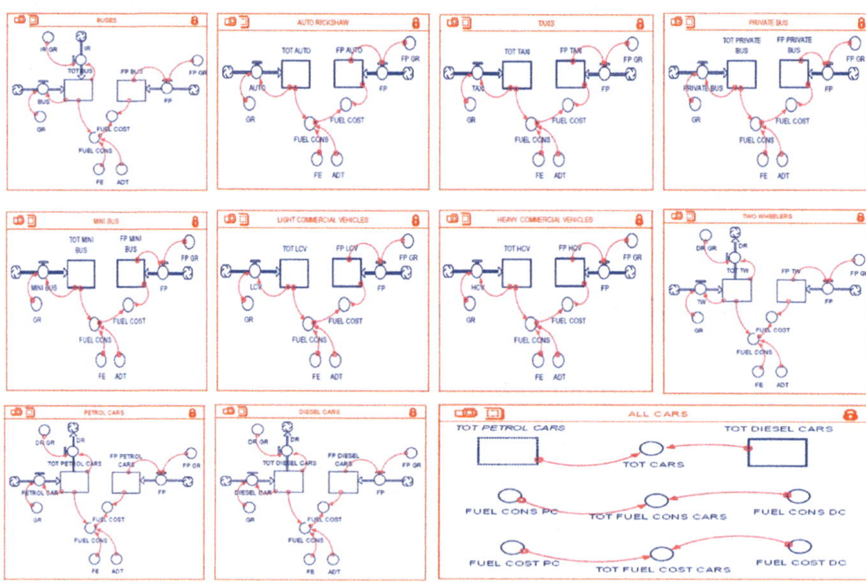

Fig. 3 SD simulation models for scenario II (Do-minimum scenario) and scenario III (Desirable scenario)

public and private modes of transport, the former has been increased to 12.63%, but the latter is limited to 6.98% and 8.24%, respectively. In the meantime, growth rates for other transport modes, including the AR, TX, LCV, HCV, PB, and MB, are slightly increased based on historical data.

The desirable scenario seeks to improve the attractiveness of public transport to optimize the structure of public transport qualitatively and quantitatively. However, total city trips cannot be covered by public transport while they can receive a considerable amount of trips. In this scenario, the growth rate values of BUS (public): TW and CR (private) have been altered in which a 20:80 modal split between public and private transport modes is assumed. In order to achieve a 20:80 modal split between public and private transport modes, the growth rate of BUS has been increased to reach 20.66%, whereas the growth rate of TW and CR is restrained from being 4.21% and 4.52%, respectively. Meanwhile, the growth rates of the other modes of transport like AR, TX, LCV, HCV, PB, and MB are assumed to be marginally incremented based on the historical data.

Additionally, it is hypothesized in this study that 30 TW and 15 CR with vehicle occupancy values 1.5 and 2.3, respectively (Chennai Comprehensive Transportation Study [16]), could replace one single BUS. Hence, a single BUS could have the capacity to accommodate a maximum of 80 persons. However, CCTS-2010 [16] reported that 'during peak hours, the buses operate with more than 100 passengers per bus indicating substantial overcrowding'.

7.2 *Forecasted Fuel Consumption*

It is well known that as the population of vehicles rises, the fuel consumption and fuel emissions also increase for different vehicle classes, and vice versa. Figure 4

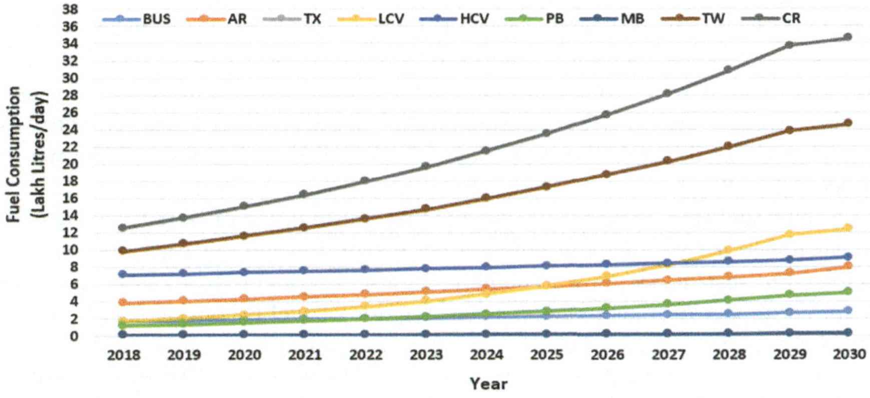

Fig. 4 Scenario I results—consumption of fuel till the horizon year

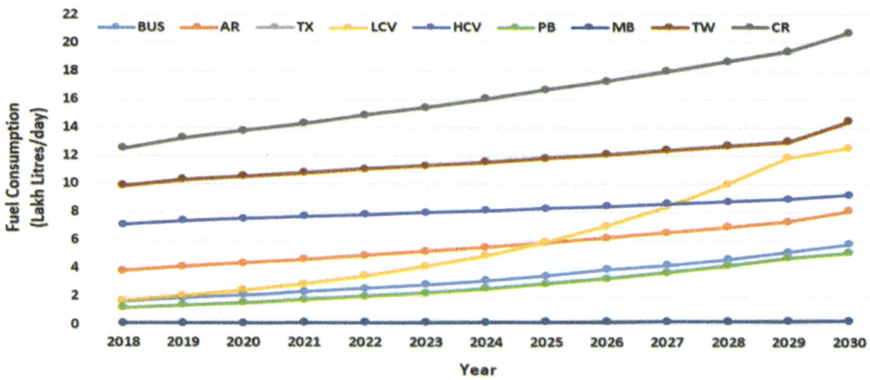

Fig. 5 Scenario II results—consumption of fuel till the horizon year

illustrates the results of scenario I (do-nothing scenario) in which the forecasted fuel consumption by each class of vehicle is represented.

The scenario of do-nothing indicates a rise in fuel consumption, following the increase in the number of vehicles in the public and private transport. CR has an expected fuel consumption of 34.02 lakh liters a day, TW has 24.11 lakh liters a day, and BUS has a forecasted fuel consumption of only 2.61 lakh liters a day over the horizon year 2030. Finally, all vehicles will consume 104.81 crore liters per day in the horizon year 2030. That will contribute to the annual fuel consumption of 382.55 crore liters annually.

In the do-minimum scenario, the changes in the share rate of different transport modes such as BUS (public): TW and CR (private) were kept at 10:90. Due to this, fuel consumption decreased, and this is shown in Fig. 5.

Adjustments in the growth rate of the number of vehicles between public and private transport have shown that fuel consumption varies accordingly for each vehicle type. The fuel consumed by CR is reduced to 20.42 lakh liters per day from 34.02 lakh liters per day (scenario I) for the year 2030. TW consumes 14.20 lakh liters per day of fuel when compared to the scenario I's 24.11 lakh liters per day in the year 2030. The fuel consumption for CR and TW is down by 39.98% and 41.10%, respectively. The fuel consumption by BUS mode of transport increases from scenario I's 2.61 lakh liters per day to 5.98 lakh liters per day in 2030, which is about 129.12%.

It can be said from these results that the fuel consumption of BUS has increased, but when we take the fuel consumption per person traveling in BUS into account compared to that of people who combinedly use TW and CR, it can be said that the former modes of transportation's consumption of fuel is considerably less than that of the latter. Compared to the outcome of scenario I, it can be seen that the total fuel consumed by all vehicles in scenario II has reduced to 308.61 Crore liters per year from 382.55 Crore liters per year in 2030, showing that all vehicles' fuel consumption in scenario II gets reduced by 73.94 Crore liters per year. Therefore, annual fuel consumption savings amount to 23.95%.

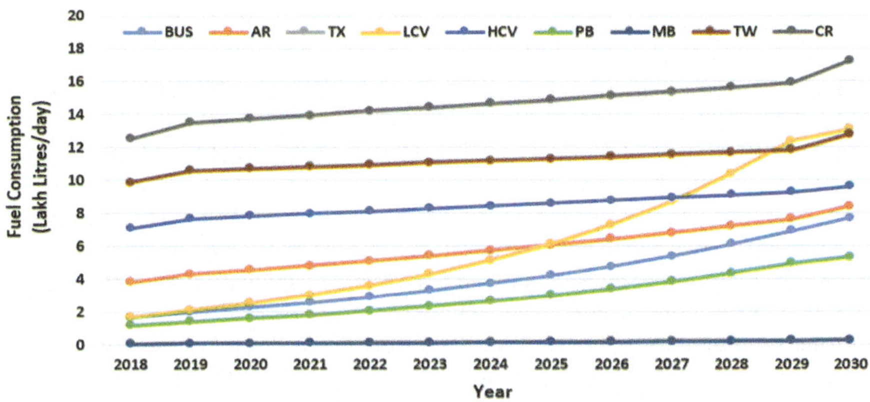

Fig. 6 Scenario III results—consumption of fuel till the horizon year

In the desirable scenario, the changes in the share rate of different transport modes such as the BUS (public): TW and CR (private) were kept at 20:80. Due to this, fuel consumption decreased, and this is shown in Fig. 6.

The CR fuel consumption reduces to 16.94 lakh liters per day in scenario III from 20.42 lakh liters per day (scenario II), and from 34.02 lakh liters per day (scenario I). The TW fuel consumption reduces to 13.01 lakh liters per day in scenario III from 14.20 lakh liters per day (scenario II), and 24.11 lakh liters per day (scenario I). The BUS fuel consumption increases to 7.98 lakh liters per day from scenario II's 5.98 lakh liters per day, and from scenario I's 2.61 lakh liters per day. The combined fuel consumption of CR and TW transport is 54.44 lakh liters per day, while that of BUS is just 7.98 lakh liters per day.

The fuel consumption in BUS has been increased, but when we take account of the fuel consumption of individuals who travel in the BUS in comparison to those who use both TW and CR, it can be assumed that the fuel consumption in the former modes of transportation is considerably lower than in the latter modes. The consequence of this scenario analysis is a combined increase. Comparing the results of scenarios II and I, with scenario III, it was found that the total fuel consumed by all the vehicles decreased to 276.44 crore-liters annually from scenario II's 308.61 crore-liters annually and to scenario I's 382.55 crore-liters annually in the year 2030. It indicates an annual fuel consumption saving of 106.11 crore-liters, which resulted in 38.38% fuel consumption reduction as compared to scenario I's fuel consumption.

7.3 Forecasted Fuel Emissions

When contemplating the do-nothing scenario, the difference in the growth rate of the number of vehicles and the fuel consumption of each type of vehicle between the public and private modes of transport has been found to have a significant direct impact on fuel emissions. Figure 7 shows this.

Figure 7 reveals that the fuel emissions of CR, TW, and BUS would hit 13.15 Gg/day, 7.89 Gg/day, and 0.82 Gg/day, respectively, if the present pattern is permitted up to the horizon year 2030. Consequently, all the vehicles could produce a total of 30.62 Gg/day of fuel emissions, which would result in a total annual fuel emission of 11,176 Gg/annum. This impact must be minimized by effective policies and preventive measures, which are to be developed and evaluated with immediate effect. Hence, the study further considers two main SD-based simulation scenario models. The following sections discuss these SD models.

In the do-minimum scenario, Fig. 8 illustrates that the fuel emissions by CR decrease to 9.30 Gg/day from scenario I's 13.15 Gg/day and the fuel emissions by TW decrease to 4.66 Gg/day from scenario I's 7.89 Gg/day while the fuel emissions

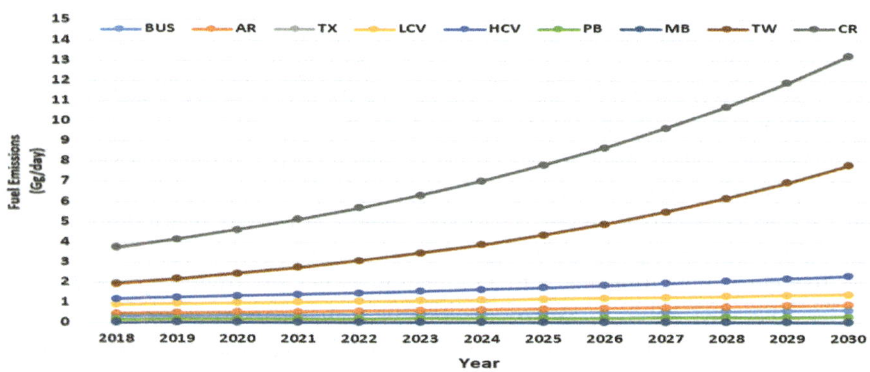

Fig. 7 Scenario I results—fuel emissions till the horizon year

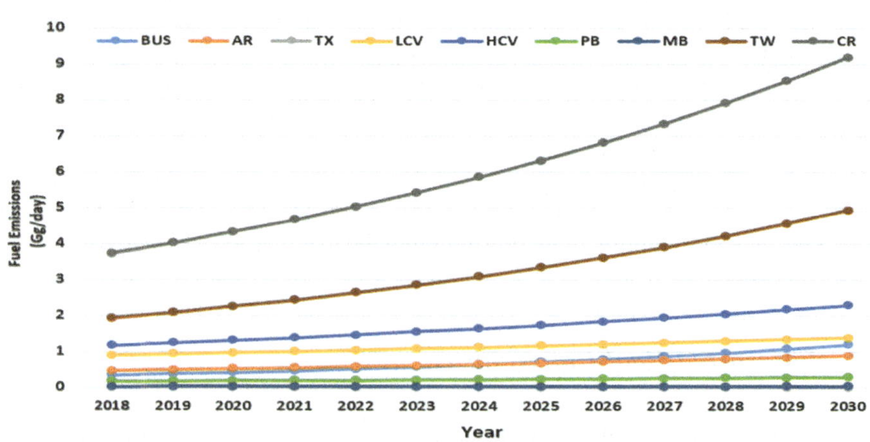

Fig. 8 Scenario II results—fuel emissions till the horizon year

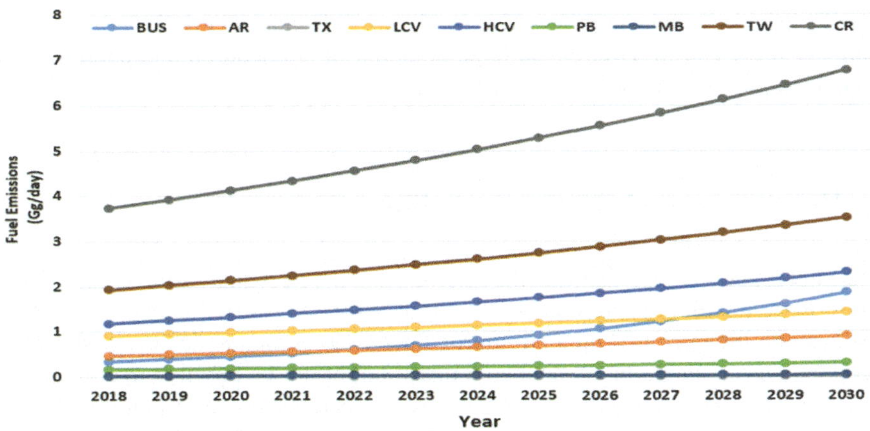

Fig. 9 Scenario III results—fuel emissions till the horizon year

by BUS increases to 1.28 Gg/day from scenario I's 0.82 Gg/day for the horizon year 2030. The private transport emissions are 12.55 Gg/day, while public transportation emissions only amount to 1.28 Gg/day.

It is evident from the above that while BUS' fuel emissions have risen, it can be said to be much lower than in the former transport mode if we consider the fuel emissions per person traveling in BUS as regards those using CR and TW combinedly. The cumulative fuel emission vehicles in scenario II have reduced to 3094 Gg/annum from scenario I's 8082 Gg/annum and from scenario II's 11,176 Gg/annum in the 2030 horizon. Therefore, fuel emissions have been reduced by 38.28%.

In the desirable scenario, the results from the contribution rate of different transport modes to fuel emissions are shown in Fig. 9.

In the desirable scenario, the fuel emission by the fuel emissions by CR decreases to 6.87 Gg/day from scenario II's 9.30 Gg/day, and from scenario I's 13.15 Gg/day. Similarly, TW decreases to 3.65 Gg/day in scenario III from scenario II's 4.66 Gg/day and from scenario I's 7.89 Gg/day, while the fuel emissions by BUS show a slight upward trend which increases to 1.88 Gg/day scenario III from scenario II's 1.28 Gg/day and from scenario I's 0.82 Gg/day. The private transport emissions are 10.52 Gg/day, while public transportation emissions are 1.88 Gg/day.

When these resulting values are compared with the results of scenarios II and I, it can be seen that the overall fuel emissions of all vehicles in scenario III dropped to 6572 Gg/annum from scenario II to 8082 Gg/annum and scenario I to 1176 Gg/annum by 2030. As a result, the annual reduction in fuel emissions relative to scenarios I and II was equivalent to 6572 Gg/annum and 1510 Gg/annum, respectively, resulting in a reduction of 41.19% and 18.68% in fuel emissions in scenario III by 2030.

As 'prepare and prevent is better than repent and repair,' promoting fuel-efficient vehicles on a larger scale can be obtained by enhancing and encouraging the use of the public transport system innovatively and systematically. Further, this scenario can bring significant amounts of annual economic savings to the people and governments by reducing energy consumption and adverse environmental effects. However, implementing this scenario requires the huge potential of cooperation and coordination from the top-down and bottom-up authorities.

8 Discussions and Recommendations

This study recommends that the increase in the BUS fleet size of the urban public transportation with greater efficiency and reliability and restriction in private transport may be implemented on a pilot basis for a zone or area to achieve sustainable development. It should be remembered that it might not be possible to enforce the desired scenario immediately. The main reason is that, because of resource and technology constraints, urban transport is difficult to make significant strides in the short term. However, the public transportation sector should be emphasized and enhanced due to its importance. Besides, 'it is believed that improving the quality of vehicles and their performance can greatly decrease fuel consumption and CO_2 emissions' [10].

Finally, this study recommends that besides implementing the conventionally applied and the recommended mitigation measures to the 4E systems—'Engineering', 'Energy,' 'Emission,' and 'Environment', calculated and strategic actions are to be taken with immediate effect considering the additional 3E's also, i.e., the Education, Ethical values, and Enforcement into the 4E systems for a prosperous and sustainable development of the city. All strategies, therefore, need tremendous cooperation and teamwork capacity from the highest and lowest governments. These policies will eventually reduce fuel consumption, fuel emissions, and environmental impacts considerably to ensure that the Sustainable Development Goals (SDG), which are adopted globally, are met.

9 Concluding Remarks

This study used SD models to understand the interrelationship between the transport system, energy system, and emission system. The SD simulation model predicted the fuel consumption and fuel emission levels when the share rate of transport was considered as business as usual or in the do-nothing scenario, 10:90 in the do-minimum scenario, and 20:80 in the desirable scenario. These policy measures will eventually lead to a considerable reduction in fuel consumption, fuel cost, and fuel emissions in the city. The study concluded that the desirable scenario gives the best results in terms of reduction in fuel consumption and fuel emissions. The presented approach and prediction results will help the transport planners and decision-makers to assess the potential needs of energy expenditure and transport infrastructure needs.

References

1. James, J., Mark, J., & Aruna, S. (2012). A review of urban energy system models: Approaches, challenges, and opportunities. *Renewable and Sustainable Energy Reviews, 16*(6), 3847–3866. https://doi.org/10.1016/j.rser.2012.02.047.
2. Litman, T. (2013). Comprehensive evaluation of energy conservation and emission reduction policies. *Transportation Research Part A: Policy and Practice, 47,* 153–166. https://doi.org/10.1016/j.tra.2012.10.022.
3. Liu, X., Ma, S., Tian, J., Jia, N., & Li, G. (2015). A system dynamics approach to scenario analysis for urban passenger transport energy consumption and CO_2 emissions: A case study of Beijing. *Energy Policy, 85,* 253–270. https://doi.org/10.1016/j.enpol.2015.06.007.
4. Wang J, Lu H, Peng H (2008) System dynamics model of urban transportation system and its application. J Trans Syst, Eng, Inform Technol 8(3):83–89
5. Han, J., & Hayashi, Y. (2008). A system dynamics model of CO_2 mitigation in China's intercity passenger transport. *Transportation Research Part D, 13,* 298–305. https://doi.org/10.1016/j.trd.2008.03.005.
6. Rentziou, A., Gkritza, K., & Souleyrette, R. R. (2012). VMT, energy consumption, and GHG emissions forecasting for passenger transportation. *Transportation Research Part A: Policy and Practice, 46*(3), 487–500. https://doi.org/10.1016/j.tra.2011.11.009.
7. Song, Y., Yao, E., Zuo, T., & Lang, Z. (2013) Emissions and fuel consumption modeling for evaluating environmental effectiveness of ITS strategies. *Discrete Dynamics in Nature and Society*. Article ID 581945. https://doi.org/10.1155/2013/581945.
8. Cheng, Y., Chang, Y., & Lu, I. J. (2015). Urban transportation energy and carbon dioxide emission reduction strategies. *Applied Energy, 157,* 953–973. https://doi.org/10.1016/j.apenergy.2015.01.126.
9. Malik, L., Tiwari, G., Thakur, S., & Kumar, A. (2019). Assessment of freight vehicle characteristics and impact of future policy interventions on their emissions in Delhi. *Transportation Research Part D, 67,* 610–627. https://doi.org/10.1016/j.trd.2019.01.007.
10. Akbari, F., Mahpour, A., & Ahadi, M. R. (2020). Evaluation of energy consumption and CO2 emission reduction policies for urban transport with system dynamics approach. *Environmental Modeling & Assessment, 25,* 505–520. https://doi.org/10.1007/s10666-020-09695-w.
11. *District statistical handbook Chennai district, 2016–2017* (pp. 85–86) (2018). Retrieved May 2020, from https://cdn.s3waas.gov.in/s313f3cf8c531952d72e5847c4183e6910/uploads/2018/06/2018062923.pdf.
12. *Ministry of Petroleum & Natural Gas, Government of India,Report of the Expert Group on a Viable and Sustainable System of Pricing of Petroleum Products.* (2010). Retrieved May 2020, from https://petroleum.nic.in/sites/default/files/reportprice.pdf.
13. Ramachandra, T. V., & Shwetmala. (2009). *Emissions from India's transport sector: Statewise synthesis.* Atmosphere Environment. https://wgbis.ces.iisc.ernet.in/energy/paper/IISc_Emissions_from_Indias_Transport_sector/index.htm.
14. Forrester, J. W. (1971). Counterintuitive behavior of social systems. *Theory and Decision, 2*(2), 109–140. https://doi.org/10.1177/003754977101600202.
15. Yuan, X. H., Ji, X., Chen, H., Chen, B., & Chen, G. Q. (2008). Urban dynamics and multiple-objective programming: A case study of Beijing. *Communications in Nonlinear Science and Numerical Simulation, 13*(9), 1998–2017. https://doi.org/10.1016/j.cnsns.2007.03.014.
16. Chennai Comprehensive Transportation Study (CCTS). (2010). *Final report executive version* (pp. 11–40). Chennai Metropolitan Development Authority. https://www.cmdachennai.gov.in/pdfs/CCTS_Executive_Summary.pdf.

Understanding Local-Level Political Economy of Urban Water Access to Poor: The Case of Mumbai

Sameer Pendharker, Subodh Wagle, and Pradip S. Kalbar

Abstract An apparent paradox in Mumbai's water scenario is the point of departure for this study. On one hand, the city is blessed with ample water resources and is governed by a financially and institutionally strong urban local body. On the other hand, in stark contrast, the poorest sections of Mumbai residing in slums of Mumbai, lack access to water. Thus, the denial of water access to the poor is the result of unequal distribution rather than of water scarcity. The study hence aims at unearthing the political economy of water operating at the slum level. To achieve this, the study draws from in-depth semi-structured interviews, as well as a few informal group discussions involving approximately more than forty respondents. The findings show that certain political actors dominate the slum-level political economy of water. Each of these actors have special interests which they try to promote through diverse activities, using different types of power which are derived from a range of sources. The local political economy dominated by local politicians and their acolytes controls alternative channels of water provisioning while ensuring denial of access to the formal water network to the poor. Understanding this political economy will help on one hand, in providing formal water access to the poor and on the other hand, in designing effective policies to regulate informal water provisioning in slum areas.

Keywords Urban water · Mumbai slums · Last mile · Political economy · Grass-root water dynamics

---

S. Pendharker (✉) · S. Wagle
Centre for Technology Alternatives for Rural Areas (CTARA), Indian Institute of Technology Bombay, Mumbai 400076, India
e-mail: sameer.pendharker@gmail.com

P. S. Kalbar
Centre for Urban Science and Engineering (CUSE), Indian Institute of Technology Bombay, Mumbai 400076, India

A. Jana and P. Banerji (eds.), *Urban Science and Engineering*, Lecture Notes in Civil Engineering 121, https://doi.org/10.1007/978-981-33-4114-2_32

1 Background and Context

1.1 Mumbai: A Water Privileged City

Any discussion on urban water starts and quite often revolves around two major issues, namely; availability of sufficient water resources and availability of funds required for the operation and maintenance of water supply systems. Mumbai is, however, an exception to this. Mumbai receives an annual rainfall of more than 2300 mm [8]. This is considerably higher than other Indian mega-cities like Bangalore (970 mm), Delhi (800 mm), Kolkata (1582 mm), Chennai (1400 mm) [6]. Besides, Mumbai is quite privileged when it comes to exclusive access to dams and water resources stored in these dams. The per capita water consumption levels in affluent areas of this mega-city are as high as 300 L per capita per day (lpcd) [7]. Bureau of Indian Standards recently deemed water supplied by MCGM as the most drinkable water among all other Indian cities [2].

The Hydraulic department of the Municipal Corporation of Greater Mumbai (MCGM) bears the responsibility to provide water to the city's residents. MCGM is endowed with surplus financial resources unlike many other urban local bodies [9].

1.2 Water Deprived Slum Dwellers in Mumbai

Despite the availability of ample water resources and robust financial condition of the MCGM, millions of people in Mumbai lack safe access to the formal water network of MCGM [5]. Most of these people who lack formal water access belong to the bottommost socioeconomic strata of Mumbai and stay in slum settlements [4]. The partial or complete denial of formal water forces people to access water from alternative sources of water. Such informal access to water is characterized by insufficient quantity and poor quality of water, as well as excessive drudgery and health, economic, and social costs to the poor [10, 11].

The alarming situation of water scarcity experienced by a huge number of families in Mumbai cannot be fully understood solely by delving into the technicalities of the water supply system, water resources engineering or municipal finances only [1, 3].

2 Research Methodology

2.1 Objective of the Study

It is clear from the above discussion that it is essential to investigate the economic and political roots of the inequalities associated with water access in Mumbai. This study aims to understand this unequal distribution of water by exploring the political economy factors, mainly those operating at the slum level.

2.2 Research Question

The broad qualitative research question investigated in this research is,

How does the politics at the level of slum settlements shape the access to water for residents of slum settlements?

2.3 Data Sources, Collection, and Analysis

This exploratory study is based on mainly the qualitative data collected from four cases represented by four slum pockets in suburban areas of the city of Mumbai. These are situated in varied terrain-conditions, in different parts of the city, and having different socio-religious-economic compositions. The study team interacted with approximately forty respondents through one-to-one, in-depth, semi-structured interviews, as well as a few informal group discussions. The method of snowball sampling was adopted to select respondents. In addition, twenty-five semi-structured interviews were conducted with other stakeholders such as local journalists, local political representatives, formal and informal water service providers, academics and researchers, and officials from MCGM. The qualitative data collected through interviews and group discussions was analyzed using the technique of thematic analysis.

3 Conceptual Framework

This qualitative study uses a political economy framework as a conceptual framework for the collection and analysis of qualitative data that views the real-life social phenomenon of denial of water access to slum dwellers as shaped by actions of a variety of actors (as shown in Fig. 1). These actions of actors are driven by the objective of protecting and promoting their respective economic, political, and other interests. While undertaking such actions, the actors involved use different kinds of powers emanating from varying sources, so as to achieve the aforementioned objective. In many cases, the informal power emanates from the weakness and vulnerabilities of the poor. Therefore, in order to understand the phenomenon of denial of access to water, the paper identifies and analyzes: (a) various activities of the powerful actors, (b) the interests driving these activities, and (c) different sources of power deployed or vulnerabilities of these actors exploited in order to affect these

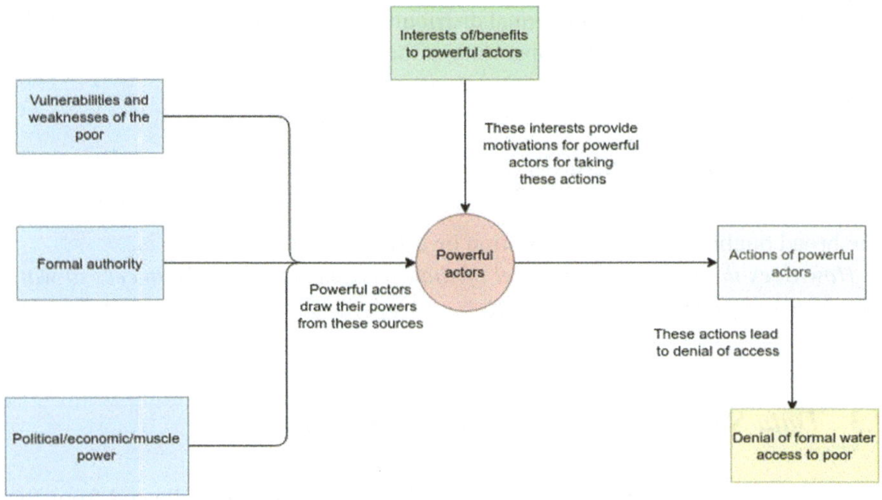

Fig. 1 Schematic of the conceptual framework

activities. Following specific research questions were used to collect the qualitative data from the field.

1. *Who are the powerful actors in the context of water access in slums?*
2. *What are the sources of power of these powerful actors?*
3. *What are the interests of these powerful actors?*
4. *What actions/activities do they undertake in order to fulfill their interests?*

Figure 1 presents the conceptual framework in the form of a schematic.

For thematic analysis of qualitative data, themes emerging from this conceptual framework were used as initial open codes.

4 Findings

The main powerful actor groups identified during the study are: (a) local politicians, (b) municipal officials, (c) private water tanker providers, and (d) resourceful residents having access to the formal water network.

The findings of the study are presented in terms of the actions undertaken by each of these four actor groups. For each of these actions, the interests driving them and the respective sources of power are presented.

4.1 Local Politicians

The most important interest of local politicians is to garner votes in elections (please refer to Table1). They garner votes by promising water or by facilitating access to water through informal sources. They ensure that the right of the poor to water access through the formal network is not often enjoyed by them so that the poor continue to rely on these politicians for their water needs. To this end, they use, misuse, mystify different rules and regulations, deny water to those who oppose or resist the activities of these politicians, and give threats of demolition of slum-dwellers' shanties if they seek help from other politicians or other political parties. They draw their power to undertake these actions from a mix of sources of two types; formal authority and informal power. Informal power emanates from different types of vulnerabilities of these poor slum dwellers such as poor people's constant need for money and their political marginalization that makes it impossible for them to fight back and resist. These findings have been detailed in Table 1.

Table 1 Interests, actions, and sources of power of local politicians

Interests of/Benefits to powerful actors	Actions affecting access to water	Vulnerabilities the poor/Formal authority
Garner votes from the electorate by promising access to water	Promise water connections during elections, but avoid delivering on the promise by giving different excuses	Formal authority as an elected representative
	Force to accept alternative (informal) ways of access to water to poor that are under their control of politicians and /or provide benefits to politicians or their acolytes	Economic and Muscle power
Maintain political control by keeping the people dependent on them for getting water	Denial of water citing different rules and regulations	Formal authority as an elected representative
	Denial of access to formal water network, saying that they bought the votes by distributing money during elections and are not accountable	Poor people's immediate and constant need for money (economic vulnerability)
Getting legal/illegal donations for elections	Denial of tenure protection by denial of access to formal water network so that poor could be evicted to clear the land for local builders	Legal Vulnerability of the poor people

Table 2 Interests, actions, and sources of power of municipal officials

Interests of/Benefits to powerful actors	Actions affecting access to water	Vulnerabilities the poor/Formal authority
Bribes	Taking hefty commission from licensed plumbers for providing formal water connections which raises the overall cost of getting a connection	Connections and/or economic exchange with local plumbers
Appeasing elected representatives (political superiors) by facilitating their anti-poor activities	Denying connections citing rules and regulations at the behest of politicians and local builders	Official authority drawn from the official post
	Inordinately delaying connections by confusing and trapping people in official paperwork	Official authority drawn from the official post
	Citing the absence of written orders from superiors to provide water connections even after High court ruling and release of GR that require them to give connections	Official authority drawn from the official post
	Asking for proof of residence prior to cutoff dates even when not required	People's lack of knowledge of rules (information vulnerability)

4.2 *Municipal Officials*

The main interests of municipal officials include earning bribes and to cater to the pressures and demands of local politicians and their acolytes. The officials take hefty commissions from licensed plumbers and if commission is not paid, they delay connections by trapping people in official paperwork, rules, and procedures. Municipal officials source their power from poor people's lack of knowledge of official procedures and their own rights besides the official authority they possess. These findings have been detailed in Table 2.

4.3 *Water Tanker Providers*

The main interest of water tanker providers is ensuring a steady sale of water in order to earn profit from the sale of water. This they do by cartelling among themselves and colluding with local politicians. They source their power over poor people from economic resources they have which they provide to elected politicians as well as from lack of other options for poor people to satisfy their water needs. These findings have been detailed in Table 3.

Table 3 Interests, actions, and sources of power of private water tanker providers

Interests of/Benefits to powerful actors	Actions affecting access to water	Vulnerabilities the poor/Formal authority
To earn profit from sale of water	Cartelling through mutual understanding with competitors	People's need for water and lack of other options to obtain water
	Colluding with local politicians, who receive regular "haftas" (bribes) from private providers in return for protection of their interests	Close relations (family and/or financial) with local politicians, as well as the exchange of economic benefits

Table 4 Interests, actions, and sources of power of formal connection holders

Interests of/Benefits to powerful actors	Actions affecting access to water	Vulnerabilities the poor/Formal authority
To earn profit out of selling water illegally to those who do not have formal water connections	By colluding with engineers, licensed plumbers, and local politicians to prevent people from getting water connections	Their slightly better of social status within the slum cluster
	By charging deposit fees from customers and refusing to return it if customers try to switch from one such connection holder to another	People's basic need for access

4.4 Formal Connection Holders

Like private water tanker providers, the interest of these formal connection holders is an economic gain in the form of additional income. These people are residents of the same slum settlements who are economically better and/or politically connected. As a result, they have been successful in getting access to the formal water network of the municipal corporation through a private, formal water connection. These connection holders sell water, which they receive from the formal network for their own consumption, to other poor people in the settlement who do not have access to the formal source of water. They charge exorbitantly for the water as compared to what they have to pay to the municipal corporation. They draw their power from their own resourcefulness, political connections, and/or social status within the community. These findings have been detailed in Table 4.

5 Conclusion

The main interests of the powerful actor groups identified in the study, fall within the ambit of economic benefits such as earning profits, bribes, additional income, or campaign funds. Only local politicians have political interest such as garnering votes. The strategies used for the fulfillment of these interests involve collusion with other powerful actors, citing rules and regulations, trapping poor people in official paperwork, asking for commissions, and cartelling. Coming to the source of power while the politicians and officials mainly make use of the formal authority they have due to their position, they also exploit the vulnerabilities of the poor people. These include their lack of knowledge about official rules and procedures, as well as the lack of options, available to them for obtaining water.

The local politicians and their acolytes control alternative channels of water provisioning, and at the same time, ensure denial of access to the poor to the formal water network. Thus, they hold poor residents of slum ransome using their formal authority and informal power. Thus, in this case, benefitting from vulnerabilities of the poor residing in these slums, political patronization, and economic exploitation of the poor by the powerful actors, all, go hand in hand. The study found an overwhelming sense of hopelessness and cynicism among slum dwellers with regards to water access, which is rooted mainly in the apathy and callousness of the state (municipal corporation) and local politicians.

The understanding gained through this study of the strategies and mechanisms used by powerful actors, their interests, and sources of their power will help devise necessary policies to effectively curb the malpractices involved and to regulate informal water provisioning in slums.

References

1. Anand, N. (2011). Pressure: The politechnics of water supply in Mumbai. *Cultural Anthropology, 26*(4), 542–564.
2. ANI (Asian News International). (2019). BIS study: Mumbai tops ranking for quality of tap water, Delhi at bottom. *The Economic Times*. https://economictimes.indiatimes.com/news/politics-and-nation/bis-study-mumbai-tops-ranking-for-quality-of-tap-water-delhi-atbottom/videoshow/72087881.cms?from=mdr.
3. Björkman, L. (2015). *Pipe politics, contested waters: Embedded infrastructures of millennial Mumbai*. Duke University Press.
4. Gandy, M. (2008). Landscapes of disaster: Water, modernity, and urban fragmentation in Mumbai. *Environment and Planning A, 40*(1), 108–130.
5. Graham, S., Desai, R. & McFarlane, C. (2013). Water wars in Mumbai. *Public Culture, 25*(1, 69), 115–141.
6. Indian climate. (2020). Indian climate: Average temperature, weather by month & weather for India. *Climate Data for Cities Worldwide–Climate-Data.org*. https://en.climate-data.org/asia/india-129/. Retrieved May 10, 2020.
7. Lewis, C. (2016). Mumbaikars get more water than they require daily, says IIT study. *The Times of India*. Retrieved April 22, 2016. https://timesofindia.indiatimes.com/city/mumbai/Mumbaikars-get-more-water-than-they-require-daily-says-IIT-study/articleshow/51936128.cms.
8. Mumbai climate. (2020). Mumbai climate: Average temperature, weather by month, Mumbai water temperature. (n.d.). *Climate data for cities worldwide–Climate-Data.org*. https://en.climate-data.org/asia/india/maharashtra/mumbai-29/. Retrieved May 10, 2020.

9. PTI (Press Trust of India). (2019). Mumbai's BMC presents Rs 30,692-crore budget for 2019–20, proposes Rs 44 crore as capital investment for BEST—India news,*Firstpost*. https://www.firstpost.com/india/mumbais-bmc-presents-rs-30692-crore-budget-for-2019-20-proposes-rs-44-crore-as-capital-investment-for-best-6027461.html
10. Subbaraman, R., Shitole, S., Shitole, T., Sawant, K., O'brien, J., Bloom, D. E., & Patil-Deshmukh, A. (2013). The social ecology of water in a Mumbai slum: failures in water quality, quantity, and reliability. *BMC Public Health, 13*(1), 173
11. Subbaraman, R., Nolan, L., Sawant, K., Shitole, S., Shitole, T., Nanarkar, M., & Bloom, D. E. (2015). Multidimensional measurement of household water poverty in a Mumbai slum: Looking beyond water quality. *PloS one, 10*(7).

Utility-Based Road Space Distribution Model for Vehicles, Pedestrian, Parking and Hawkers

Solanki Ghosh and R. N. Datta

Abstract The vibrant and multifunctional nature of the Indian roads makes it unique and challenging for effective traffic management. But, in today's urban planning models, road space allocation is done based on automobile requirements. Hence, there is a need to develop a methodology for optimizing the allocation of competing road space. This study aims at calculating the utility of each competing use at any particular space and place based on the socio-economic and functional fabric of the area. Three road stretches in Kolkata, India, are selected and a primary survey of 300 samples is conducted. Utility maximization theory is applied to maximize the road space utilization. Utility is assigned to every competing uses under the constraints of right of way, level of service requirement and mode demand. The outcome of the proposed methodology is in terms of the percentage of actual demand to be met while allocating road space to different competing use, to maximize the social and economic benefits. This methodology helps achieve optimal allocation of road space for varying kinds of demand that are characteristic of Indian urban streets.

Keywords Road space · Utility maximization · Pedestrians, hawkers, social choice

1 Introduction

In Indian cities, significant numbers of roads and streets have multifunctional use. This vibrant nature of the Indian roads makes it unique and challenging for effective traffic management. Conflict for space happens between various kinds of users like cars, public transport, slow moving vehicles, pedestrians, hawkers, parking and roadside shops. Demand for space for these users also differs greatly with changing adjacent land use and socio-economic condition of the area. In general, the prime

S. Ghosh (✉)
Indian Institute of Technology, Bombay, Mumbai, India
e-mail: solankiarchi@gmail.com

R. N. Datta
Indian Institute of Technology, Kharagpur, Kharagpur, West Bengal, India

A. Jana and P. Banerji (eds.), *Urban Science and Engineering*, Lecture Notes in Civil Engineering 121, https://doi.org/10.1007/978-981-33-4114-2_33

objective of allocating space for movement on Indian streets is biased towards automotive modes, mainly cars. Should it be so? The 2011 Census of India shows that only 4.7% of India's population has ownership of scooter/car/jeep/van. Majority of our population still avail public transport, bicycles, two-wheelers and even walking as the mode of travel. Therefore, there is a need to study and develop a methodology for the effective utilization of road space by different users. Now the question remains that, in which part of the city, we should give more importance to the car and where importance is to be given to pedestrian and other modes' movement? How much space do we allow for the hawkers? Where on-street parking should be allowed and where it should be restricted? How does adjacent land use and socio-economic composition of the area accounts for the above decisions?

Hence, there is a need to develop a methodology for optimizing the allocation of competing road space, demanded by motorized traffic, non-motorized vehicular traffic, pedestrians, parking and vendors with respect to social, economic and functional demands. The aim of the study is to develop a methodology for optimizing the allocation of road space, to competing motorized traffic, non-motorized vehicular traffic, pedestrians, parking and vendors with due consideration to their functional demand and socio-economic needs. The next section of this study focuses on various literature to arrive at the factors which contribute to the utility of any mode with respect to its surroundings. Section 3 outlines the survey procedure and its outcomes. In Sect. 4, a conceptual framework has been proposed for optimizing the allocation of road space, to competing motorized traffic, non-motorized vehicular traffic, pedestrians, parking and vendors. In Sect. 5, the application and validation of the proposed conceptual framework has been done. This paper is concluded with some limitations of the study.

2 Literature Review

There are several types of schemes/concepts that talk about reallocation of road space. These concepts use various concepts of equity to reallocate space. Living Streets is one such concept where priority is given to pedestrians and cyclists, and create safe places for people to walk, cycle, play and meet friends [1]. Shared streets is another such concept which introduces an additional 'place' function for urban road space in addition to its basic functions of 'mobility' and 'access' [2, 3]. Complete Streets give emphasis to the value of safe access for all users, not just automobiles. [4]. Land use planning can contribute to a reduction in kilometers driven by influencing the spatial structure area. Dense and mixed-use development helps to keep walking and cycling attractive [5]. Most analyses identify income and automobile ownership as primary determinants for explaining differences in mode choice [6, 7]. Both variables are closely correlated as an increase in incomes makes owning and maintaining a car more feasible. Additionally, higher incomes increase the opportunity costs of travel time, and thus private motorized vehicles become more attractive. On average, business owners overseas overestimate car uses by their customers by approximately

20% and underestimate walking trips to local shopping areas by 13–19% [8, 9]. For some land use, like roadside shopping areas, there is a demand for pedestrians. But with the absence of proper pedestrian facilities, people walk in a very unsafe environment, adding to the congestion in the road by spill over. From the literature study, the following factors are identified as factors which are most likely to affect the demand for road space by competing uses like motorized traffic, non-motorized vehicular traffic, pedestrians, parking and vendors:

- Land use
- Socio-economic
- Modal split.

3 Survey Procedure and Data Collection

3.1 Survey Procedure

Three road stretches, Hatibagan, College Street and Saltlake 3rd Avenue, had been identified in Kolkata Metropolitan Area having diversity in land use, socio-economic conditions and modal split. Then detailed physical characteristics survey had been done to collect information on right of way, lane widths and footpath width, vendors' encroachment, pedestrian traffic spill over and street parking. A questionnaire survey was conducted with visitors and/or shoppers, shop owners and/or hawkers and inhabitants to understand their demographic variance, spending patterns, mode choice, needs and preferences. The survey procedure is indicated in Fig. 1 and Table 1.

4 Introduction to Study Area

4.1 Identification of the Stretches

Three arterial road stretches of Kolkata, India, namely, Hatibagan, College Street and 3rd Avenue of Saltlake have been selected for the study. The selection has been done mainly on the basis of diversity of land use, socio-economic conditions and modal split in these stretches. Table 2 describes the physical characteristics of the chosen road stretches. College street road stretch, famously known as 'Boi Para' and is dotted with small and big bookstores and kiosks. It has many educational institutions like Presidency University, Hindu school, etc. and the famous Medical College. Along with the bookstalls and institutes, the entire street feels like an extended campus. The selected stretch runs from Madan Mohan Burman street (Bata junction) in the north to B B Ganguly Street (Bowbazar junction) in the south.

Hatibagan road stretch is known for its 100 years old Hatibagan market. This stretch is a predominantly commercial area with some public buildings, mostly

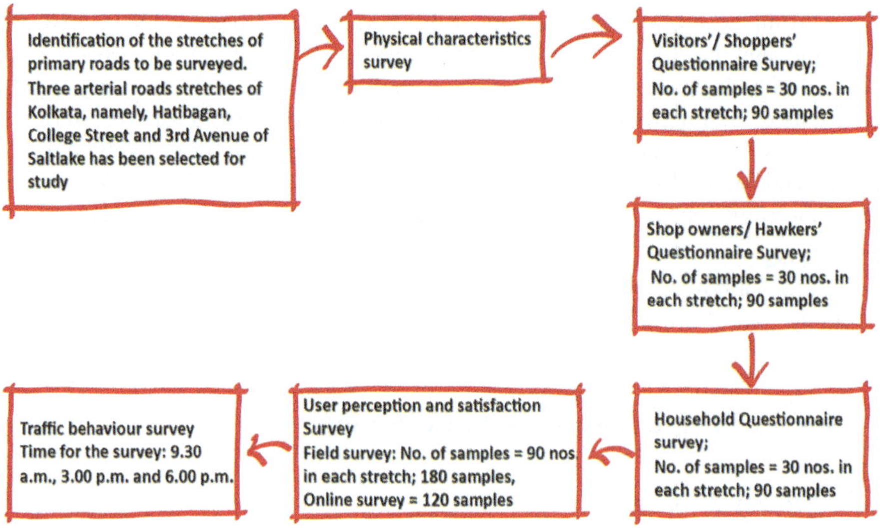

Fig. 1 Flow diagram of survey

theatres like Star Theatre. The vendors' market here is one of the vibrant and economically thriving markets of the city. It stretches from Shyambazar 5 point crossing in the north to Hedua crossing in the south. Saltlake is a satellite township of Kolkata and 3rd avenue is one of the arterial roads of this township. Major adjacent land use is public and semi-public. It also has considerable residential land use. City centre is one major commercial establishment along this road. It also abuts the Karunamoyee Bus Terminal.

4.2 Volume Count Survey

The existing traffic condition of the three road stretches is being summed up in the following Table 3. Survey nodes A2 and B1 has v/c ratio more than 1, and thus need immediate attention. Detailed analysis of PCU count has been done in the following section.

5 Conceptual Framework

The aim of the study is to develop a methodology for optimizing the allocation of road space, to competing uses with due consideration to their functional demand and socio-economic needs. All competing uses have been categorized into 4 categories; Motorized Traffic (MT), Non-Motorized Traffic (NMT), Pedestrians (Ped)

Table 1 List of data collected

Survey	Secondary source	Information to be derived
Physical characteristics survey	Map: urban land use Calcutta City; NATMO GIS road map of Kolkata: KMDA	Land use map Land use mix percentage Detailed dimensions Spill over from footpaths
Visitors'/shoppers' survey		Modal split Change in modal choice with change in land use, income, etc. Parking characteristics % expenditure by various income groups, mode of transport Informal economic activity Socio-economic survey for the floating population
Shop/Hawkers' survey		Informal and formal economic activity characteristics Busiest hours of activity No. of customers Modal split of the customers Change in modal choice with change in land use, income, etc.
Socio-economic survey	Census of India, 2011, 2001 and 1991	Population, HH nos, HH income Change in modal choice with change in income, vehicle ownership, etc. % expenditure by various income groups, mode of transport Daily transportation habits
User perception and satisfaction survey		User perception of what should be the optimum level of service provided User satisfaction from existing services Transportation habits
Traffic behaviour survey	Comprehensive mobility plan, Kolkata metropolitan area, August, 2008; KMDA	Traffic volume Travel time Speed distribution of the traffic Parking accumulation Public transport system Para-transport system (transport department)

and Parking & Hawkers (PH). From the literature review, we can see that the utility of the first three categories, MT, NMT and Ped, is dependent upon Land Use, Socio-economic factors like gender and income, and the characteristics of the mode like mode cost and travel time. The utility of parking and hawkers is seen as a function of

Table 2 Physical characteristics of the road stretches

Characteristics	College street	Hatibagan	Saltlake
Length (m)	1.2	1.5	3.4
Right of way (m)	15	18	30
Fee on on-site parking	Present	Present	Present
Designated lane	Not present	Not present	Not present
Public transport system	Bus service present, auto-rickshaw service not present, tram service present	Bus service present, auto-rickshaw service not present, tram service present	Bus service present, auto-rickshaw service present, metro to come
Adjacent dominant land use	Commercial and Institutional	Commercial and recreational	Public, Residential and Commercial
Bus stops	3	2	4

Table 3 Volume Count Survey Results (Source: Primary survey)

Stretch	Time	Pcu	Flow rate Pcu/hr	V/C	Pedestrian volume
College street	Morning	101	1508	0.75	154
	Evening	107	1605	0.80	206
Hatibagan	Morning	93	1388	0.93	43
	Evening	84	1266	0.84	70
Saltlake	Morning	143	2152	0.60	385
	Evening	149	2242	0.62	383

the economic benefits it brings to the area. The following Fig. 2 outlines the conceptual framework to arrive at a maximization equation. The utility of each category is explained in Eq. 1 and 2 below

$$Ua = Social\,Cost * (Transport\,Space\,Utility + Socioeconomic\,Utility - mode\,cost) \quad (1)$$

where, a represents MT, NMT and Ped;

$$U_{PH} = \text{Percentage of Economic Revenue per Rs.1000 revenue of permanent shops} \quad (2)$$

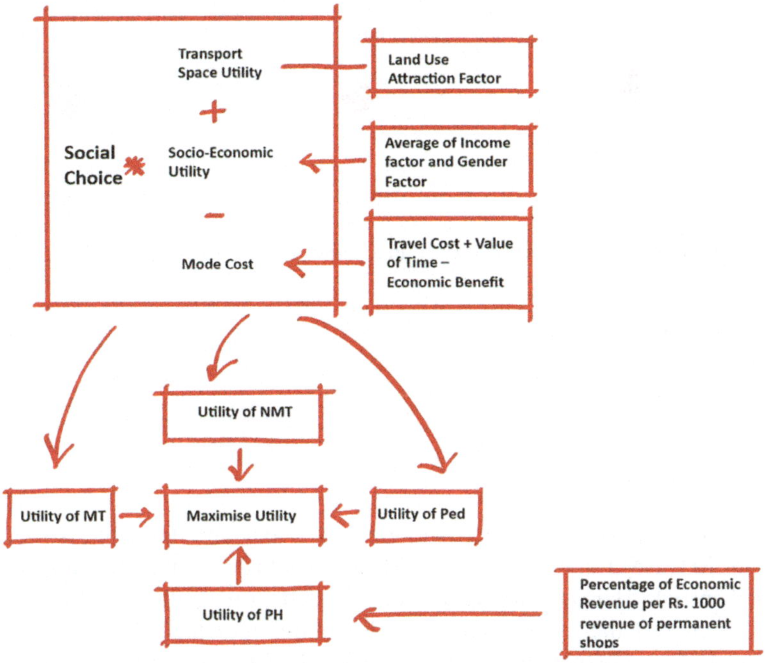

Fig. 2 Conceptual framework flow diagram

5.1 Social Choice

Social choice is obtained from Soft Multi-Criteria Analysis of Users' preferences data. The Users were asked to rate each factor (Table 4) in terms of value and impact. Value is the rating given to each factor in terms of much that factor influences in mode choice of the user and the impact represent how much each category fares well in each of these factors.

Table 4 Impact and value matrix (more numbers of Xs and +s higher is the impact or value respectively)

Value		MT	NMT	Ped
+++	Door to door travel is possible	xx	xx	xxxx
+	More comfortable than other alternatives	xxx	xxxx	xx
++	Less costly than other alternatives	xx	xx	x
++	Safe to travel	xx	xx	xx
+++	Time of travel is less	xxx	xxx	x
++	Non-availability of other alternatives	xx	xx	x

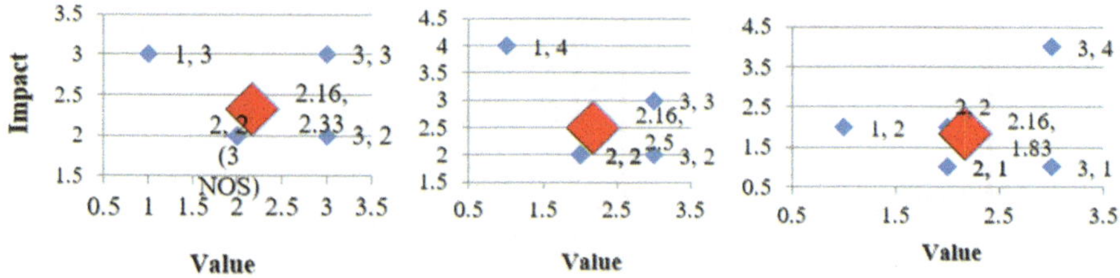

Fig. 3 Centroid for categories MT, NMT and Ped, respectively

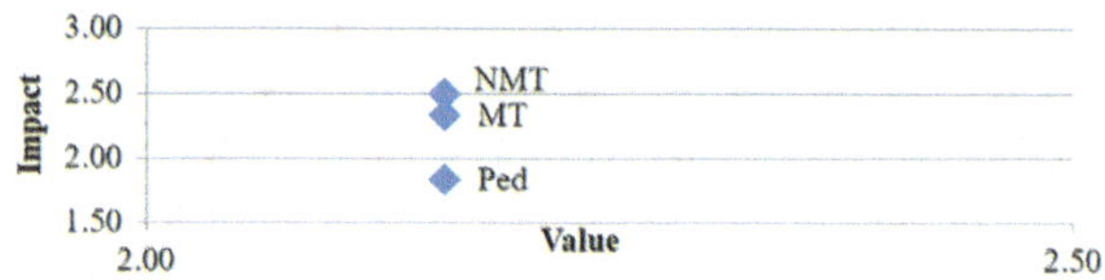

Fig. 4 Comparison of centroids

For every category, the impact is plotted on Y-axis and the value is plotted on X-axis for MT, NMT and Ped (Fig. 3). The centroid is obtained from all the points is plotted in red. These centroids are then plotted in the following Fig. 4. These centroids represent the social choice. Because in this case, the value of X-axis is the same for all three alternatives, the value of Y-axis is taken as the social choice for respective alternatives. This method used is soft multi-criteria analysis [10].

5.2 *Transport Space Utility (L)*

User preference for mode of transport for various Land use is identified from the survey. For finding user preferences, the total modal frequency for the trip generated for different kinds of land use and total modal frequency for trips attracted for different kinds of land use have been identified from the primary survey. Weightage **Li** is given by adding the frequency and then normalizing them. This weightage will be used as a factor for land use utility (**L**).

$$\mathbf{L} = \sum \mathbf{Li} * \%\ \mathbf{of\ Land\ use\ I} \tag{3}$$

where Li is the Land use attraction factor for land use i (Table 5).

Table 5 **a** Land use attraction factor **b** Transport space utility for the three stretches

	MT	NMT	Ped
Commercial	5.11	0.71	4.18
Institutional	5.91	1.45	2.64
Residential	5.56	1.37	3.07

	Hatibagan	College street	Saltlake
MT	445.61	460.52	352.00
NMT	88.10	98.62	83.05
Ped	296.28	260.86	184.95

5.3 Socio-economic Utility

The mode choice frequency for various socio-economic factors is calculated. Weightage is given by the weighted sum method (sg and si). This weightage will be used as a factor for socio-economic utility (Table 6).

$$\begin{aligned} \mathbf{S} = \{ & (\mathbf{sgm} * \%\mathbf{of\,male} + \mathbf{sgf} * \%\mathbf{of\,female}) + \\ & (\mathbf{si1} * \%\mathbf{of\,income1} + \mathbf{si2} * \%\mathbf{of\,income2} + \\ & \mathbf{si3} * \%\mathbf{of\,income3} + \mathbf{si4} * \%\mathbf{of\,income4} + \mathbf{si5} * \%\mathbf{of\,income5})\} \end{aligned} \quad (4)$$

5.4 Mode Cost (C)

Mode Cost is calculated using the following equation.

$$\text{Mode Cost} = \text{Travel cost} + \text{Value of time} - \text{Economic benefit} \quad (5)$$

Travel cost factor is calculated as travel cost per km/summation of travel cost per km of all modes * 100. Value of time is taken from 'Road User Cost Study in India' by CRRI [11] (Table 7).

Table 6 **a** Income factor **b** gender factor **c** socioeconomic utility

Income	MT	NMT	Ped
0-5000	0.29	2.57	7.14
5000-10000	2.90	0.32	6.77
10000-25000	4.71	0.59	4.71
25000-50000	7.45	0.36	2.18
>50000	8.75	0.14	1.11

	MT	NMT	Ped
Male	6.16	1.43	2.41
Female	5.26	0.00	4.74

	Hatibagan	College street	Saltlake
MT	575.48	618.92	592.61
NMT	64.27	62.60	65.38
Ped	360.26	318.48	342.02

Table 7 Mode cost

	MT	NMT	Ped
Mode cost	55.68	50.61	– 6.28

Table 8 Utility of PH

Utility	Hatibagan	College street	Saltlake
PH	0.21	0.8	0.8

Table 9 Total Utilities of all categories

Utility	Hatibagan	College street	Saltlake
MT	5.86	6.18	5.06
NMT	0.66	0.72	1.72
Ped	3.16	2.78	2.86
PH	0.21	0.8	0.8

5.5 Utility for Category PH

Utility of category PH is calculated as percentage of economic revenue to total revenue. The normalized utility of category PH can be summarized as follows (Table 8).

5.6 Final Utilities of All Categories

See (Table 9).

5.7 Maximizing Equation—Proposals

Maximizing equation for each category arrives in the following format:

- Max U = UMT * XMT + UNMT * XNMT + UPed * XPed + UPH * XPH
- Now, we have to maximize the total utility U. Given the constraints, XMT, XNMT, XPed and XPH <= Demand and XMT * AMT + XNMT * ANMT + XPed * APed + XPH * APH = Road width, Here, AMT, ANMT, APed and APH are per unit area required (Table 10).

In Hatibagan and College street, Category NMT traffic should be completely restricted. Hawkers and parking can be provided only on one side of the road. In all cases, the proposed demand for motorized traffic is lessened considerably. In Salt Lake, due to the presence of residential land use, the requirement of Non-motorized traffic is considerable and it demands the provision of a dedicated lane.

Table 10 Final results of maximization

	Hatibagan	College street	Salt Lake
Demand MT	1900	2500	3300
Proposed MT	1850	1050	2200
Demand NMT	700	500	1300
Proposed NMT	0	0	300
Demand Ped	1300	2100	2500
Proposed Ped	1300	2100	2500
Demand PH	240	205	335
Proposed PH	200	100	100

6 Conclusion

The proposed methodology achieves optimal allocation of road space for varying kinds of demand that are characteristic of Indian urban streets. For existing streets, where the constraints are more pronounced, the methodology helps to determine a workable allocation of space to varying uses on their socio-economic priorities. The basic form of the proposed methodology is an optimization model that aims to maximize the utility of road space. The utility function has three components, viz. transport space utility, socio-economic utility and mode cost as a negative utility which have coefficients based on social choice. The objective function is maximized under the constraints of the level of service limit, available road width and lane capacities. The methodology has been applied to three road stretches having different land use characteristics in Kolkata. Application of the methodology on the three stretches indicates that there are scopes for improvement of the existing conditions of movement and road space usage. For green field developments, the methodology can be effectively applied to determine proper road space allocation by drawing data from similar existing urban places. Given the constraints of time and resource, the study had to be restricted to some limitations. Cultural and political factors are not considered. The impact of the solution on the greater part of the city is considered negligible for this study. Climatic and environmental factors are also not considered.

References

1. Bain, L., Gray, B., & Rodgers, D. (2012) *Living streets: Strategies for crafting public space.* John Wiley & Sons.
2. Polus, A., & Craus, J. (1988). Planning considerations and evaluation methodology for shared streets. *Transportation Quarterly, 42*(4).
3. Karndacharuk, A., Wilson, D. J., & Dunn, R. (2013). Evaluating shared spaces: Methodological framework and performance index. *Road & Transport Research: A Journal of Australian and New Zealand Research and Practice,* 22(2), 52.
4. Litman, T. (2015). *Evaluating complete streets.* Victoria Transport Policy Institute.

5. Wang, Y., Monzon, A., & Di Ciommo, F. (2015). Assessing the accessibility impact of transport policy by a land-use and transport interaction model–The case of Madrid. *Computers, Environment and Urban Systems, 49,* 126–135.
6. Cameron, I., Kenworthy, J. R., & Lyons, T. J. (2003). Understanding and predicting private motorised urban mobility. *Transportation Research Part D: Transport and Environment, 8*(4), 267–283.
7. Buehler, R. (2011). Determinants of transport mode choice: A comparison of Germany and the USA. *Journal of Transport Geography, 19*(4), 644–657.
8. Sustrans, U. K. (2006). Shoppers and how they travel. *Livable Neighborhoods Information Sheet.*
9. Feng, T., & Timmermans, H. J. (2014). Trade-offs between mobility and equity maximization under environmental capacity constraints: A case study of an integrated multi-objective model. *Transportation Research Part C: Emerging Technologies, 43,* 267–279.
10. Nijkamp, P., & Soffer, A. (1979). *Soft multicriteria decision models for urban renewal plans.*
11. Swaminathan, C. G., & Kadiyali, L. R. (1983). Road user cost study in India. *Journal of the Indian Roads Congress, 44*(1).

Assessment of Urban Foodbowl Using Spatiotemporal Analysis of Urban Growth

Shahana Usman Abdulla and Bimal Puthuvayi

Abstract Drastic transformations in land use often impact the sustaining capacity of a city, the major impact being the one on the productive agricultural land available in an urban area, termed as the "urban foodbowl". This study tries to analyze the transitions in the productive agricultural land of the urban area of Palakkad, a city situated in Kerala, India, due to the impact of land cover changes. Landsat images obtained from remote sensing data were used to map the land cover of the study area at four time points, i.e., 2000, 2005, 2015, and 2020, which were then subjected to supervised classification into categories of built, water, open areas, plantations, forests, and agricultural land. The transition of agricultural land to other land uses was determined. Then, the factors most likely to have influenced were used to analyze the transitions measured earlier using logistic regression. An increase of 2,335,452 ha (500%) in the built-up area was found over a period of 20 years. The analysis revealed an increasing trend of transition from agricultural land to built-up land use which ranged from 1050.3 ha (2000–05) to 3540.24 ha (2015–20). The factor analysis revealed the influence of states of neighborhood cells, distances to the roads, and urban centers to be higher on the transition of agricultural land to other land uses. This study brings our attention to the impacts of the transformation of the urban landscape on the foodbowl which imposes challenges to the sustainability of food production in urban areas.

Keywords Urban land change · Urban foodbowl · Transition analysisn

1 Introduction

Understanding urban growth spatiotemporally is important for landscape and urban development planning [1]. The changes in land use and land cover occur due to urbanization caused by unplanned and uncontrolled urban sprawl which leads to change in nature, destroys green cover, and pollutes the water resources [2]. Over

S. U. Abdulla (✉) · B. Puthuvayi
Research Scholar, National Institute of Technology, Calicut, Kerala, India
e-mail: shahana.usab@gmail.com

A. Jana and P. Banerji (eds.), *Urban Science and Engineering*, Lecture Notes in Civil Engineering 121, https://doi.org/10.1007/978-981-33-4114-2_34

the last 20 years, many urban areas have experienced dramatic growth, as a result of rapid population growth and as the world's economy has been transformed by a combination of rapid technological and political change. Around 3 billion people—virtually half of the world's total population—now live in urban settlements. And while cities command an increasingly dominant role in the global economy as centers of both production and consumption, rapid urban growth throughout the developing world is seriously outstripping the capacity of most cities to provide adequate services for their citizens. [3]. The capacity to feed people in urban areas depends mainly on the impact of rapid urban growth on the highly productive agricultural region on the city fringe termed as the "Urban Foodbowl" [4].

The urban population of Kerala, one among the southern states of India, is about to cross the 50% mark with a high rate of urbanization and if the trend continues, Kerala is most likely to be 100% urban in the future [5]. Palakkad is one among the major rice producing districts of Kerala. The conversion of paddy fields into residential and commercial plots has been going on at a rapid pace in Kerala, India, since the 1980s. The high density of population, inflow of remittance incomes from migrant workers, and the fast growth of the services sector have created a high demand for land in the state [6].

This study tries to analyze the impact of urban growth on the urban foodbowl of Palakkad which would be helpful for future predictions and policy interventions. The transition of agricultural land to other land uses have been analyzed using remote sensing data and GIS techniques. The factors influencing these transitions were analyzed using logistic regression to find the predominant causes of transition.

2 Materials and Methods

2.1 Study Area

We examined the spatiotemporal growth of the urban area of Palakkad, located in Palakkad district, Kerala, India (Fig. 1a, b), from 2000 to 2020. Palakkad district is one of the main granaries of Kerala and its economy is primarily agricultural engaging more than 44 per cent of the workers. The district shares 20.45% of the state agricultural area (Fig. 1c), expressing that agriculture is the main sector of economy of the district [7].

Palakkad shows an increase of 10.47% in urbanization, from 13.62% in 2001 to 24.09% in 2011 [4]. The Palakkad town spreading over an area of 26.6 km^2 has agricultural land both inside and along the fringes of the town. In order to find the impact of urbanization on the urban foodbowls of Palakkad, the study area was delineated.

Images from Landsat 7 ETM + (Jan 2000 and Jan 2005) and Landsat 8 OLI/TIRS (Jan 2015 and Dec 2019) (Fig. 2a) with one scene (path 144, row 53) covering the study area were acquired. The study area was then delineated with a buffer of 15 km

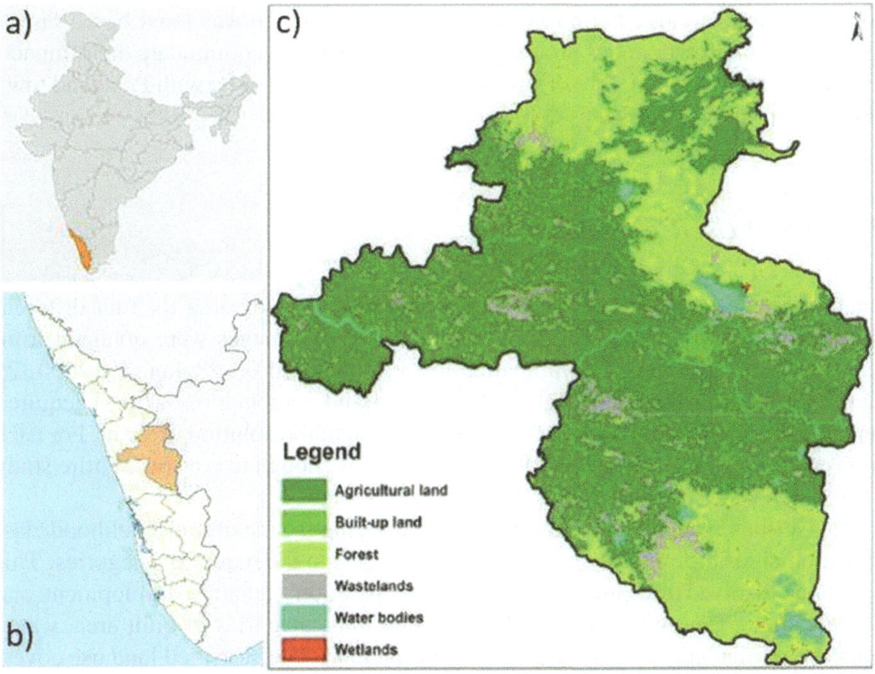

Fig. 1 Location of the study area: **a** Location of Kerala in India. **b** Location of Palakkad district in Kerala. **c** Land use of Palakkad district (*Source* State Urbanization Report, 2012)

Fig. 2 Location of study area: **a** Landsat 8 OLI/TIRS (144, 53), Jan 2015 **b** Delineated study area showing Palakkad town (*Source* USGS)

from the administrative boundary of Palakkad town which was fixed based on the distribution of growth nodes around the town in order to accommodate their impacts on the urban growth (Fig. 2b). Thus, a total area of 8, 04, 330 ha with Palakkad town and its surrounding areas was delineated.

2.2 *Land Use/Cover Mapping*

Landsat images were used to map the land use of the study area for four different time points being 2000, 2005, 2015, and 2020. These images were obtained from Landsat 7 ETM + (Band combination—4,3,2) acquired on 28 January 2000, 25 January 2005, and from Landsat 8 OLI/TIRS (Band combination—7,6,4) acquired on 13 January 2015, 27 January 2020, all with a spatial resolution of 30 m. For each time point, one scene with path 144 and row 53 was enough to cover the entire study area.

A pixel-based supervised classification employing the maximum-likelihood classification algorithm was used to classify the images to the required categories. This technique involved three main steps: sample preparation, signature development, and classification. The images were then classified into categories of built area, water, open areas, plantations, forests, and agricultural land. The classified land use covers were then overlaid with Google Earth imagery for accuracy.

2.3 *Transition Analysis*

This analysis was done in ArcGIS in which the transitions of agricultural land to other land uses of built, plantation, and open areas during the periods 2000–05, 2005–15, and 2015–20 were determined. For this, an ArcGIS model (Fig. 3) was created which used a raster calculator to take the paddy layer during the first year of transition and then calculate its transition to other land uses of the second year of transition using "&" operation with the required target values of each layer.

2.4 *Factor Analysis*

The factors here taken are those which could most probably have an influence on the land use transitions calculated. Based on the study area, the factors selected were neighborhood, distance to junctions, distance to first order urban centers, and aspect (Table 1).

For the analysis, a sample of 1 lakh pixels was taken using sampling by random points. Firstly, the transitions calculated earlier were extracted to these sample points. Then the neighborhood (3 × 3) pixels of each sample point for each land use were

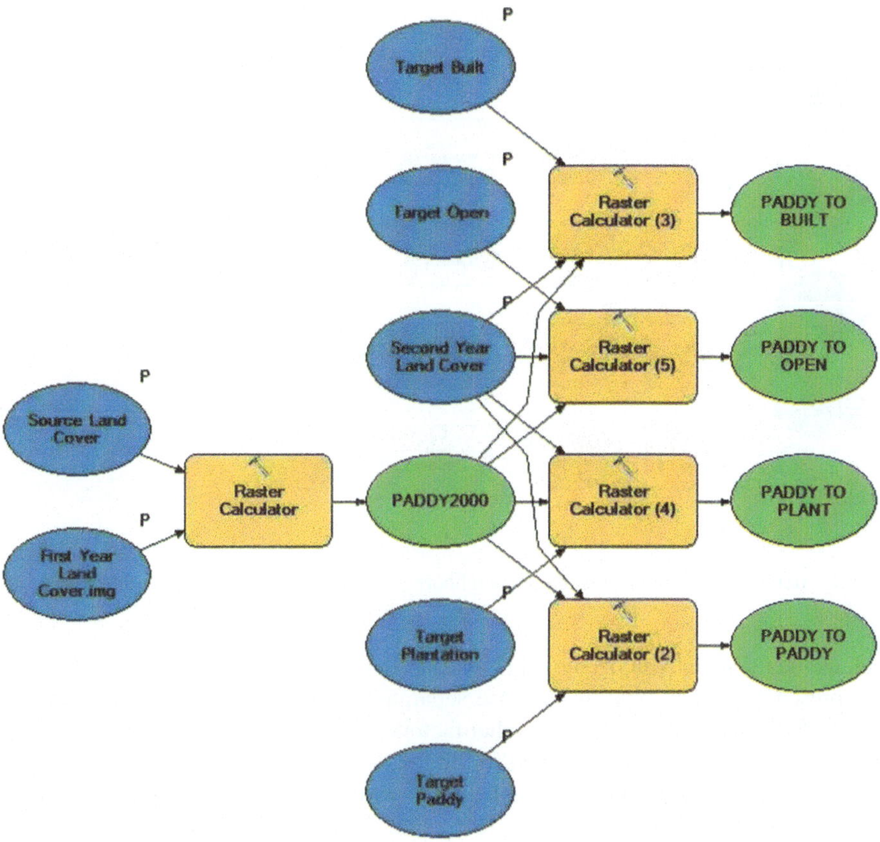

Fig. 3 ArcGIS model for calculating transitions of agricultural land to other land uses

Table 1 Driving factors and their descriptions

Representative factors	Description
Cellular automata	Represents the built and plantation land use present in 3 × 3 neighborhood cells of each sample point
Distance to first order urban centers	Represents access to urban facilities
Distance to second order urban centers	Represents access to urban facilities
Distance to junctions	Represents access to nearby growth nodes
Distance to built-up areas	Represents proximity to built environment
Distance to roads	Represents access to transport facilities
Population density	Represents the spread of population
Slope	Represents the terrain of the study area
Aspect	Represents the terrain of the study area

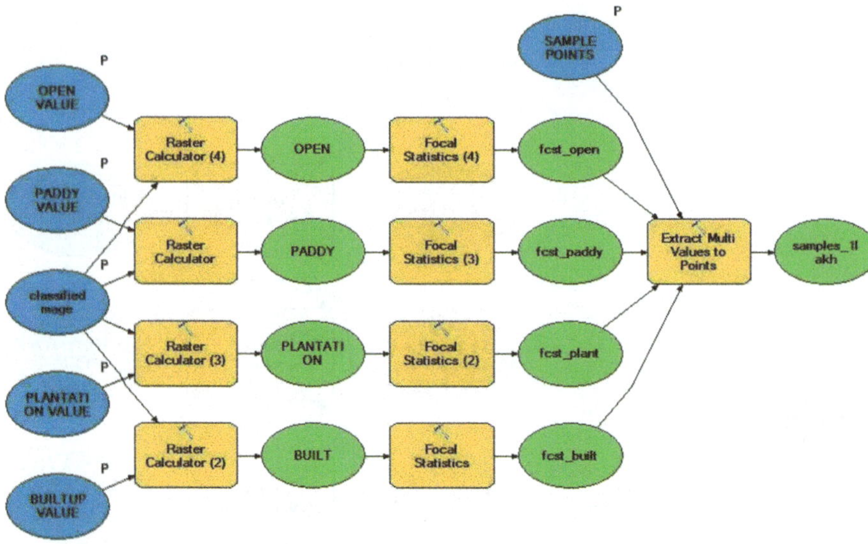

Fig. 4 ArcGIS model for calculating the neighborhood land use for the sample points

calculated using an ArcGIS model (Fig. 4) utilizing focal statistics which calculated the presence of each type of land use separately extracted to each of the sampled points. Similarly, data of all of the other factors were extracted to the sampled points. The whole procedure was followed for all the three time periods taken.

The data extracted from the sample points are then exported to STATA for further analysis to find the influence of the factors on the transition of agricultural land to other land uses. The calculated transition of agricultural pixels to built-up area, open area, plantations, and which remained agricultural are taken as dependent variables and the rest of the factors are taken as independent variables for analysis. These are then subjected to logistic regression from which the odds ratio gave the influence each of the representative factors had on the transition of agricultural land to built-up, open and plantation areas. The factors were also checked for multi-collinearity and the redundant factors were removed based on their correlation. Logistic regressions and correlations were run using coding in STATA. The analysis was done to the datasets of all the three time periods taken; 2000–05, 2005–15, and 2015–20.

3 Results and Discussion

3.1 Transition Analysis

Our results show a total loss of agricultural land being 14,940.45 ha during 2000–2005, 15,303.33 ha during 2005–15, and 9405.09 ha during 2015–20 in the study

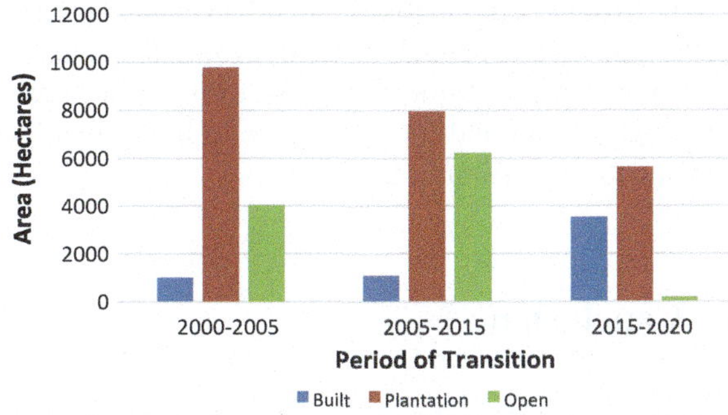

Fig. 5 Loss of agricultural land to other land uses

area (Fig. 5). The losses were measured to land uses of built area, plantation, and open area. Out of the losses considered, transition to built-up land use seems to be at 1050.03 ha (7%) during 2000–05 and 1109.34 ha (7%) during 2005–15, whereas it gets catapulted to 3540.24 ha (38%) during 2015–20. Transition to an open area is found to be at 4077.99 ha (27%) during 2000–05, 6233.04 ha (41%) during 2005–15, and 212.67 ha (2%) during 2015–20. The highest transition of agricultural land was to plantations which were found to be 9812.43 ha (66%) during 2000–05, 7960.95 ha (52%) during 2005–15, and 5652.18 ha (60%) during 2015–20.

The results clearly prove the practice of conversion of agricultural land to other land uses as it is evident from the transition of agricultural cells to other land uses over a period of time. The loss of the urban foodbowl is majorly attributed to plantations, which mostly is related to the nature of the study area where the conversion of paddy fields to coconut plantations so that the area could be further used for other purposes is prevalent. Loss of agricultural land to built-up areas is showing an increasing trend over the years. Urban growth may be attributed as the major reason behind such transitions. Since the loss of urban foodbowl to other land uses has been substantiated, further analysis of factors has been done to find the influence behind the transitions.

3.2 *Factor Analysis*

The logistic regression gave the odds ratio which showed the odds of a factor influencing a particular land use transition. During 2000–05, the aspect, distances to the urban centers and junctions were found to have a higher influence on the transition of agricultural land to built-up areas. But during 2005–15, the transition to built-up areas is found highly dependent on the built neighborhood indicating an increase in the urban growth during that particular period in the area. During 2015–20, again the aspect and distances to nearby facilities become dominant in influencing transition to built-up areas. These indicate the nature of terrain and access to urban and transport facilities as the major reasons for the transition of agricultural land for constructing buildings, in turn, indicating an increasing trend of urbanization.

The distances to junctions and urban centers are found to have a higher influence on the transition to plantations. This again substantiates the conversion of paddy lands to plantations to be further used for construction, a prevalent practice in Kerala. And the rough terrain and long distances to urban facilities have resulted in less or no transition of agricultural land.

4 Conclusion

This study examined the impact of urban growth on the foodbowls of the urban area of Palakkad, known as the "rice bowl" of Kerala, India, from 2000 till 2020, using remote sensing data and GIS techniques. The transition analysis revealed a loss of urban foodbowl showing an increasing trend over the course of years. The major losses were found to be attributed to plantations and built-up areas. The transition to plantations can in-turn be attributed to the emerging built-up areas as it is a practice prevalent in Kerala to convert them to plantations first and then to built-up areas. The factor analysis revealed an influence of the access to urban, transport facilities, and terrain as the major factors influencing the transitions. Also, the nature of the neighborhood of agricultural land has also resulted in its conversion majorly to built-up areas. This study provided insight into the decline of urban foodbowls upon which the food production is dependent, especially in a state like Kerala. This further results in the deterioration of resources and urges us to depend more on other states for food supply and survival. Analyzing such a decline would help us combat further losses and take necessary policies for retaining a sustainable food production system. The findings of the study could further be used to model the future land use transitions and their probable impacts.

Acknowledgements We acknowledge Dr. Althaf S., Assistant Professor, School of Management Studies, National Institute of Technology, Calicut, for sharing his expertise in the analysis of datasets using STATA.

References

1. Subasinghe, S. (2016). Spatiotemporal analysis of urban growth using GIS and remote sensing: A case study of the Colombo metropolitan area, Sri Lanka. *International Journal of Geo-Information, 5*(197), 1–19.
2. Al-shalabi, M. (2013). Modelling urban growth evolution and land-use changes using GIS based cellular automata and SLEUTH models: The case of Sana'a metropolitan city, Yemen. *Environmental Earth Science, 70*, 425–437
3. Cohen, B. (2006). Urbanization in developing countries: current trends, future projections, and key challenges for sustainability. *Technology in Society, 28*, 63–80.
4. Sheridan, J., Larsen, K., & Carey, R. (2015). *Melbourne's foodbowl: Now and at seven million*. Victorian Eco-Innovation Lab: The University of Melbourne.

5. State Urbanisation Report. (2012). Department of town and country planning, Government of Kerala. https://townplanning.kerala.gov.in/town/wp-content/uploads/2018/12/SUR.pdf. Accessed on 06/05/2020.
6. Paddy Cultivation in Kerala. https://ras.org.in/paddy_cultivation_in_kerala. Last accessed 2020/5/8.
7. District Spatial Plan Palakkad. (2011). Department of Town and Country Planning, Government of Kerala.

Understanding the Implications of the Loss of Peri-Urban Arable Land—A Case of Pune Metropolitan Region

Amruta Garud and Bakul Rao

Abstract Land, one of the non-renewable resources, is primarily used as a commodity to cater to food, shelter, and other infrastructure demands. Urbanization has resulted in unmonitored conversions of agricultural lands to non-agricultural (NA) activities in the peri-urban and rural areas, which were once fertile lands. These arable parcels of land have now been exposed to the haphazard urban sprawling. The study aims to map non-agricultural conversions temporally and understand the implications on the local food security and livelihood. A ground-level data extracted from 7/12 forms have been used to estimate the loss of land and the resulting impacts. The Pune Metropolitan Region in India has been considered as a case for investigation. It has recorded a total of 320 NA conversion cases for 140 villages between 2015 to 2019. The results show a cumulative loss of 4739 ha arable land, of which 1146 ha, 716 ha, and 403 ha under cereals, legumes, and vegetables, respectively. It is also observed that 80% of the land lost was under rainfed and 13% under irrigated agriculture. In just four years, it has been estimated that around 22,716, 74,627, and 319,952 people have lost their local source of land for cereals, legumes, and vegetable production, respectively. The potential local livelihood loss is estimated at around 411 agricultural labourers per year. The research concludes that every hectare of land converted to non-agricultural use has direct and much severe implications at the grassroots level, especially on the local or regional food security and agricultural livelihood.

Keywords Land-use conversions · Peri-urban · Loss of arable land · Food security · Local livelihood · Metropolitan regions

---

A. Garud · B. Rao (✉)
Centre for Technology Alternatives for Rural Areas, Indian Institute of Technology Bombay, Mumbai, India
e-mail: bakulrao@iitb.ac.in

A. Garud
e-mail: amrutagarud.iitb@gmail.com

A. Jana and P. Banerji (eds.), *Urban Science and Engineering*, Lecture Notes in Civil Engineering 121, https://doi.org/10.1007/978-981-33-4114-2_35

1 Introduction

Agriculture is one of the most essential and exploited provisioning services of the land resource. As the global population continues to rise and require ever-greater quantities of food [12], dependency on the agriculture sector has increased than ever, and the constrained access to the land resource has further widened the gap between food production and consumption [15]. As urban ecosystems themselves have less land available for agriculture, the peri-urban lands are susceptible to take a burden of urban sprawling [1, 15, 16]. The Ministry of Rural Development, Government of India, in 2013, released the National Land Utilisation Draft Policy [11] to address the growing concern due to increasing pressure on land resources and sustainable development challenges therein. Further, the policy has widened the ambit of the term *land* by defining it as things attached to the earth or permanently fastened to anything attached to the earth and which also includes benefits to arise out of the land. The said benefits from the land resource are the offshoots of the ecosystem's provisioning, regulating, supporting, and cultural services [5]. The provisioning ecosystem services primarily depends on agricultural land cover (for example, cropland versus pasture) and management practices. Hence, the study of arable lands becomes significant.

For India, though the seventh-largest country in the world, land resource management is critical as India has over 17% of the world's population living in 2.4% of the world's geographical area. Therefore, India can no longer afford to neglect current malpractices or inefficient planning and management of land resources. In metropolitan regions, the dependency on the rural ecosystem for agriculture increases as urbanization takes a heavy toll on local land-use practices. The impact of urban sprawling on the rural ecosystem is noticeable as one travels outward from the city limits. The ramification is paramount in the peri-urban area of the city, which once was the most fertile arable land catering to the city's demands. As cities grow, urbanization further outspread this belt of fertile land, leaving resources crunch and an increase in resource miles in the future. So how does this arable land gets snatched? The gimmick we are interested in here is the conversion of native agricultural lands to non-agricultural use, otherwise referred to as NA conversion in India. This aspect of land-use conversions and its implications on the local food and livelihood system has not been adequately addressed in the current regional planning practices and researches in India, making this research a valuable contribution.

2 Literature Review

A considerable number of studies have concluded that global warming and increased concentration of greenhouse gases impacting temperature trends, as a significant implication of land-use conversions [4, 15]. Also, some studies have precisely underlined the implications on the micro-nutrients and soil profile [7, 10, 13]. The changing land-use practices such as deforestation and tillage, resulting in a net loss of soil carbon to the atmosphere. A study in China has found that the land-use conversion may affect the soil organic carbon concentration distribution in the soil profile [2]. As most of the literature has focused on the environmental implications of land-use conversions, there is a need to focus on the socio-economic implications as well. How do we estimate these socio-economic implications? A study by Etter et al. [3] emphasized a spatially explicit understanding of existing and predicted land cover changes and knowledge of their underlying drivers. It was also evident from the literature review that analyses of multiyear time series of the land attributes, their fine-scale spatial pattern, and their seasonal evolution have led to a broader understanding of land cover change [7, 17]. Based on the literature insights and data availability, a suitable methodology has been developed for further investigation.

3 Study Area

The Pune Metropolitan Region (PMR), identified for this study, was notified in 2014, under the Maharashtra Regional Town Planning Act of 1966. Compared to other metropolitan regions in India, the PMR shows substantial loss of fertile arable land to urbanization, hence makes a suitable case for investigation. The PMR consists of 9 blocks (see Fig. 1a) of the Pune district of which Mulshi, Maval, and Haveli blocks are included totally; and Bhor, Shirur, Velhe, Purandar, Khed, and Daund are partially included. The PMR consist of 845 villages, 13 census towns, 8 municipal councils, and 2 municipal corporations, amounting to the overall population of 7.33 million as of 2011. The mix of urban–rural population in this region is 42–58%. The PMR is known for its highly fertile land and quality of produce, especially vegetables, because of geographic and climatic suitability.

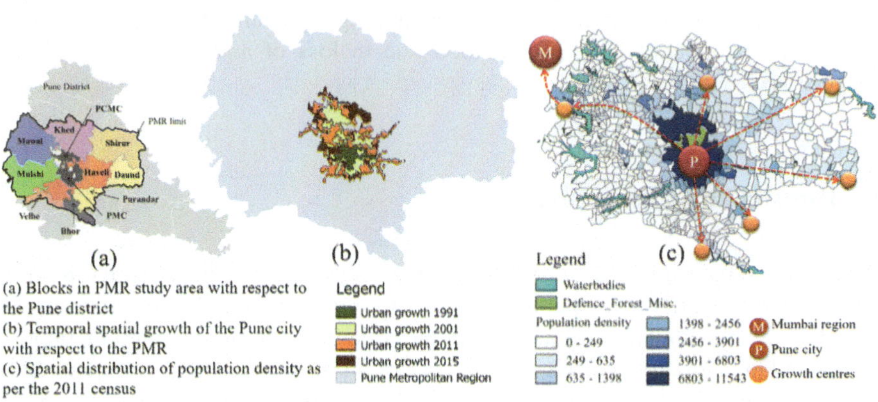

Fig. 1 A key map showing blocks and administrative boundary of PMR (**a**), temporal, spatial growth of Pune city (**b**) and village wise spatial distribution of population density for PMR (**c**), *Source* modified using the google earth historical imagery tool and Census of India 2011

4 Data and Method

The NA conversion is a commonly known but crucial process as several factors, stakeholder approvals, and high conversion fees are involved. The procedure of converting land-use to non-agriculture comes under the State governments in India, as land is a state subject. In Maharashtra, the conversions are monitored in the purview of the Maharashtra Land Rules (conversion of use to of land under non-agricultural Assessments) 1969. In this research, the land-use conversion data for the past four years has been used to establish the research problem spatially and numerically. In Maharashtra, the land record form, commonly known as 7/12 form, is a combined registration of Record of Right and Registration of crops (Land Record office, no date). At the grassroots level, the land record officer, commonly known as Talathi in Maharashtra, is responsible for maintaining the land record register for their respective jurisdictions. The compiled raw data for the PMR was acquired from the Pune Metropolitan Region Development Authority (PMRDA). The data was in a tabular format in which village wise non-agricultural land use conversions, commonly known as NA conversions, recorded during 2015–2019 were listed. For every NA conversion listed, a form number 7/12 was acquired from the state government's land record website. For this study, a total of 320 NA conversion cases recorded for 140 villages of PMR are considered for further investigation. From the 7/12 form, for every conversion case, data on agriculture type, crop, and season-wise area under cultivation and its irrigation status were categorically tabulated for further descriptive analysis.

In this research, based on the data available and literature review, a suitable methodology has been developed for estimating the potential implications. The socio-economical implications have been studied here and estimated using the following four steps. (1) estimating the loss of arable land under cultivation under cereals, legumes, and vegetables food groups, (2) estimating the loss of agricultural produce therein for the identified food groups, (3) estimating the number of people losing a potential local source of production, (4) estimating a potential number of marginal workers losing potential local livelihood due to loss of arable land to land-use conversions.

5 Results and Discussions

The results of the study can be broadly grouped depending on the potential implications for food security and local livelihood. The summary of the analyzed year wise and block-level cumulative data of NA conversions have been presented in Table 1. The cumulative loss to land-use conversion, for all the food groups, is around 4739 ha of arable land from December 2015 to April 2019. The studied data has shown that the Haveli block, which encompasses the Pune city, shows maximum loss of arable land around 40.66% of the total recorded land-use conversions and Bhor being the lesser of all the blocks with 4.34%. Mawal and Mulshi blocks are the parts of the

Table 1 Block-wise cumulative loss in hectares from Dec 2015 to April 19 compiled from the land-use conversion data

Blocks	Area of the block (A)	2015		2016		2017		2018		2019		Block total	
		Area (a)	% of A	Area (a)	% of A	Area (a)	% of A	Area (a)	% of A	Area (a)	% of A	$\sum$ Area (a)	% of $\sum$ (a) over $\sum$ yr total
Bhor	25,766	5.5	0.02	57.4	0.22	12.3	0.05	125.7	0.49		0.00	200.9	4.24
Daund	53,955		0.00	114.6	0.21	178.0	0.33	130.1	0.24	20.5	0.04	443.2	*9.35*
Haveli	91,835	0.4	0.00	658.7	0.72	750.4	0.82	334.6	0.36	182.9	0.20	1927.1	**40.66**
Khed	94,967		0.00	219.5	0.23	143.0	0.15	15.6	0.02	22.7	0.02	400.6	*8.45*
Maval	109,636	80.5	0.07	110.1	0.10	60.7	0.06	160.8	0.15	5.5	0.01	417.6	*8.81*
Mulshi	100,396	5.3	0.01	366.0	0.36	90.3	0.09	267.0	0.27	5.9	0.01	734.5	*15.50*
Purandhar	32,177		0.00	9.9	0.03	100.1	0.31	272.5	0.85	2.2	0.01	384.8	*8.12*
Shirur	110,539		0.00	173.1	0.16	32.6	0.03	24.8	0.02		0.00	230.6	***4.87***
Year Total	619,271	91.6	0.01	1709.3	0.28	1367.4	0.22	1331.1	0.21	239.7	0.04	4739.1	

Note The coding used is bold—higher; italic—moderate; bolditalic—lower
This summary includes other food groups like flowers, fruits, grass, etc. as well

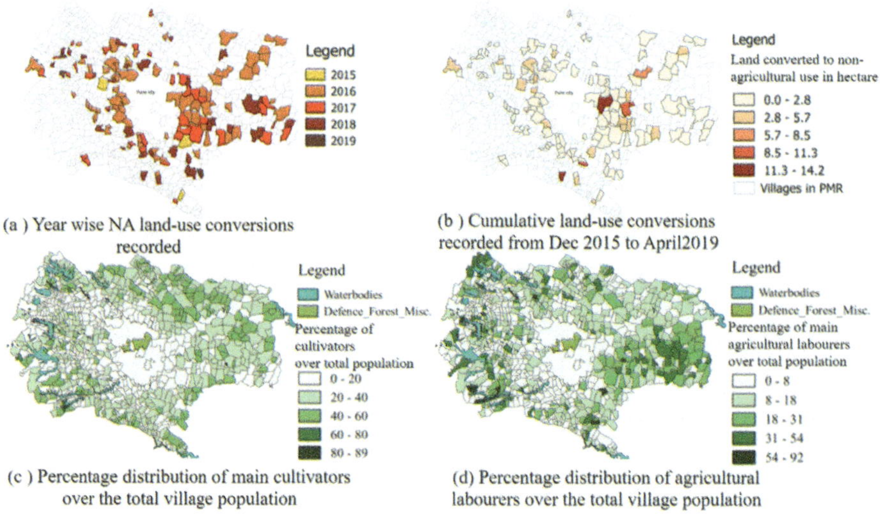

(a) Year wise NA land-use conversions recorded

(b) Cumulative land-use conversions recorded from Dec 2015 to April2019

(c) Percentage distribution of main cultivators over the total village population

(d) Percentage distribution of agricultural labourers over the total village population

Fig. 2 Spatial distribution of (**a**) year wise and (**b**) cumulative land-use conversions recorded between December 2015 to April 2019 and village working population percentage of (**c**) main cultivators and (**d**) agricultural labourers as per Census 2011 in the PMR

western ghats, that is more undulating and suitable for paddy fields and vegetables, also having functional connectivity to both the Pune city and its primate city Mumbai, have recorder loss of 8.81% and 15.50%, respectively. Daund, Khed, Purandhar, and Shirur, blocks with less undulating topography and less fragmented fields, are suitable for wheat and horticulture and have recoded 9.35%, 8.45%, 8.12%, and 4.87%, respectively. These blocks also show a majority of the percentage of workers engaged as agricultural labourers (See Fig. 2). These prime agricultural lands, sufficing the local demand, show a considerable loss of arable land. Further, the spatial distribution of the land-use conversions shows a spatial correlation with the transport corridors, proximity to the urban centers, comparatively less undulating topography, and geographically more significant villages (Refer Figs. 1c, 2b).

5.1 Cropping Season and Food Group Wise Estimation of Loss of Arable Land

While studying agriculture within a region, it is crucial to understand the prevalent cropping seasons, cropping pattern, irrigation type. The loss of area under cultivation can be estimated for three prevalent cropping seasons in Maharashtra, namely Kharif, Rabi, and Summer crops. The Kharif from June to October, Rabi from November to March, and Summer from April to June. In the PMR region, which is blessed with right agro-climatic conditions, the maximum loss of arable land is under Kharif

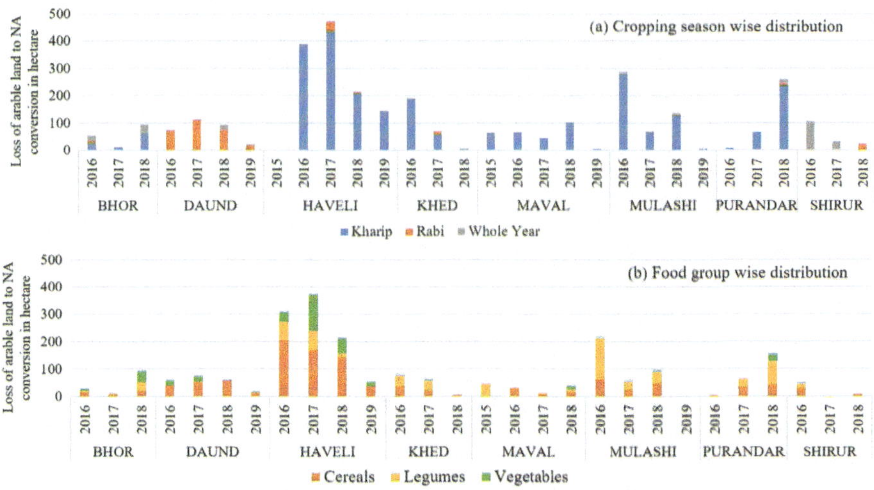

Fig. 3 Block wise loss of arable land to NA land-use conversion and concerning (**a**) cropping season and (**b**) food groups, recorded between December 2015 to April 2019, in the PMR *Note* Data only for food groups in focus has been presented. The data for fruits, flowers, grazing land, and grass has not been considered in the food group wise study.

cultivation, in which almost 80% of the total conversion has taken place (see Fig. 3a). The crops in the Kharif season are primarily rainfed crops; thus, it can be said that the loss of rainfed fields is higher compared to the irrigated crops. The Haveli block shows the highest loss of Kharif and Daund block with the highest loss of Rabi season crops. Another way of looking at these land-use conversion losses is that there is a loss of major food groups. The loss of crop is grouped under three groups, namely *Cereals*,*Legumes*, and *Vegetables*, as they form a central part of the local diet (see Fig. 3b). The percentage losses for cereals, legumes, and vegetables are 24.19%, 15.12%, and 8.51%, respectively, and together with a total of 47.81% of the total recorded land-use converted area. The Haveli block has shown a maximum loss of area under cereals and vegetable cultivation.

5.2 *Implications on Local Food Security*

The aspect of food security considered here in this research is in terms of 'access to a local source of production for catering average consumption demands' of the region. Hence, two prime questions answered include; first, how much produce would have been produced if the lands were not converted, and second, with that produce, to what extent the consumption demand would have been catered? The loss of agricultural produce due to the land conversions is calculated by multiplying the area of the loss of agricultural land and the average productivity of the respective crop (Refer Table 2). The average productivity for cereals is 2.1 tonnes/ha, for legumes is 1 tonne/ha,

Table 2 Block-wise estimations of implications on food security derived from Table 1

Column Head	(a) Total NA converted area (Ha)				(b) Production Loss (tonnes/ha)				(c)People losing a local source of production			
Notes					(b) = (a) * Avg. food group productivity by the local government				(c) = (b)/Avg. food group wise per capita per annum consumption from NSSO			
Blocks	C	L	V	BT	C	L	V	BT	C	L	V	BT
Bhor	41.06	43.51	51.61	***136.18***	86.2	43.5	825.7	*955.5*	814	4,532	40,960	*46,306*
Daund	178.9	5.01	44.57	*228.48*	375.7	5	713.1	*1093.9*	3,546	522	35,374	*39,442*
Haveli	551.66	153.28	246.72	**951.66**	1158.5	153.3	3947.5	**5259.3**	10,933	15,966	1,95,809	**2,22,709**
Khed	64.54	76.33	5.91	***146.78***	135.5	76.3	94.6	***306.4***	1,279	7,951	4,692	***13,922***
Maval	57.57	65.57	14.45	***137.59***	120.9	65.6	231.2	***417.7***	1,141	6,831	11,468	***19,440***
Mulshi	135.48	227.06	8.33	*370.87*	284.5	227.1	133.2	*644.8*	2,685	23,652	6,608	*32,945*
Purandhar	79.36	123.49	27.97	*230.82*	166.7	123.5	447.6	*737.7*	1,573	12,863	22,201	*36,637*
Shirur	37.6	22.18	3.58	***63.36***	79	22.2	57.2	***158.4***	745	2,310	2,839	***5,894***
Column Total	**1146.16**	*716.42*	***403.14***	2265.74	*2406.9*	***716.4***	**6450.2**	9573.6	***22,716***	*74,627*	**3,19,952**	4,17,295

Annotations used – C: Cereals; L: Legumes; V: Vegetables; BT: Block total
The colour coding used is bold—higher; italic—moderate; bolditalic—lower

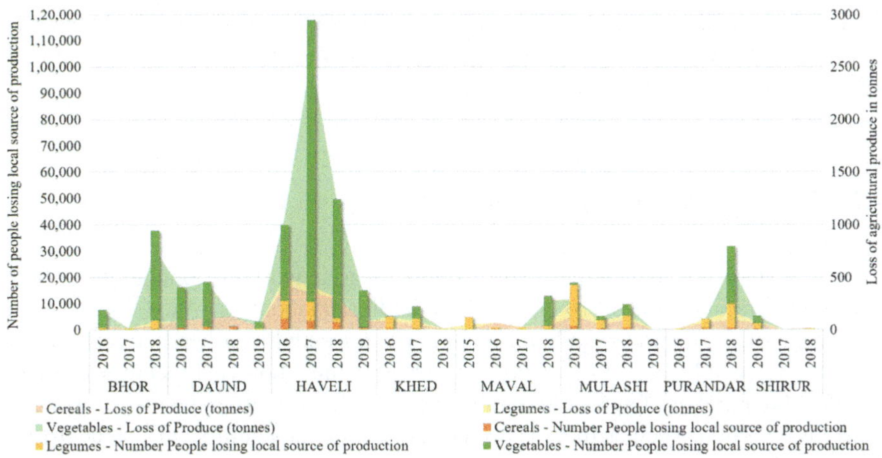

Fig. 4 A temporal block wise estimation of loss of produce and the potential number of people losing a local source of production

and vegetables are 16 tonnes/hectares, respectively, and the same is considered for the calculation purpose, and results have been presented in Fig. 4. The average productivity estimates are taken from the Agriculture Contingency Plan for Pune District [9]. While estimating overall production, the two inherent assumptions made here are—first, that total NA converted land recorded was being cultivated, and there is no fallow land; second, there is no productivity enhancement since 2011, and the same average productivity estimates can be considered for calculations. The study shows that the overall estimated produce loss is higher for vegetables, which is 6450 tonnes, compared to cereals (2406 tonnes) and legumes (716 tonnes), and this potential nearby local source of cultivation, over four years, has been lost. This also means increasing dependency on arable land elsewhere to cover the lost produce as consumption and productivity remain the same. Loss of local sources also means an increase in the travel distance to produce perishable vegetables.

On the consumption part, the per-annum-per-capita average consumption estimates for cereals as 0.096 tonnes, legumes as 0.011 tonnes, and vegetables as 0.019 tonnes given by the National Sample Survey Office, Government of India, in its NSS 68th Round (2014) are considered. The research assumes these per capita averages are as per the NSSO sampling strategy for the Pune division, represent close to the reality, and reliable figures for further estimation. Further, the estimates for the number of people losing the potential source of local production are derived by dividing the estimated loss of agricultural produce with per-annum-per-capita average consumption estimates (see Fig. 4). In just four years, it is estimated that around 22,716 people for cereals, 74,627 people for legumes, and 3,19,952 people for vegetables (total of 5.53% of the total PMR population) have lost their local source of land for production. As the Haveli block has the highest amount of NA converted land, it can be said that the potential of the block to suffice consumption demands locally has been compromised. In case of cereals and legumes, local dependency seems to be lesser than vegetables, as perishable items are more dependent on the cold chain systems, which are much less in India.

5.3 *Implications on Local Livelihood*

As per Census 2011, in the Pune district, out of the total working population, 21.90% are cultivators and almost 10.24% are agricultural labourers. As a considerable percentage of the district population is in agriculture and its allied activities as the primary source of local livelihood, loss of agro activities shows a direct impact on the considerable number of agricultural labourers. While estimating the loss of working hours, assumptions such as there is no mechanization happening and a standard 8-h shift have been made.

The labour requirements for cereals as 1098 h/ha, legumes as 683 h/ha, and for vegetables 1834 h/ha, derived from a suitable study for non-mechanized farming [2], have used to determine per kg working hours required. This number divided by average of 189 days of agriculture labour [14] in a year and 8 h. per day to estimate per year loss of agriculture employment due to loss of arable land. Refer Fig. 5 for block wise and Table 3 for food group wise loss of local livelihood.

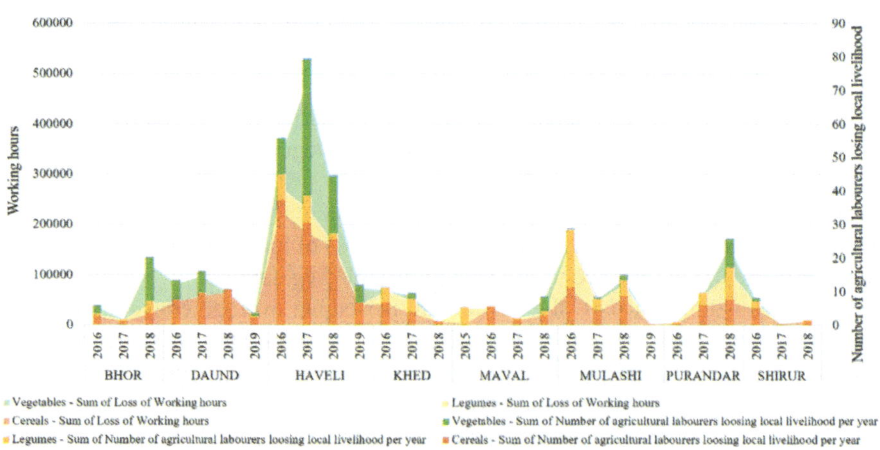

Fig. 5 Block-level understanding of the implications of NA land-use conversions on agriculture livelihood

Table 3 Block-wise estimations of implications on local livelihood due to land-use conversions in the PMR

Column Head	(d) Loss of working hours				(e) Number of marginal agricultural workers losing local livelihood per year			
Notes:	(d) = (a) * Avg. food group wise working hours required				(e) = ((d)/ 189/8/4) Avg.189 days/year of agriculture labour; 8 h. daily working hours; 4: total period for which data considered			
Blocks	C	L	V	BT	C	L	V	BT
Bhor	45,082	29,718	94,652	*169,451*	7	5	16	*28*
Daund	196,436	3423	81,744	*281,604*	32	1	14	*47*
Haveli	605,717	104,687	452,484	**1,162,888**	100	17	75	**192**
Khed	70,861	52,132	10,842	***133,835***	12	9	2	***22***
Maval	63,212	44,787	26,502	***134,501***	10	7	4	***22***
Mulshi	148,759	155,080	15,271	*319,110*	25	26	3	*53*
Purandhar	87,135	84,343	51,303	*222,782*	14	14	8	*37*
Shirur	41,285	15,146	6559	***62,991***	7	3	1	***10***
Column total	**1,258,489**	***489,316***	*739,357*	2,487,162	**208**	***81***	*122*	411

Note For reference (a), see Table 2. Annotations used—C: Cereals; L: Legumes; V: Vegetables; BT: Block total. The coding used is bold—higher; italic—moderate; bolditalic—lower

6 Conclusions

From the implications studied and geospatial understanding, it can be concluded that the urban sprawl occupies fertile parcels of the surrounding arable land by changing their land-use. The NA method is one of the dominant tools of land-use conversion exploited by the commercial developers; hence, the rate of NA conversion can be used as an indicator while arguing the loss of agriculture and threat to the local food security. It is evident that the blocks which are closer to the metropolitan city, like Haveli, Maval, Bhor, and Mulshi, show vast land getting converted. Overall, at a regional level, though the rate of NA conversions may not seem alarming, at a village level and block level, it shows higher implications. Hence, it can be argued that the intensity of the implications of per unit land-use conversions is highest at the grassroot level, which gets dispersed as the expanse increases. The PMR study has shown that arable land parcel, which is a non-fragmented parcel, rainfed, under cereals cultivation, has access to transport corridor, and is in proximity to an urban area, is more vulnerable to conversion. The implications on local food security and livelihood may not be alarming, currently, due to NA conversion, but as urbanization continues, the peri-urban areas are going to outspread, making more arable land vulnerable. For this purpose, reasonable restrictions on the acquisitions and conversions of such vulnerable arable land should be enforced for ensuring local agricultural sustainability.

Acknowledgements The research presented is a part of the ongoing Ph.D. at the Centre for Technology Alternatives for Rural Areas IIT Bombay, India. We are thankful to the Pune Metropolitan Development Authority for giving access to the land-use conversion data.

References

1. Badami, M. G., & Ramankutty, N. (2015). Urban agriculture and food security: A critique based on an assessment of urban land constraints. *Global Food Security. Elsevier, 4*, 8–15. https://doi.org/10.1016/j.gfs.2014.10.003.
2. Chen, L., et al. (2007). Effect of land use conversion on soil organic carbon sequestration in the loess hilly area, loess plateau of China. *Ecological Research, 22*(4), 641–648. https://doi.org/10.1007/s11284-006-0065-1.
3. Etter, A., et al. (2006). Regional patterns of agricultural land use and deforestation in Colombia. *Agriculture Ecosystems and Environment. Elsevier, 114*(2–4), 369–386. https://doi.org/10.1016/j.agee.2005.11.013.
4. Fall, S., et al. (2010). Impacts of land use land cover on temperature trends over the continental United States: Assessment using the North American Regional Reanalysis. *International Journal of Climatology Royal Meteorological Society, 30*(13), 1980–1993. https://doi.org/10.1002/joc.1996.
5. Hardelin, J., & Lankoski, J. (2018). *Land use and ecosystem services, OECD food, agriculture and fisheries papers*. Paris: OECD Publishing. https://doi.org/10.1787/c7ec938e-en.
6. Ibarrola-Rivas, M. J., Kastner, T., & Nonhebel, S. (2016). How much time does a farmer spend to produce my food? An international comparison of the impact of diets and mechanization. *Resources, 5*(4). https://doi.org/10.3390/resources5040047.
7. Lambin, E. F., Geist, H. J., & Lepers, E. (2003). Dynamics of land -use and land cover change in tropical regions. *Annual Review of Environment and Resources, 28*(1), 205–241. https://doi.org/10.1146/annurev.energy.28.050302.105459.
8. Land Record office (no date). *GoM, Government of Maharashtra*. Available at: https://bhulekh.mahabhumi.gov.in.
9. Maharashtra, G. of (2011). *State: Maharashtra agriculture contingency plan for district: Pune*. Available at: www.kvkbaramati.com.
10. Mercedes, G., & Montenegro, L. (2005). Globalwarming and tropical land-use change: greenhouse gas emissions frombiomass burning, decomposition and soils in forest conversion, shifting cultivation and secondaryvegetation, pp. 115–158.
11. MoRD (2013). *National Land Utilisation Policy -framework for land use planning & management*. Available at: https://smartnet.niua.org/sites/default/files/resources/draft_national_land_utilisation_policy_july_2013.pdf.
12. Muir, J. F. et al. (2010). Food security : the challenge of feeding 9 billion people. 327, pp. 812819.
13. Murty, D., et al. (2002). Does conversion of forest to agricultural land change soil carbon and nitrogen? A review of the literature. *Global Change Biology, 8*(2), 105–123. https://doi.org/10.1046/j.1354-1013.2001.00459.x.
14. Padhi, K. (2007). Agricultural labour in India-a close look. *Orissa Review, 63*(7/8), 23–28.
15. Seto, K. C. et al. (2012). Urban land teleconnections and sustainability. https://doi.org/10.1073/pnas.1117622109.
16. Seto, K. C. et al. (2017). Sustainability in an urbanizing planet. *114*(34), 8935–8938. https://doi.org/10.1073/pnas.1606037114.

17. Tavares, A. O., Pato, R. L., & Magalhães, M. C. (2012). Spatial and temporal land use change and occupation over the last half century in a peri-urban area. *Applied Geography, 34*, 432–444. https://doi.org/10.1016/j.apgeog.2012.01.009.

Optimizing Building Fenestration Design for Daylight and Energy Savings in Low-Income Housing: Case of Mumbai, India

Nikhil Kumar and Ronita Bardhan

Abstract The low-income housing of dense metropolitan cities focuses solely on occupancy maximization, thus significantly neglecting indoor livability parameters energy-saving potential and daylight efficiency. The existing unit designs lack rational insight on fenestration detailing window size and location, which, when modified, impacts the indoor daylight levels and its corresponding energy demands. This study unveils a comparison of two scenarios, one considering the single building without context, and the second case considers the contextual effect of surrounding urban geometry on daylight performance. This work bridges the existing literature gap in connecting fenestration design through an optimization route. While most optimization studies deal with early design stage strategies, this study looks into the post-construction stage using retrofit and refurbishment strategies. A systematic parametric design framework was adopted for rating fenestration designs based on illuminance indicators like Daylight Autonomy (DA), Daylight Factor (DF), and Usable Daylight Index (UDI). The simulations were performed at the neighborhood level considering the impacts of nearby hyper-dense urban geometry on a given building for hot and humid climate. The results suggest that out of various fenestration design parameters, the window size was observed to be the most significant in affecting indoor daylight levels. An increase in window size from 10% WWR to 30% WWR significantly impacted daylight metrics like DA and UDI. Compared to the building without context, the contextual simulations showed that the lower floors received deficient levels of daylight, whereas the upper floors maintained the right amount of constant daylight for both the scenarios.

N. Kumar (✉) · R. Bardhan
Centre for Urban Science and Engineering, Indian Institute of Technology Bombay, Mumbai 400076, India
e-mail: nikhil.89.kumar@gmail.com

R. Bardhan
e-mail: ronita.bardhan@gmail.com

R. Bardhan
Behaviour and Building Performance, Department of Architecture, University of Cambridge, Cambridge CB3 9DT, UK

A. Jana and P. Banerji (eds.), *Urban Science and Engineering*, Lecture Notes in Civil Engineering 121, https://doi.org/10.1007/978-981-33-4114-2_36

Keywords Daylight · Fenestration · Refurbishment · Optimization · High occupancy buildings

1 Introduction

The current multi-rise low-income hyper-dense buildings have less degree of freedom for designing fenestrations that would provide ample daylight, visual communication with energy savings. Hence, daylight designing through architectural intervention has become a significant contributor to a sustainable living environment in modern deep-plan buildings. Windows, which critically influence indoor daylight, also characterize energy use, lighting performance, and extent of visual communication in a building [1].

Most of the daylight studies have been conducted at places with higher latitude. In such places, the sun is always on one side of the sky hemisphere, which gives an intuitive understanding as to which face will receive better daylight. In the case of the tropics, the sun shifts its position for a part of the year. Sunshine duration is much longer. The context of the building becomes much more significant in the case of the tropics. Henceforth there is a clear gap in studying the daylight for buildings in the tropical climate and under contextual conditions.

The most common metric used for quantifying daylight in architecture is Daylight Factor (DF). DF is calculated under overcast sky conditions; hence the location of the building, orientation, location of the window does not impact the results of the simulation. DF is dependent on the geometry of the building. Furthermore, the illuminance of the interior is calculated depending on the illuminance of the exterior. This leads to unreliability as DF ignores the clear sky conditions, whereas Daylight Autonomy (DA) is a dynamic metric dependent on the sky conditions, location of the window, orientation, location of the building. Defined as a minimum illuminance threshold, which is met for a percentage amount of time in a year.[2]. Useful Daylight Illuminance (UDI) is another daylight metric that gives a threshold of illuminance (minimum to maximum). UDI is an efficient metric for optimization purpose as it high amounts of daylight exposure does create glare and other side effects as well. Henceforth UDI gives an optimum range of lux level, which can be used for optimizing window design [3]. In this study, we use a simple graphical optimization method, with some constraints subjected to DA and DF. There has been the application of different optimization algorithms in several daylight studies, though using a graphical method helps us to study the interaction between the various metrics used for optimization.

This study aims to determine the importance of window size and location for improving daylight in a residential unit. A large number of daylight studies have been done, but they miss out on considering the context of the building. As the surroundings majorly impact the daylight available for the building. The main objective is to obtain the most optimal window design for different floor heights of the building. It has been achieved by performing a sensitivity analysis with Daylight Autonomy (DA) as the measurement metric and optimizing the Useful Daylight Illuminance (UDI) with respect to Daylight Factor (DF) and DA. The novelty of the study lies in the idea of increasing the daylight by using generative design for optimum window size and location for low-cost housing present in Mumbai. The used methodology can be used pre-construction. As well as the same methodology is applicable in case of retrofit and refurbishment of the neighborhood.

Section 2 describes the methodology, which includes simulation settings, measurement parameters, study area, description of the proposed window designs, and optimization. Section 3 consists of the results of sensitivity analysis and optimization. Section 4 concludes the study.

2 Method

2.1 Simulation Settings and Performance Indicators

Daylight incidence at residential units on different floors was simulated. The units were considered empty with no furniture or shading. A sensor grid was generated at the height of 850 mm from the floor. Equidistant sensors were placed on the grid where the distance between the two adjacent sensor points was kept at 500 mm. The reflectance values of 0.7 for the ceiling, 0.2 for the floor, 0.5 for interior walls, and 0.3 for surrounding outside buildings were considered. The model geometry was created in Rhinocerous 5.0, and Diva for Rhino was used for daylight simulations [4]. The simulation parameters are shown in Table 1, and the occupancy was considered for daytime only between 8 AM and 6 PM, amounting to a total of 3650 h in a year.

Table 1 DIVA simulation parameters

Parameter	Value
Ambient bounces (ab)	7
Ambient divisions (ad)	1000
Ambient super samples (as)	20
Ambient resolution (ar)	300
Ambient accuracy (aa)	0.1
Limit reflection (lr)	6
Specular threshold (st)	0.15
Specular jitter (sj)	1
Limit weight (lw)	0.004
Direct jitter (dj)	0
Direct sampling (ds)	0.2
Direct relay (dr)	2
Direct pretest density (dp)	512

The daylight performance of different window sizes and designs was studied by using three daylight performance indicators, which are defined as follows.

Daylight factor: The average daylight factor (DF_{av}) is the ratio between the exterior sky illuminance on an overcast day and illuminance inside a building at the sensor point. Generally, the DF should be 2% or above for activities [5]. For this study, DF is considered as a constraint where its value is to be more than or equal to 2%.

Daylight Autonomy 300 lx: Daylight Autonomy (DA) is the percentage of occupied times in a year during which a minimum illuminance (300 lx in this case) can be met by daylight alone in the residential unit. The Illuminating Engineering Society (IES) recommends a target illuminance of 300 lx for typical residential buildings [5] and a minimum DA of 50% of the occupied times of a year. The minimum DA is considered to be more than or equal to 50%.

Useful Daylight Illuminance: Useful Daylight Illuminance ($UDI_{100–2000\ lx}$) is the percentage of occupied times in the year during which a specific range of illuminance (100–2000 lx in this case) [5] can be met by daylight alone. Lux levels higher than 2000 can cause glare, whereas Lux level less than 100 lx is not suitable for working conditions. $UDI_{100–2000\ lx}$ is considered as the main factor and is to be maximized for all the cases in this study.

2.2 *Study Area and Case Explanations*

The selected study area is low-income group tenement housing made for Project Affected Population (PAP) in the city of Mumbai. The housing design represents a cookie cutter prototype of low-income housing in Mumbai. A single building consists of Ground +7 floors with commercial space on the ground floor. All the above floors are residential space. Each floor has 13 single room tenement units abutting a corridor. A single unit consists of a multipurpose space that includes kitchen area, a toilet, and a bathroom. The total area of a single unit is 21.25 m^2 containing one window in the multipurpose space and one window in the toilet. For the study window present in the multipurpose space is considered for daylight. Figure 1 shows the building location plan, 3D of the neighborhood used for simulation in contextual condition, 3D building plan for single building simulations, typical floor plan, and floor plan of a single unit.

2.3 *Parametric Generative Design*

A total of 140 scenarios of window design with varying location and size were considered in the study by varying the Window to Wall Ratio (WWR) from 10, 15, 20, 25, and 30% and change in design and location of window represented by Type A, Type B, Type C, and Type D. In Fig. 2, window designs are illustrated. Type A

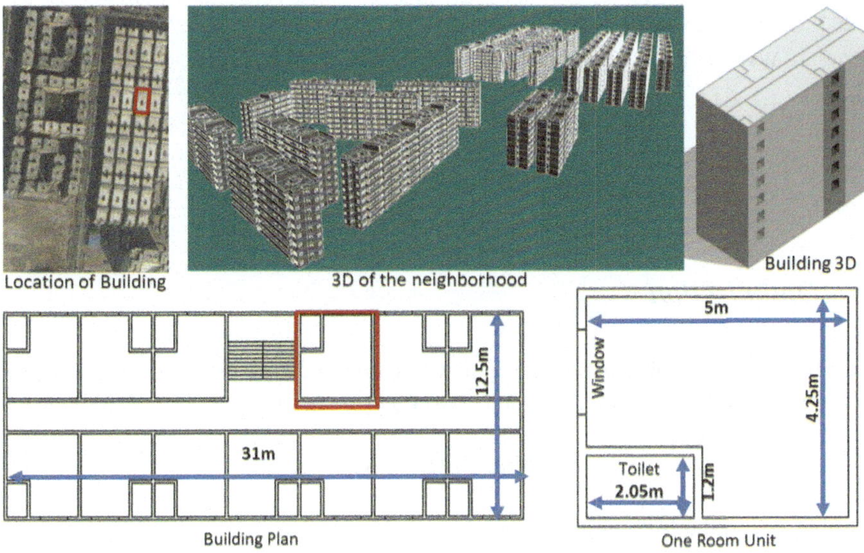

Fig. 1 Building location, typical floor plan, and 3D model used for simulation

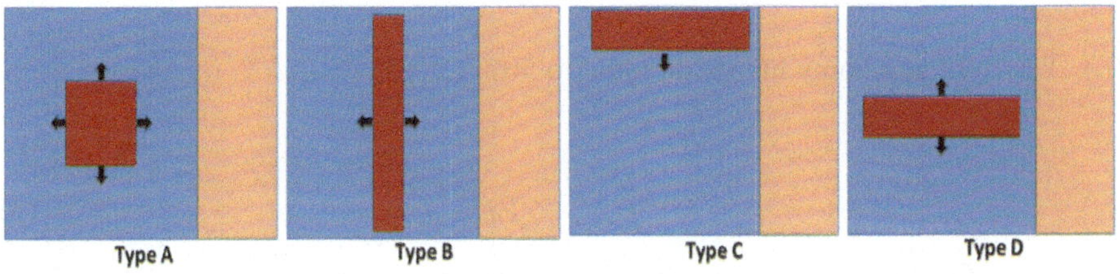

Fig. 2 Location and direction of increase in window size

has a square window located in the center of the exterior wall, which increases in all directions with an increase in the WWR. Type B represents a rectangular window that covers the vertical exterior wall and expands horizontally with an increase in WWR. Type C is a horizontally long window starting from the ceiling level and extends down with the increase in WWR. Type D has a horizontally long window located at the center of the exterior wall, which can be extended up and down proportionally with the increase in WWR. The arrow shows the direction of an increase in window size with an increase in WWR. Table 2 shows the exact sizes of the windows in the building model.

Table 2 Sizes of the windows used for simulation

	WWR 10 (m)	WWR 15 (m)	WWR 20 (m)	WWR 25 (m)	WWR 30 (m)
Type A	1.2 × 1.2	1.4 × 1.4	1.6 × 1.6	1.8 × 1.8	2 × 2
Type B	0.5 × 2.8	0.7 × 2.8	0.9 × 2.8	1.2 × 2.8	1.4 × 2.8
Type C	2.5 × 0.5	2.5 × 0.8	2.5 × 1	2.5 × 1.3	2.5 × 1.6
Type D	2.5 × 0.5	2.5 × 0.8	2.5 × 1	2.5 × 1.3	2.5 × 1.6

2.4 Optimization

The optimization problem for daylight was formulated by using the function for maximizing the UDI Eq. (1). The objective function is to

$$MaxUDI_{100-2000} \tag{1}$$

Subjected to the following constraints:

(i) A daylight factor higher or equal to than 2% Eq. (2)

$$DF \geq 2\% \tag{2}$$

(ii) Daylight autonomy higher or equal to 50% Eq. (3)

$$DA \geq 50\% \tag{3}$$

The graphical optimization method was used to find optimum solutions. Using a graphical method helps us to understand the trends of DA, DF, and UDI for different window location and their WWR. To find the optimum solution DF and DA were paired with UDI. The optimum solutions were generated, and the best possible window location and its WWR were found. In the case of similar results, the smaller WWR was preferred. Some floors had multiple solutions.

3 Sensitivity Analysis

3.1 Comparison Between Single Building vs Building in Context

The DA was simulated for a standalone building and under two contextual conditions with varying the aspect ratio, it was done by altering the distance between the buildings. Two different context scenario entailed 6 m (current condition) to 10 m

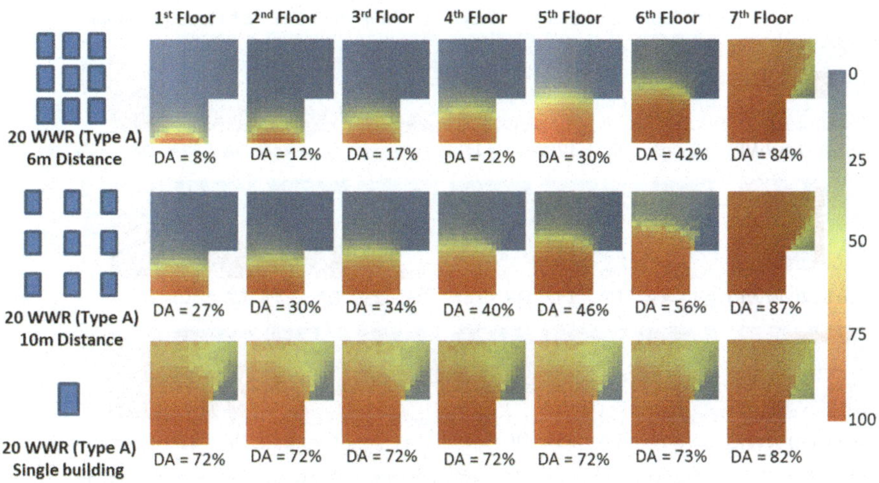

Fig. 3 Daylight autonomy values for standalone building and in context (6, 10 m)

(proposed condition) between the buildings. The window size was set at 20% WWR for Type A. This resembles the window size at the study site. In Fig. 3, the variation of DA for a standalone building ranged between is 72 and 73% from 1st floor to 6th floor and 82% on the 7th floor. In case when the building is surrounded with other buildings at a distance of 10 m, the DA on the 1st floor is 27%, which linearly increased to 87% on the 7th floor. There is a significant difference of 31% in the DA of the 7th floor to that of the 6th floor. This is majorly due to reduced access to daylight on the lower floors caused due to mutual shading from surrounding buildings. When the distance between the buildings was set to 6 m, the DA reduced further and was found to be below 50% till the 6th floor. The 7th floor in all the cases received similar amount of daylight with DA values ranging between 82 and 87%. The amount of daylight reaching the building is dependent on the context of the building. Henceforth it can be said that the context of a building is essential for correct daylight simulations as it represents the ground reality.

3.2 *Impact of Change in Window Design and Location.*

The simulations were conducted by changing the window position in the residential units, as mentioned earlier. The results for 20% WWR with four different window positions were shown in Fig. 4. In Type A the DA for 1st floor was 8% and gradually increased to 84% on the 7th floor. For Type B the DA from 1st floor to 3rd floor was similar to Type A. Then, there was an increase in the DA from 4th floor to 7th floor. It can be said that window Type B performs better than Type A, where the incidence daylight linearly increases with the floor. In Type C the DA on 1st floor was 7%, and

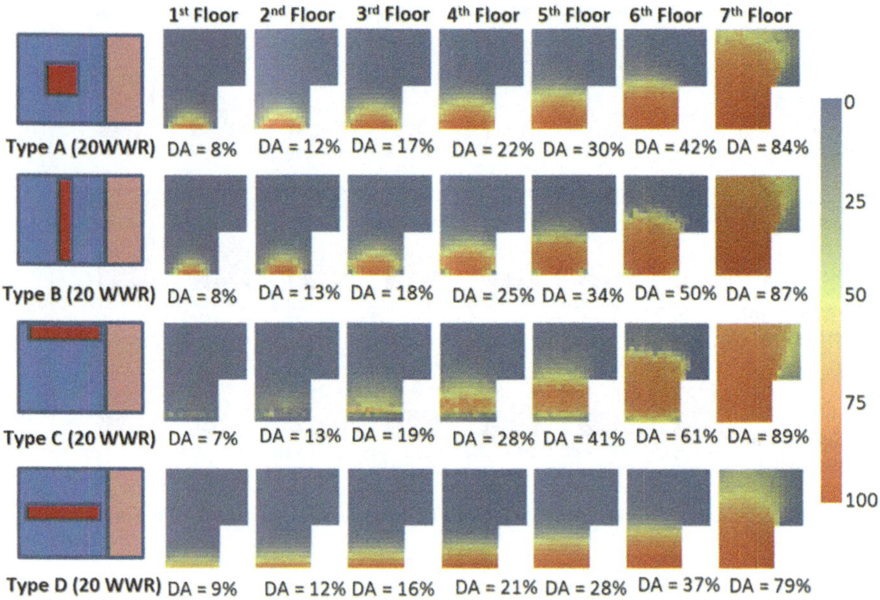

Fig. 4 Daylight autonomy of units with a change in the building design

89% for the 7th floor, which is highest in all scenarios. Type C performs the best at higher floor levels. In the case of Type D, the DA on 1st floor was 9%, which was highest, and the DA on 7th floor is 79%, which was lowest among the four window designs. Type C performs well at the top floors and Type D on the 1st floor. Hence window design does play an essential role in increasing daylight in the residential unit, but the increase in daylight remains inconsequential.

3.3 *Increase in WWR.*

A set of simulations were conducted for Type A window design. Where WWR was increased from 10 to 30% by an interval of 5%. The maximum WWR was set at 30%, as the building structure does not allow for a higher WWR. Figure 5 demonstrates the DA percentages of the residential unit at each floor with varying WWR. WWR is directly proportion to DA. 4% DA was found in the residential unit of 1st floor with 10% WWR, which gradually increased to 52% DA on the 7th floor. The maximum 92% DA was realized in the 7th floor residential unit with a window size of 30% WWR.

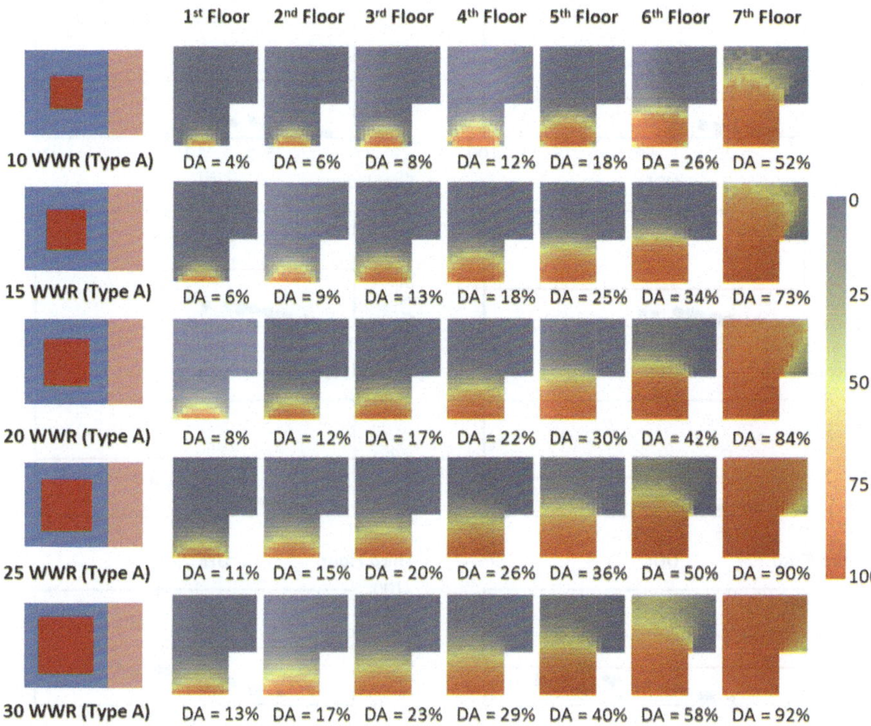

Fig. 5 Daylight autonomy of units with an increase in the size of the window

3.4 Optimization

The simulation results for the 1st to 7th floor have been displayed in Fig. 6. The scatter plots demonstrate the maximum UDI in each case concerning DF and DA. The minimum DF and DA was not achieved from the 1st floor to the 4th floor. Henceforth the window design with maximum UDI has been considered as the best possible window design solution. The maximum UDI achieved for the 1st floor was 41 for 30% WWR with Type C window. For 2nd floor, the maximum UDI was 51 for 30% WWR Type C window. 3rd floor had maximum UDI of 63% for WWR 30% Type C. For 4th floor the maximum UDI was 83 for 30% WWR Type C and 25% Type C, we opted for 30% WWR Type C as the value of DA and DF was higher in this particular window design. For the 5th floor, the maximum UDI was 91 for 25% WWR and Type C window design. The DA was 49% and DF_{avg} 1.7. For 6th floor, the maximum UDI was 96 for 20% WWR Type C window design, the DA was 61%, and DF_{avg} was 2.1. The maximum UDI for 7th floor was 100 for 15% WWR Type C the DF_{avg} was 2.3 and DA was 83%. Window design was universal in all the results; hence it can be said that Type C design performs better than the rest for daylight within the residential unit. From 1st floor to 4th floor, the optimal window size is

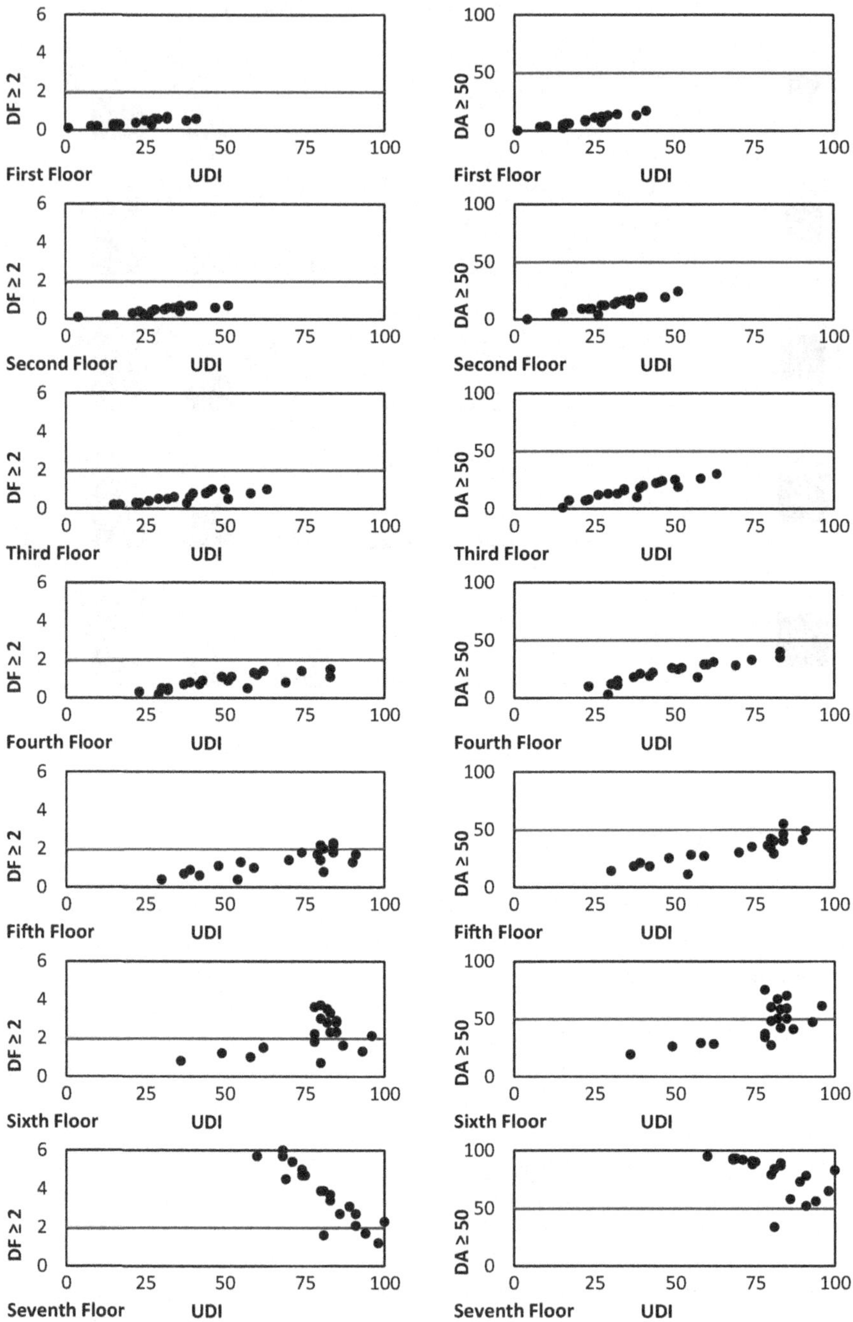

Fig. 6 Floor wise scatter plot of daylight autonomy and daylight factor against useful daylight illuminance

30% WWR with Type C window design. From 5th floor to 7th floor, the size of the window reduces by 5% WWR per floor, though the window design remains same.

4 Conclusion

A simple graphical optimization approach was used for the assessment of window design and location in this study, to look for the optimum window size and location for different floors of a small rise apartment building, in a contextual scenario. The significant findings from the research are:

This research shows that there is less variation of daylight across different floors in standalone buildings. Though the results changed when the context was considered. An increase in distance between the buildings leads to an increase in daylight in the residential units, an incremental increase in the distance escalates the daylight significantly. Due to changes in window design, the location of the high DA zone inside the residential unit changes, but the amount of DA remains approximately similar and is generally below 25% till the 3rd floor. In the case of upper floors 4th to 7th, the change in window design impacts the DA significantly. With the increase in WWR, the daylight increases, but the amount of daylight reaching the lower floors is quite less, and thus increasing the WWR does not impact the daylight much, whereas for the upper floors increasing WWR leads to discomfort and low UDI. Using optimization, the best possible WWR and window location has been recorded, and it can be suggested to have different window sizes and designs for different floors in a particular building.

However, it was noted that though the study is specific to a single building and hence contextual differences can lead to varied results depending on the orientation and design of the building. There is a need to study the additional aspects related to windows like natural ventilation, thermal comfort, energy saving, and the view from the window, which can lead to better window positioning. Despite the limitations, this study presents a new lookout at window design in apartment buildings and can help improve building performance at design stage and at refurbishment stage as well.

References

1. Acosta, I., Campano, M. Á., & Molina, J. F. (2016). Window design in architecture : analysis of energy savings for lighting and visual comfort in residential spaces. *Applied Energy, 168,* 493–506.
2. Acosta, I., Varela, C., Molina, J. F., Navarro, J., & Sendra, J. J. (2017). Energy efficiency and lighting design in courtyards and atriums: a predictive method for daylight factors. *Applied Energy, 211,* 1216–1228

3. Futrell, B. J., Ozelkan, E. C., & Brentrup, D. (2015). Optimizing complex building design for annual daylighting performance and evaluation of optimization algorithms. *Energy and Buildings, 92,* 234–245.
4. Bardhan, R., & Debnath, R. (2016). Energy for Sustainable Development Towards daylight inclusive bye-law : Daylight as an energy saving route for affordable housing in India. *Energy Sustainable Development, 34,* 1–9.
5. Mangkuto, R. A., Rohmah, M., & Asri, A. D. (2016). Design optimisation for window size, orientation, and wall reflectance with regard to various daylight metrics and lighting energy demand: A case study of buildings in the tropics. *Applied Energy, 164,* 211–219.

Author Index

A. Jana and P. Banerji (eds.), *Urban Science and Engineering*, Lecture Notes in Civil Engineering 121, https://doi.org/10.1007/978-981-33-4114-2

The manufacturer's authorised representative in the EU is Springer Nature Customer Service Centre GmbH, Europaplatz 3, 69115 Heidelberg, Germany. If you have any concerns regarding our products, please contact ProductSafety@springernature.com

Printed and bound by CPI Group (UK) Ltd, Croydon, CR0 4YY
30/09/2026
02225295-0015